New models in geography

Volume II

TITLES OF RELATED INTEREST

The carrier wave
P. Hall & P. Preston

City and society
R. J. Johnson

The city in cultural context
J. A. Agnew *et al.* (eds)

Collapsing time and space
S. Brunn & T. Leinbach (eds)

Cost-benefit analysis in urban and regional planning
J. A. Schofield

Exploring social geography
P. Jackson & S. Smith

Gentrification of the city
N. Smith & P. Williams (eds)

Geography and gender
Women & Geography Study Group (IBG)

Geography of gender in the Third World
Janet Henshall-Momsen & Janet G. Townsend (eds)

High tech America
A. Markusen *et al.*

Ideology, science and human geography
Derek Gregory

Inferential statistics for geographers
G. B. Norcliffe

International capitalism and industrial restructuring
R. Peet (ed.)

Localities
Philip Cooke (ed.)

Location and stigma
C. J. Smith & J. Giggs (eds)

London 2001
P. Hall

Method in social science
Andrew Sayer

Place and politics
J. A. Agnew

The political economy of owner-occupation
Ray Forrest, Alan Murie & Peter Williams

The political economy of rural areas
P. J. Cloke

The politics of the urban crisis
A. Sills, G. Taylor & P. Golding

Policies and plans for rural people
P. J. Cloke (ed.)

Population structures and models
R. Woods & P. Rees (eds)

The power of geography
M. Dear & J. Wolch (eds)

The power of place
J. A. Agnew & J. S. Duncan (eds)

The price of war
N. Thrift & D. Forbes

Production, work, territory
A. Scott & M. Storper (eds)

Race and racism
P. Jackson (ed.)

Regional dynamics
G. Clarke *et al.*

Regional economic development
B. Higgins & D. Savoie (eds)

Rural land-use planning in developed nations
P. J. Cloke (ed.)

Rural settlement and land use
Michael Chisholm

Shared space: divided space
M. Chisholm & D. M. Smith (eds)

Silicon landscapes
P. Hall & A. Markusen (eds)

State apparatus
G. Clark & M. Dear

Technological change, industrial restructuring and regional development
A. Amin & J. Goddard (eds)

Urban and regional planning
P. Hall

Urban problems in Western Europe
P. Cheshire & D. Hay

Western sunrise
P. Hall *et al.*

New models in geography

The political-economy perspective

edited by

Richard Peet & Nigel Thrift

Clark University & University of Bristol

Routledge
Taylor & Francis Group
LONDON AND NEW YORK

First published 1989 by Academic Division

2 Park Square, Milton Park, Abingdon, Oxon OX14 4RN
711 Third Avenue, New York, NY 10017, USA

Routledge is an imprint of the Taylor & Francis Group, an informa business

First issued in hardback 2016

British Library Cataloguing in Publication Data

. New models in geography: the political-economy perspective.
1. Human geography. Mathematical models

I. Peet, Richard, *1940-* II. Thrift, Nigel, *1949-*

304.2'0724

ISBN 978-0-04-445421-2 (pbk)
ISBN 978-1-138-18044-4 (hbk)

Library of Congress Cataloguing-in-Publication Data

New models in geography: the political-economy perspective/edited by Richard Peet & Nigel Thrift.
p. cm.
Bibliography: p.
Includes index.
ISBN 0-04-445420-1. — ISBN 0-04-445421-X (pbk.)
1. Anthropo-geography. 2. Geography, Economic. 3. Geography, Political. I. Peet, Richard. II. Thrift, N. J.
GF41.N48 1989 89-5261
304.8—del 9 CIP

Typeset in 10 on 11 point Bembo by Computape (Pickering) Ltd, North Yorkshire

Contents

Foreword

It seems such a long time ago, another age – yet it is a mere twenty-odd years since the original *Models in Geography* was published. It is an even shorter time since the first tentative steps were taken towards an alternative formulation of what might constitute a geographical perspective within the social sciences. It all seemed very daring at the time, and it began with critique and with an eager reading of basic texts. But what came to be called the political-economy perspective has progressed with remarkable speed and energy to generate its own framework of conceptualization and analysis, its own questions and debates.

The papers in these two volumes are witness to the richness and range of the work which has developed over this relatively short period within the political-economy approach. Moreover, from being a debate within an institutionally-defined 'discipline of geography', to introducing into that discipline ideas and discussions from the wider fields of philosophy and social science and the humanities more generally, it has now flowered into a consistent part of enquiries that span the entire realm of social studies. Not only has 'geography' increasingly become an integral part of the study of society more widely, but a geographical perspective is contributing to, as well as learning from, that wider debate.

The political-economy approach has been of central importance in this move. Indeed, debate within political-economic approaches to geographical studies has reflected, in its different phases, that reintegration of geography within social sciences. The form of this integration is still an issue today, but it is striking how many of the chapters in these volumes, while often talking about quite distinct empirical areas of concern, document in broad outline a similar trajectory on this issue.

The path has not always been smooth. There have been difficult and sometime confusing debates, which have involved the reformulation of questions as well as of answers. Many of the longer-running (and in the main continuing) debates are reflected in these papers; again it is striking how different authors in distinct fields frequently agree on which discussions have been of central importance. Perhaps most fundamental to the reformulation of geography's place within the social sciences has been the thoughtful and productive debate (productive in the sense that it really has moved on and has made progress from stage to stage) concerning the relation between the social and the spatial, and whether it is in any case an impossible dichotomy which should be dissolved. (Maybe we ought to be conducting a similar debate about the relation between the social and the equally difficult concept of the natural?) That debate is documented here from a number of angles. It is also clear that, even if we have understood a few things better, there are still important issues unresolved. There is still debate over exactly what it means to say that space makes a difference. Is it that particular time-space contexts trigger the realization of causal powers embedded in the social or, alternatively, do not trigger them? That is one position persuasively argued here. Or is it that some social phenomena cannot be adequately conceptualized without some degree of spatial content? And how does this tie in to the notion locality effects? Whatever 'the answers' are, it is an important debate which connects directly to the philosophical foundations of the subject and which links geography inextricably to other areas of social science.

The relation between theoretical and empirical work has also been a consistent preoccupation, both in terms of the priority that should be given to one or the other and in terms of the relation between them. A number of the papers here document the debate and wrestle with the problem. The importance of the contribution of realism to the discussion is evident, even if its usefulness, and even its form, is not always agreed. Many of the papers call for the development of 'middle-level' theories or concepts. One fascinating thing here is how different authors, from the evidence presented in these volumes, seem to mean quite distinct things by these terms. There is clearly debate here which could perhaps be addressed more directly.

Much of the early political-economy writing within geography grew up within studies of uneven development and of industrial geography. Perhaps for that reason, but also reflecting the contemporary character of political economy more widely, it had a heavily economic, at times economistic, bent and one which often accompanied a greater attention to structures than to agency. Once again there is agreement in a very wide set of the papers in this collection that this is changing. The newer research, which focuses on cultural forms and on representation and interpretation, is breathing a different kind of life into the debates. It also, very importantly, promises to help us establish closer links between areas of geographical studies which might have become too separated. But here, too, the debate is not finished; indeed in geography it has only just begun.

There are also debates between the contributors to these volumes which the careful reader will detect, but which are not addressed directly. There are contrasting understandings of the meaning of basic terms, such as 'theory'; I suspect that there are also variations in what people would include under the rubric 'political economy approach'! And I am sure that the meaning, if not the use, of the term 'model' would make a lively topic of debate! But the persistence of debate should not be seen as deeply problematical, nor necessarily as a weakness. It certainly proves the political-economy perspective is alive; over two decades it has moved from being a few isolated voices to being one of the major influences on the current development, and richness, of human geography. That implies responsibilities too, as well as a pat on the back for our collective achievement. It means pursuing these debates in a constructive, friendly and un-pompous (if there is such a word) manner; it means writing in a way which is accessible to participants outside the immediate discussion. The contributions to these volumes, in my view, achieve that aim.

Another thing many of them achieve very well is to set the development of these discussions in their (also developing) historical contexts, both societal and theoretical. Sometimes we have clearly been guilty of bending the stick too far, but often that has been the importance of stressing particular arguments at particular moments. The emphasis in the early years on social causes at the expense of the spatial dimension is a case in point. Arguments are not developed in vacuums. Today both social and theoretical contexts are presenting new challenges. Similarly, a particular focus for research may be appropriate, even urgently needed, in some contexts, without any implication that it should be a priority in some absolute, eternal, sense. The current emphasis on locality research in the UK is, in my opinion, a case in point. Amidst a flood of writing on national structural change (in which, for example, 'the end of the working class' figured prominently) it was among other things important to point out that, and to analyze how, the picture varied dramatically between different parts of the country; how some of the social dynamics in which people were caught up were often quite different from what one might divine from a national picture. This focus also gelled with questions of theory and methodology which had been raised at the same time. But none of it means to say that a focus on locality research will always and everywhere be important.

Which raises another point: the political import and impact of our work. How much difference has it made that the political-economy perspective has blossomed academically? I think the editors of these volumes are right to say in their introduction that here the record is mixed. There have been other shortcomings, too, inevitably. Some are mentioned by contributing authors, but two in particular struck me. There is a UK-US focus to much of the work, which is reflected not only in the object of study, but also in tendencies to universalize from their particularities, and also to be less than aware of academic work going on in other parts of the world. And there is still a million miles to go before the full impact of the feminist critique has been taken on board.

Nonetheless, what these volumes incontrovertibly establish is the enormous progress which has been made since those early days. They are something of a monument to years which have been energetic, full of debate, and often fun. What is more, the possibilities now opened up by geography's more fruitful relations with the rest of social science mean that the future looks set to be equally productive.

Doreen Massey

Introduction

The publication of *Models in geography* (Chorley & Haggett 1967) presaged a sea change in the practice of Anglo-American geography. Since that date, the practice of geography has changed again. A set of new models – based upon a political-economy perspective – now peppers the geographic landscape. This book provides a summary of the nature of these models, their spirit, and purpose.

The new models often took their original inspiration from Marx and Marxism.[1] That original inspiration has by now been overlain with many other layers of influence, so that this book reports on what is an increasingly diverse body of work, but one which still holds to the critical vision of society which was at the heart of Marx's project. Of course, these new political-economy models do not form the only approach to geography, but they have certainly been influential in the subject over the past few years. Their influence can be measured in three ways.

First, there is the quality of the work that the models have generated; on this count, they can surely be judged a success. As the following pages record, the approach has generated a flood of substantive theoretical and empirical work in geography, ranging all the way from class to culture, from gentrification to geopolitics, from restructuring to the urban–rural shift.

A second gauge of the influence of an approach is its ability to move outside narrow disciplinary boundaries and influence other disciplines. On this count again, the new political-economy models can surely be judged a success. Geographers who subscribe to these models now feature regularly in books and journals the length and breadth of social science, where not too many years ago it would have been very difficult to find any work by geographers at all. Geographers have also had notable success in participating in certain debates in the social sciences as a whole, for example, on subjects such as realism, structuration theory, deindustrialization and industrial restructuring. Geograhers using political-economy models are also disseminating their work to a wider audience.[2] As a result geography is now surely held in greater respect in the social sciences.

Finally, there is a judgement to be made about the practical import of the political-economy models in terms of active intervention. Here, the record is mixed. But the responsibility is greater because the political economy approach does, after all, encapsulate an avowedly critical approach to society (Johnson 1986). Yet, in a world where millions of people are dying in famines or war, where more millions live in acute poverty and fear, and where there is an ecological crisis of grave proportions, it is surely important to hold on to that emancipatory vision. Here, at the cutting edge of capitalism, much new thinking and ideological face work remains to be done.

Organization of the book

The book is split into two volumes, each consisting of four parts. Both volumes have a common introduction and first introductory part. Subsequently, volume 1 consists of the second, third and fourth parts of the book while volume two consists of the fifth, sixth and seventh parts. Each part of the book except the introductory part is prefaced with an introduction written by one of the editors.

The first introductory part provides essential background to the book. It sketches the history of political-economic models in geography and their chief characteristics. In addition, the changing fortunes of the original models in geography are documented. The second part of the book is devoted to the natural environment. It is true to say that the natural environment has received less attention than its due from geographers interested in the new models, although there are signs that these omissions are now being righted. The third part is concerned with models of the geography of production. These have been at the hub of the new models in geography and they are therefore given considerable attention. The fourth part considers models of the state and politics in all their manifestations. The fifth part explores the struggle to provide political-economic models of the city in a time of considerable social and economic change. The sixth part is concerned with models of civil society, ranging from gender through race to landscape and locality. The seventh and final part of the book moves to the links currently being forged between political-economic models and social theory. Appropriately, the book ends on an open note.

Most of the chapters have two emphases. The first consists of a review of the work of the past 20 years. A good chapter tells the reader what main ideas have developed, in which order, and where they fit in terms of the changing social structures. But we are also concerned with a second emphasis: where the political-economy approach is going. Most of the review-type chapters conclude with prospects for future research and several are almost exclusively concerned with expanding the frontiers of political-economic theory. The book, then is intended as both retrospect and prospect.

As in any edited collection, there are omissions which we have not, because of pressures of time, circumstance and (most especially) space, been able to rectify. To a degree we have tried to minimize these omissions by pointing to them in the introductions to each part of the book. Nevertheless, there are still omissions of which we are particularly conscious, especially in four areas of work. The first of these is the Third World. Some of this work is documented in these volumes, but not enough. A second area omitted concerns the socialist countries. It is striking how little work in political-economic geography has been directed towards the socialist countries – a case, perhaps, of capitalism becoming an obsession. The third omission concerns historical geography. One of the most important elements of political-economic models is their sensitivity to the importance of history, so that most work of this kind includes a strong sense of change and process. Hence we have not included a specific section on historical geography. Suffice it to say that a book which went beyond the contemporary era was likely to become monumental in size. The final omission is of physical geography. Clearly, unlike the original *Models in geography*, this is not a book that includes the work of physical geographers. This is chiefly because, whether for good or ill, in the years since the publication of the original *Models in geography* human and physical geography have drifted further apart (Johnston 1983). Human geography now lies firmly in the camps of the social sciences and the humanities. There are encouraging signs of a renewal of the *entente cordiale* between human and physical geography (Peake & Jackson 1988), but as yet they hardly constitute sufficient grounds for an integrated volume.

In what follows we have tried to retain the initial sense of criticism and excitement about new approaches to new and old topics which pervaded the early 'radical geography' while also displaying the more sophisticated work of recent years, which no longer needs to criticize the conventional to establish a position. Here is what we have done, with hints at how we felt; there is where we are going. The struggle continues.

Richard Peet
1989 Nigel Thrift

Note

1 Marx was mentioned in the original *Models in geography* by Hamilton, Harries & Pahl, if only in passing. That epitaphs should never be written can be seen in the examples of the return to popularity in the late 1980s of Talcott Parsons and Althusser.

References

Chorley, R. J. & P. Haggett, (eds) 1967. *Models in geography*. London: Methuen.
Johnston, R. J. 1983. Resource analysis, resource management, and the integration of human and physical geography. *Progress in Physical Geography* **7**, 127–46.
Johnston, R. J. 1986. *On human geography*. London: Edward Arnold.
Peake, L. & P. Jackson 1988. The restless analyst: an interview with David Harvey. *Journal of Geography in Higher Education* **12**, 5–20.

Acknowledgments

Richard Peet would like to thank ETC (United Kingdom) for a grant enabling him to visit Britain in 1988 for consultation with Nigel Thrift. He also acknowledges the comradely support of several generations of students at Clark University and the tolerance of his faculty peers for views which must frequently appear extreme and doctrinaire. Finally, for her warm companionship during the editing of this book, thanks to Kathy Olsen. Nigel Thrift thanks Lynda, Victoria, and Jessica for their continuing forbearance.

Both editors wish to acknowledge Roger Jones at Unwin Hyman for his encouragement and patience during the course of an enterprise which sometimes seemed to have no end.

We are grateful to the following individuals and organizations who have kindly given permission for the reproduction of copyright material:

Figure 9.1, reprinted with permission of Peter Kennard; Figure 10.1 reproduced from *Private Eye* (issue 643); parts of Chapter 11 by David Slater, © David Slater 1989, reprinted from *Territory and state power in Latin America*, by permission of Macmillan.

Contributors

Contributors to this volume are shown in bold type.

Sophia Bowlby, Lecturer, Department of Geography, University of Reading, Reading, RG6 2AB, UK.

Lata Chaterjee, Associate Professor, Department of Geography, Boston University, Boston, Massachusetts 02215, USA.

Gordon Clark, Professor, School of Urban and Public Affairs, Carnegie-Mellon University, Pittsburgh, Pennsylvania 15213, USA.

Martin Clarke, Lecturer, School of Geography, University of Leeds, Leeds, LS2 9JT, UK.

Paul Cloke, Reader, Department of Geography, Saint Davids University College, Lampeter, Dyfed, SA48 7ED, UK.

Philip Cooke, Reader, Department of Town Planning, University College of Wales, Cardiff, CF1 3EU, UK.

Stuart Corbridge, Lecturer, Department of Geography, University of Cambridge, Cambridge, CB2 3EN, UK.

Stephen Daniels, Lecturer, Department of Geography, University of Nottingham, Nottingham, NG7 2RD, UK.

Simon Duncan, Lecturer, Department of Geography, London School of Economics, London, WC2A 2AE, UK.

Jacque Emel, Assistant Professor, Graduate School of Geography, Clark University, Worcester, Massachusetts 01610, USA.

Ruth Fincher, Department of Geography, University of Melbourne, Parkville, Victoria 3052, Australia.

Jo Foord, Principal Policy and Research Officer, Borough of Southwark, London, UK.

Derek Gregory, Professor, Department of Geography, University of British Columbia, Vancouver V6T 1W5, Canada.

Peter Jackson, Lecturer, Department of Geography, University College London, London, WC1H 0AP, UK.

R. J. Johnston, Professor, Department of Geography, University of Sheffield, Sheffield, S10 2TN, UK.

Helga Leitner, Associate Professor, Department of Geography, University of Minnesota, Minnesota 55455, USA.

Jane Lewis, Economic Development Unit, Borough of Ealing, London, UK.

John Lovering, Research Fellow, School for Advanced Urban Studies, University of Bristol, Bristol, BS8 4EA, UK.

Linda McDowell, Senior Lecturer, Faculty of Social Sciences, Open University, Milton Keynes, MK7 6AA, UK.

Suzanne Mackenzie, Assistant Professor, Department of Geography, Carleton University, Ottawa K1S 5B6, Canada.

Doreen Massey, Professor, Faculty of Social Sciences, Open University, Milton Keynes, MK7 6AA, UK.

Timothy O'Riordan, Professor, School of Environmental Sciences, University of East Anglia, Norwich, NR4 7TJ, UK.

Richard Peet, Professor, Graduate School of Geography, Clark University, Worcester, Massachusetts, 01610, USA.

Geraldine Pratt, Assistant Professor, Department of Geography, University of British Columbia, Vancouver, British Columbia V6T 1W5, Canada.

Erica Schoenberger, Assistant Professor, Department of Geography and Environmental Engineering, Johns Hopkins University, Baltimore, Maryland 21218, USA.

Eric Sheppard, Professor, Department of Geography, University of Minnesota, Minneapolis, Minnesota 55955, USA.

David Slater, CEDLA, 1016 EK Amsterdam, The Netherlands.

Neil Smith, Associate Professor, Rutgers University, New Brunswick, New Jersey 08855, USA.

Edward Soja, Professor, Graduate School of Architecture and Urban Design, University of California, Los Angeles, California 90024, USA.

Nigel Thrift, Reader, Department of Geography, University of Bristol, Bristol, BS8 1SS, UK.

John Urry, Professor, Department of Sociology, University of Lancaster, Lancaster, LA1 4YL, UK.

Alan Wilson, Professor, Department of Geography, University of Leeds, Leeds, LS2 9JT, UK.

Part I
NEW MODELS

1 *Political economy and human geography*

Richard Peet & Nigel Thrift

Introduction

Since the publication of the original *Models in geography* (Chorley & Haggett 1967) some 20 years ago, human geography has changed dramatically. It has matured theoretically, it is more directly oriented to social problems, and it has achieved an awareness of politics without sacrificing its advance as a 'science'. This transformation can be traced to the emergence, and the widespread acceptance, of a new set of models which have a common root in the notion that society is best understood as a political economy.

We use the term 'political economy' to encompass a whole range of perspectives which sometimes differ from one another and yet share common concerns and similar viewpoints. The term does not imply geography as a type of economics. Rather economy is understood in its broad sense as social economy, or way of life, founded in production. In turn, social production is viewed not as a neutral act by neutral agents but as a political act carried out by members of classes and other social groupings. Clearly, this definition is influenced by Marxism, the leading class-orientated school of critical thought. But the political-economy approach in geography is not, and never was, confined to Marxism. Marxism was largely unknown to early radical geographers. Humanists and existentialists, who had serious differences with Marxism, have definitely been members of the political-economy school. At present, there are several critical reactions to Marxism, particularly in its stucturalist form, which nevertheless remain broadly within the political-economy stream of geographic thought. So, while political economy refers to a broad spectrum of ideas, these notions have focus and order: political-economic geographers practise their discipline as part of a general, critical theory emphasizing the social production of existence.

A number of themes related to the development and present contents of this school of thought are examined in this introductory chapter. We begin by tracing, in barest outline, the history of radical or critical geography. We then consider the development of the structural Marxist conception of society in the 1970s and early 1980s which provided the chief guiding theoretical influence over this development. We follow by noting some of the critical reactions to this conception in the discipline in the mid-1980s which have strongly influenced the current direction of the political-economy approach. Finally, we conclude with a statement of the present position of political-economic geography in the late 1980s. It is important to note that the chapter makes no claim to be all inclusive, noting every byway that the political-economy approach has taken. Rather, we will examine a few of the more important *theoretical* debates that have taken place in and around the political-economy approach to human geography since it first became of consequence.

The development of a political-economy approach

The critical anti-thesis to the thesis of conventional geography developed unevenly in time and space, so unevenly, indeed, that its various phases have frequently emerged independently rather than in linked sequence. Each phase had its distinct character, its own unique reaction to the events of its time. Each phase was also a particular reaction to themes in conventional explanation of geography at the time. Here we examine three of these phases in the recent development of conventional, geographic, thought and their critical counterparts: environmental determinism and its anarchist and Marxist critics; areal differentiation and its (limited) opposition; and, in more detail, conventional quantitative–theoretical geography and the radical geography movement.

Environmental determinism and its critics

It has been argued that modern geography first emerged as a justification for the renewed Euro-American imperial expansion of the late 19th century (Hudson 1977, Harvey & Smith 1984, Peet 1985b, Stoddart 1986). The need to explain Euro-American dominance compounded with the biological discoveries of Darwin, and Spencer's ideology of social Darwinism, to produce an explanation of social conquest cast in terms of the varying natural qualities and abilities of different racial groups. In the new modern geography this took the particular form of environmental determinism: differences in humans' physical and mental abilities, and in the level of their cultural and economic potential and achievement, were attributed to regionally differing natural environments. Euro-American hegemony was the natural, even god-given, consequence of the superior physical environments of Western Europe and North America.

Social Darwinism, and its geographic component environmental determinism, were opposed by the anarchist Russian geographer Kropotkin (1902). Kropotkin agreed that interaction with nature created human qualities, but differed on what these might be. As opposed to the social Darwinists' theory of inherent competitiveness and aggression as behaviours suggesting capitalism and imperialism as the natural modes of human life, he argued for co-operativeness and sociability as the natural bases for an anarchist form of communism. Only in the 1920s did Wittfogel (1985), a Marxist with geographical interests and training, criticize the environmental thesis from a position opposed to the direct natural causation of inherent human characteristics. For Wittfogel, human labour, organized in different social forms, moulded nature into the different material (economic) bases of regional societies. These in turn were the productive bases of different human personalities and cultures; that is, humans made themselves, rather than were made by nature. Yet Wittfogel remained within the environmental tradition by concluding that nature differentially directed the development of regional labour processes. Specifically, he argued that the climatically determined need for irrigation in the East (India, China) yielded a line of social development greatly different from that followed by rainfall-fed agriculture in the West (Wittfogel 1957). Hence, entirely different kinds of civilization developed in East and West.

Kropotkin and Wittfogel both achieved political and intellectual notoriety outside geography, but they were peripheral to the main lines of development of the discipline. Conventional geography tended to stand firm in support of the current social order. This was certainly one of the reasons for its widespread adoption in schools and universities.

Areal differentiation and its opponents

The 30 years between the late 1920s and the late 1950s must be characterized as the period of conventional geography's retreat from its position as a science of the origins of human nature, in the light of internal and external critiques of environmental determinism. Possibilism, a leading school of thought of the time, was so vague a formulation of environmental causation as to preclude systematic, theoretical, or even causal generalizations. In the United States, geography turned into areal differentiation (Hartshorne 1939, 1959): the description of the unique features of the regions of the Earth's surface. Critical reactions to this extremely conservative position, which began to surface in the 1940s and 1950s, were muted by the rampant anti-communism of the Cold War. Some regional geographies carried isolated, critical statements. The Lattimores' (1944) regional history of China, for example, says of late 19th-century United States foreign-policy makers that they 'did not propose a cessation of imperialist demands on China; they merely registered a claim of "me too"'. (A few years later Lattimore (1950, p. vii, Harvey 1983, Newman 1983) found himself labelled 'the top Russian espionage agent in this country' by US Senator McCarthy.) Hartshorne's conception of geography as a unique integrating science which, however, precluded generalization in the form of universal laws, also began to be opposed on theoretical grounds. Schaefer (1953) mildly proposed instead that geography explains particular phenomena as instances of general laws. In reply Hartshorne, philosopher-general of geography at the time, had merely to label Schaefer's criticisms 'false representation' to dismiss them. Hartshorne commented on a brief (and critical) mention of Marx in Schaefer's article:

> Whether the analysis of Karl Marx is sound, few readers of the *Annals* would be competent to judge. They should be competent to judge the appropriateness of including the analogy [between Marx and the geographer Hettner] in a geographic journal (Hartshorne 1955, p. 233).

After such broadsides, criticism was limited to less directly political arenas in the purely quantitative 'revolution' (Burton 1963) of the late 1950s and early 1960s.

Quantitive theoretical geography and the radical geography movement

We must leap into the late 1960s to find a widespread critical *and* political geography continuously responding to social crises and conventional geography's analysis of them. **Radical geography** originated as a critical reaction to two crises of capitalism at that time: the armed struggle in the Third World periphery, specifically United States involvement in the Vietnamese War, and the eruption of urban social movements in many cities, specifically the civil rights movement in the United States and the ghetto unrest of the middle and late 1960s in the United States, Great Britain, and elsewhere. Conventional geography's response to these momentous events lacked conviction, in more ways than one.

However, in the late 1960s some geographers already active in broader sociopolitical movements began to turn their attention inwards, towards their own discipline. The Detroit Geographical Expedition, led by William Bunge (Horvath 1971), used its conventional geographical skills on behalf of the black residents of the city's ghettos. At Clark University, in Worcester, Massachusetts, the radical journal *Antipode* began publication in 1969, carrying articles on socially relevant geographic topics (Peet 1977a).

But, it soon became apparent that conventional geographic theories and methodologies were inappropriate for a more relevant geography. The search for an alternative theoretical approach is exemplified by the intellectual biography of radical geography's leading theorist. David Harvey (1969) had previously written a conventional treatise on geographical methodology, but in the early 1970s began exploring ideas in social and moral philosophy – topics neglected in his earlier work. The journey took him through a series of liberal formulations, based on social justice as a matter of eternal morality, to Marxism with its analysis of the injustices built into specific societies; and from an interest in material reality, merely as the place to test academic propositions, to the transformation of capitalist society through revolutionary theory (Harvey 1973, pp. 9–19, 286–314). Harvey's journey was made by many other young radical geographers in the 1970s. For a few years in the early part of the decade radical geography explored, still from a liberal-geographical perspective, the many social injusticies of advanced capitalism (Peet 1977a). But increasingly, as the 1970s wore on, and environmental crises and economic recession were added to political problems of the 1960s critical liberal formulations were found lacking and radical geographers increasingly turned to the analysis of Marx.

The mid-1970s saw a flowering of radical culture in geography celebrated by the publication of *Radical geography* (Peet 1977b). Here radical geographers critically examined almost every geographic aspect of life in modern capitalism: the geography of women, the ghetto, the mentally ill, housing, rural areas, school busing, planning, migrant labour, and so on. The period was notable for a series of increasingly sophisticated critiques of conventional geography by Anderson (1973), Slater (1973, 1975, 1977), and Massey (1973). A series of exegetical writings (e.g. Harvey 1975) explored areas of Marx's writing most applicable to geographical issues. The growing interest in Marxism was broadened to include a comprehension of social anarchism (Breibart 1975, Galois 1976). The geographical expeditionary movement, which had spread to the Canadian cities of Toronto and Vancouver and over the Atlantic to London, was joined in 1974 by the Union of Socialist Geographers, which organized leftist faculty and students in the discipline. In the late 1970s *Antipode* published issues on the environment and anarchism which, in retrospect, were the last bursts of colour in the fall of its 1960s-style radicalism (Peet 1985a).

The radical geography movement changed again in the 1980s. In general, it became more sober and less combative for at least four reasons. First, the mainstream of Marxist thought was subjected to a number of more or less powerful critiques. Second, the disciplining effect of the 1979–83 economic recession and a greater knowledge of existing socialist countries made revolutionary politics a less certain quantity. Third, the laid-back academic style of the 1970s was replaced by the narrower professionalism of the 1980s. Finally, some of the Young Turks who had battled against the human geography establishment now found themselves part of it.

Yet, such a momentum had been built up in the 1970s that Marxist and related scholarship continued to flourish in geography. For example, major works were published by Harvey (1982, 1987a & b) and Massey (1984). In some areas of research, such as industrial geography, views influenced by Marxism had become engrained (e.g. Massey 1984, Massey & Meegan 1986, Peet 1987, Scott & Storper 1986, Storper & Walker 1988), and even in the last bastion of the traditionalist approach, cultural geography, Marxism and other interpretations of political economy were accepted as at least one valid viewpoint (e.g. Cosgrove 1985, Cosgrove & Jackson 1987). New journals such as *Society and Space*, founded in 1983, were still springing up, and important collections, such as *Social relations and spatial structures* (Gregory & Urry 1985) have continued to appear.

Thus, the political-economy approach to human geography now stretches through more than two decades. It has survived counterattack, critique, and economic and professional hard times, and has matured into a leading and, for many, *the* leading school of contemporary geographic thought.

The history of the approach can be roughly split into phases. The first phase, the 1970s and early 1980s, covers a period when structural Marxism was particularly influential. The second period, beginning in the late 1970s but peaking in the mid-1980s, sees a greater diversity of concerns, especially the relative potency of social structure and human agency, realism, and the study of localities. Finally, the latest period, the late 1980s, finds such issues as postmodernism and its critiques coming to the fore.

The 1970s and early 1980s: structural Marxism

The most dramatic event in the intellectual Odyssey of the political-economy approach was the turn to Marxism in a discipline in which, as Hartshorne's earlier remarks suggest, the very mention of Marx had certainly been unusual, and sometimes even anathema. Not only did geographers now read Marx, they were influenced by a particularly powerful version of Marxism, the structural ideas of Louis Althusser and his followers. To appreciate Althusser's version of Marxism, however, we must first briefly outline some of the basic theses of Marx and Engels themselves.

Marx on social and natural relations

Marxism is simultaneously politics and science. The political purpose of Marxism is social transformation on behalf of the oppressed people of the world. Communism proposes that power be placed in the hands of the workers and peasant masses in the belief that economic and political democratization will produce a higher order of society and a new kind of human being (Peet 1978–9). This proposal does not stem from utopian optimism alone. It results from a whole way of knowing the world, the science of existence called **dialectical materialism**.

Dialectics is a way of theoretically capturing interaction and change, history as the struggle between opposites, with a conception of long-term dynamics in the form of non-teleological historical laws. **Materialism** proposes that matter precedes mind, consciousness results from experience, and experience occurs primarily in the material reproduction of life. Combining the two, dialectical materialism analyzes societies in terms of modes of production, the struggle within them of the forces and relations of production, and the succession of modes of production through time towards the eventual achievement of a society characterized by high levels of development, socialized ownership of the means of production, economic democracy, and freedom of consciousness within a system of social responsibility. In the following paragraphs we emphasize the geographical aspect to the Marxian idea that social production is fundamental to human existence. By 'geographical' we mean an emphasis on the social transformation of nature followed by non-geographer Marxists (Schmidt 1971, Timpanaro 1975) as well as Marxist geographers (Burgess 1978, Smith & O'Keefe 1980, Smith 1984). We shall follow Marx directly rather than interpreting his interpreters.

Marx's view of the human relation to nature was fundamentally different from that of the classical economists. Smith and Ricardo began their analysis of production and exchange with the individual already formed by nature. Marx begins with production by individuals who form their personalities as they transform nature through the labour process. For Marx, the fact that all people are involved in broadly similar natural and social relationships makes possible a discussion of human nature in general side by side with a set of abstract (transhistorical) analytical categories (Horvath & Gibson 1984). Thus Marx always regards nature as the 'inorganic body' of the human individual, the source of the means of continued existence and locational context in which life unfolds. Matter is always exchanged between the inorganic and the organic bodies, described by Marx (1976 p. 209) as the 'universal condition for the metabolic interaction between man and nature, the everlasting nature-imposed condition of human existence . . . common to all forms of society in which human beings live'. During this necessary interaction, humans develop themselves as particular kinds of social beings. The distinguishing feature of this history is an increasingly conscious direction of labour and natural relations by human subjects.

However, Marx spends little time at the transhistorical level of analysis – elaborating the production of life in general – preferring a more concrete, historical understanding. Implicit in the above relation with nature is a second relation essential to life: the social relation among people, especially co-operation in the labour process: 'All production is appropriation of nature on the part of an individual within and through a specific form of society' (Marx 1973 p. 87). Here the essential analytical category is the *property relation*. At the dawn of human history, nature was communally owned; over time, parts of it became the private property of certain individuals; communism envisages a return to the social ownership of nature at a high level of economic development.

In this interpretation, **mode of production**, the central category of Marxian analysis, appears 'both as a relation between the individuals, and as their specific active relation to inorganic nature' (Marx 1973, p. 495; see also Godelier 1978). These relations form the economic structure of society, the foundation on which arises 'a legal and political superstructure, and to which correspond definite forms of social consciousness' (Marx 1970, p. 20). The type and level of social and natural relations correspond with a 'specific stage in the development of the productive forces of working subjects' (Marx 1973, p. 495).

Development of the productive forces fundamentally changes the economic structure and, through it, the entire society. However, this productive forces–social-relations framework should be understood as a very general conception for long-term historical analysis. The productive forces make up a structure of limitations and probabilities within which class struggles, resulting from opposition to the prevailing relations of production, actively bring about social change. Marx provides at least two accounts of the historical succession of modes of production: a 'broad outline' in which 'the Asiatic, ancient, feudal and modern bourgeois modes of production may be designated as epochs marking progress in the economic development of society' (Marx 1970, p. 21) and a more complete version (Marx 1973, p. 471–514) in which universal primitive communism decomposes into classical antiquity (based on slavery) and then feudalism and capitalism in the West and an Asiatic mode in the East. In this second version, we see the potential for an historical and geographical theory. History may be interpreted as the development and interaction of regional social formations characterized by different modes of production, each mode being further characterized by dominant social relations, including the social relation to nature.

Structural Marxism

The particular version of Marxism that was dominant in the West in the 1960s and for much of the 1970s grew in the fertile intellectual and political soil of France in the postwar years. The orthodox Marxism of the French Communist Party took the Stalinist position that all human natural history could be replicated in the scientific laws of dialectical materialism. A critical reaction to this notion in the late 1950s led many West European intellectuals towards a new synthetic form of Marxism, drawing on diverse systems of non- and neo-Marxist thought. One source was the existential and phenomenological ideas developed by Merleau-Ponty and Sartre in postwar France, particularly their critique of Stalin's insistence on the unity of the natural and human worlds. This unity, they claimed denies the specificity of the human being – her social and creative potential, his subjectivity in the historical process – and thus destroys Marxism as a theory of revolutionary self-emancipation. In opposition to Stalin's iron laws of history, Merleau-Ponty and Sartre proposed a subject-centred history with lived experience as the source of consciousness.

Stalinism, however, was not the only theoretical tradition that saw subjectivity as constitut*ed* rather than constitut*ive*. The various functionalist and structuralist streams of thought emerging from 19th century biology and sociology saw the human being made by her social milieu. In the late 1950s and early 1960s, intellectual attention (particularly in France) shifted from existential phenomenology towards structuralist ideas developed in linguistics, anthropology, and psychology. In the structural linguistics of Saussure, a coercive sign system bestows meaning on the speech of the subject. In Levi-Strauss's anthropology, the meaning of history is imparted not by historical actors but by the totality of rule systems within which actors are located. And in Lacan's psychology, the phases by which Freud's human individual achieves identity are reinterpreted as stages in the subjection of the personality to the authority of culture (Benton 1984, Callinicos 1976, 1985, Elliott 1987).

The French philosopher Althusser (1969), responding critically to Sartre and more positively to structuralism, reworked Marx's theoretical schema and analytical categories. For Althusser, as for Stalin, Marxism was indeed science. But in contradiction to Stalin's *direct* economic and technical determination, determination by the economy was, for Althusser, a thesis of the *indirect* causal relations between elements of society – relations, however, which he theorized in abstraction from actual history. In Althusser's formulation, 'non-economic' elements, such as consciousness and politics, were relatively autonomous in an *overdetermined* social structure (i.e. one in which there are diverse elements interacting one with another). For Althusser, society was a complex 'structure in dominance', yet human beings were bearers, rather than makers, of social relations.

The details of this structuralist position were elaborated by Althusser's collaborator Etienne Balibar (Althusser & Balibar 1970). Balibar argued that mode of production, the central category of Marxism, had two distinct roles: as a principle for identifying periods of history and as a means of conceptualizing the relationship between the economic, political, and ideological 'levels' of societies. In the second, synchronic role, mode of production assigned each social element its place in a hierarchy of dominance and subordination. Economic class relations (between owners and workers) always determine the structure of society in the last instance. But determination takes an indirect form; the economic level assigns to the non-economic levels their place in a hierarchy of dominance and the kinds of connection or articulation between them. Historical materialism so conceived became a theory of connections or articulations between, and the dynamic of, the main social elements. As such, structural Marxism claimed a status as a true, theoretical science (Benton 1984 p. 115).

Structural Marxism in geography

Marxist theorists influenced by Althusser subsequently applied this version of science to a range of problems, many of particular interest to geography, such as the structures of precapitalist societies (Meillasoux 1981, Terray 1972, Hindess & Hirst 1975) the historical transition and articulation of modes of production (Rey in Wolpe 1980), the state (Poulantzas 1975, 1978), and critical analysis of culture, ideology and consciousness. This work, with its potential to yield a theory of regional social structures, thought of as particular interconnected modes of production, went only a limited way before being replaced by more diverse 'post-structuralisms' as the leading frontier of Marxist and neo-Marxist geography. Structuralist geographers became bogged down defining details of space and nature, rather than applying the broad conceptions of mode of production and social formation to geohistorical development. We cannot, therefore, report on a sophisticated structuralist geography with a rich history of conceptualization and application. We shall, however, follow one line of theoretical development that has received sustained attention: the connection between society and space, or social structure and spatial structure.

Society and space

The most direct importation of structuralist ideas into geography was undoubtedly Manuel Castells's *The urban question* (1977; originally published in French in 1972). Castells saw the city as the projection of society on space: people in relation one with another give space 'a form, a function, a social signification' (Castells 1977, p. 115). The theory of space, he insists, is an integral part of a general social theory. For this theory, Castells turns to Althusser's conception of modes of production and their constituent elements, for instance:

> To analyse space as an expression of the social structure amounts, therefore, to studying its shaping by elements of the economic system, the political system and the ideological system, and by their combinations and the social practices that derive from them (Castells 1977, p. 126).

Under capitalism, the economic system is dominant and is the basic organizer of space. (By economic Castells means economic activities directly producing goods located at certain places, activities that reproduce society as a whole such as housing and public services, and exchange activities such as communications and commerce.) The political system organizes space through its two essential functions of domination/regulation and integration/repression. The ideological system marks space with a network of signs, with ideological content. Over and above this the social organization of space is determined by each of the three instances: by structural combinations of the three, by the persistence of spatial forms created in the past, and by the specific actions of individual members of social and spatial groups. As an Althusserian, Castells believed that the analysis of space first required abstract theorization of the mode of production and then concrete analysis of the specific ways structural laws are realized in spatial practice (Castells 1977, Ch. 8).

But when it came to empirical research, Castells's own formulation of the urban question did not turn on production directly (i.e. the city as a locus of production and class struggle between workers and owners in the factory). Rather he turned to consumption or the 'reproduction of labor power', and the increasing intervention of the state in such areas of *collective consumption* as housing and social services. Through state intervention, Castells argued, collective consumption is made the political arena for the struggles of the urban social movements its deficiencies produce. The reasons for Castells's diversion from production to consumption were not made clear, for he shared with Althusser an imprecise mode of expression and an argumentative style. Even so, his ideas were extremely influential in work on urban services, politics, and social movements in the 1970s, particularly in France, but also in Britain and the United States (Castells 1977, pp. 465–71).

Writing in France in the 1960s and 1970s, Castells was immediately exposed to Althusserian thought; indeed, we can call *The urban question* a direct, if idiosyncratic, application of structural Marxism to urban space. But in the Anglo-American world, structural Marxism has always been more eclectic, especially in geography. David Harvey's *Social justice and the city*, written at about the same time as Castells's book, used a concept termed 'operational structuralism', drawn from Piaget (1970), Ollman (1971), and Marx directly, rather than Althusser. It emphasized the relations between the constituent elements of governing structural change. Elements such as social classes, frequently in contradiction, force changes in society. For Harvey (1973, p. 290), structure is defined as a 'system of internal relations which is being structured through the operation of its own transformation rules'. Contradictions occur between structures as well as within them. For example, Harvey says that the political and ideological structures have their own contradictions and separate revolutions, as well as being in contradiction with the economic base of society. But some structures are regarded as more basic than others. Thus Harvey follows the Marxist view that the reproduction of material existence forms the starting point for tracing the relations within society. So for Harvey (1973, p. 302): 'Any attempt to create an interdisciplinary theory with respect to a phenomenon such as urbanism has perforce to resort to the operational structuralist method which Marx practices and which Ollman and Piaget describe.'

Harvey's (1973, p. 205) approach to the city was quite general: 'some sort of relationship exists between the form and functioning of urbanism . . . and the dominant mode of production.' Cities are economic and social forms capable of extracting significant quantities of the social surplus created by people. For Harvey, the central connections lay between the mode of economic integration in history, (whether reciprocity, redistribution, or market exchange), the subsequent creation of social surplus, and the form of urbanism. The transformation from reciprocity to redistribution precipitated a separated urban structure with, however, limited powers of inner transformation. Born of a contradiction between the forces and relations of production, the city functioned as a political, ideological, and military force to sustain a particular set of relations of production – especially property rights. The movement from this early political city to a commercial city based on market exchange was interpreted, following Lefebvre (1972), as an inner transformation of urbanism itself. The industrial city resulted from a reorganization of the industrial forces of production. But urbanism is not simply created by the forces and relations of production; it both expresses and fashions social relations and the organization of production.

Even so, urbanism is channelled and constrained by forces emanating from the economic base and, to be understood, it has ultimately to be related to the reproduction of material existence. Thus, industrial society dominates urbanism, by creating urban space through the deployment of fixed capital investments, by disposing of products through the process of urbanization, and by appropriating and circulating surplus value through the device of the city. Cities, for Harvey, are founded on the exploitation of the many by the few. Therefore:

> It remains for revolutionary theory to chart the path from an urbanism based in exploitation to an urbanism appropriate for the human species. And it

> remains for revolutionary practice to accomplish such a transformation (Harvey 1973, p. 314).

Such was the optimistic tenor of the times! Such is the conclusion inherent in structural Marxism: changing any part of society, such as the city, involves changing the relations of production that guide human development.

Other works written in the 1970s saw a more direct 'one-way' relation between mode of production and the social organization of space. For example, Buch-Hansen & Nielsen (1977) coined the term 'territorial structure' to refer to the totality of production and consumption localities, their external conditions, and the infrastructures linking the whole together. Again, the most crucial determinant of territorial structure was the mode of production, both the forces of production that directly form the material contents of space and the relations of production that condition development of the productive forces in space. The social and political superstructure, which has some independence from the productive base, also makes and transforms territorial structure. In turn, however, territorial structure also conditions the further development of production.

Harvey's major work, *The limits to capital* (1982), declared its intention of steering a middle course between spatial organization as a mere reflection of capitalism and a spatial fetishism which treats the geometric properties of space as fundamental. But in reality, Harvey emphasized the first tendency, extending the analytical categories of Marx's *Capital* (use and exchange value, competition, etc.) to the explication of spatial organization through an extended theory of crisis. Although he no longer employed the explicitly structuralist language of parts of *Social justice and the city*, Harvey nevertheless integrated social and spatial structure in a 'landscape that has been indelibly and irreversibly carved out according to the dictates of capitalism'. This position was carried to its extreme by Harvey's student, Smith (1984), who termed the connections between society and environment 'the production of space' and 'the production of nature'. For Smith, the transformation of environment by capitalism implies an end to the conceptual division between the natural and the social:

> In its constant drive to accumulate larger and larger quantities of social wealth under its control, capital transforms the shape of the entire world. No god-given stone is left unturned, no living thing is unaffected. To this extent the problems of nature, of space and of uneven development are tied together by capital itself. Uneven development is the concrete process and pattern of the production of nature under capitalism . . . . There can be no apology for the anthropomorphism of this perspective: with the development of capitalism, human society has put itself at the centre of nature, and we shall be able to deal with the problems this has created only if we first recognize the reality (Smith 1984, p. xiv).

With this recognition of uneven development as a structural imperative of capitalism, as a necessary outcome of the unfolding of capital's inherent laws, the structural approach to space reached its ultimate conclusion.

It should be clear from the above discussion that while there were obvious parallels between Castells and Harvey, Marxist geography in the English-speaking world was influenced by Althusser only indirectly; a much broader and more fluid conception of 'structure' and of 'structuralism' was employed. By the early 1980s, this conception was broadening ever further. One index of this change is the work of Doreen Massey. Massey's key work, *Spatial divisions of labour* (1984) expresses the transition she had made from using geographical space as a passive surface, expressing the mode of production, to a conception of space as an active force. As she put it 'geography matters' too.

> The fact that processes take place over space, the facts of distance, of closeness, of geographical variation between areas, of the individual character and meaning of specific places and regions – all these are essential to the operation of social processes themselves. Just as there are no purely spatial processes, neither are there any non-spatial processes. Nothing much happens, bar angels dancing, on the head of a pin . . . . Geography in both its senses, of distance/nearness/betweenness and of the physical variation of the earth's surface (the two being closely related) is not a constraint on a pre-existing non-geographical social and economic world. It is constitutive of that world (Massey 1984, p. 52).

As a Marxist, Massey emphasized variation in the social relations of production over space. Thus as social classes are constituted in places, so class character varies geographically. But Massey went on to argue that the social structure of the economy necessarily develops in a variety of local forms which she termed 'spatial structures of production'. The archetype she developed was the hierarchy of functions of the multi-locational company, different stages in production (organization, research, assembly, parts-making) being assigned in different combinations to various regions, although other ways of conceptualizing spatial structures are possible. Massey maintained that spatial structures not only emerge from the dictates of corporate initiative but are also established and changed through political and economic battles on the part of social groups (i.e. through class struggle). In turn, spatial structures, through differential employment possibilities, create, maintain, or alter class and gender inequalities over space. Her main point was that 'spatiality is an integral and active condition' of the production process (Massey 1984, p. 68).

The mid-1980s: the structure–agency debate, realism and locality

Before the 1980s began, Althusser's influence had already sparked off a furious debate throughout the social sciences about the relative contributions of economic structure and human agency to the making of history. The 'structure–agency' debate, as it has come to be known, is documented in a whole series of responses and counter-responses to E. P. Thompson's original critique of Althusser, *The poverty of theory* (1978), which ranged widely across all social science, taking in life, the universe, and everything along the way (see, for example, Anderson 1980, 1983).

Human geography's version of this debate was prefigured in the work of

Gregory (1978), but did not fully take off until the exchanges that followed Duncan & Ley's critique of structural Marxism in 1982 (see Duncan & Ley 1982, Chouinard & Fincher 1983). Like the debate in the social sciences as a whole, human geography's version of the structure–agency debate was wide ranging, but, in particular, it intertwined three themes; the relative importance of structure and agency, and how they might be reconciled in a single approach; the efficacy of a realist methodology; and the importance of localities. However, in essence, the structure–agency debate in human geography was multifaceted because of the several rather different impulses that fuelled it. What were these impulses? Five of them seem particularly relevant.

Human geography

First of all, the debate came about because of the peculiar circumstances of human geography. Almost alone amongst other human sciences, in the 1950s human geography still had a poorly developed base in social theory. One only has to compare such work as C. Wright Mills's (1959) *The sociological imagination* with the extant human geography books of the period to grasp the differences in range and depth. Thus, as Urry points out in this volume, when Marxism began to have an influence on the subject in the 1960s and into the 1970s it was successful in part because there was so little in the way of social theory in human geography with which it had to contend. The discourses of Marxism and social theory were almost synonymous.

Therefore the structure–agency debate was in many ways a parade of traditions of social theory well known in the social sciences but which hitherto had received scant attention in human geography, traditions such as phenomenology, symbolic interactionism, even hermeneutics (Gregory 1978). Marxism provided both the space for these different traditions to be introduced into the subject and, at the same time, a suitable theoretical orthodoxy, in its structuralist form, against which to battle (see Billinge 1978, Ley & Samuels 1978, Duncan & Ley 1982).

Of course, the Marxist tradition in human geography did not remain unchanged in the face of the assaults that were mounted upon it. The different traditions encouraged a more eclectic approach to political economy. There were parallels here with what had been happening in the Marxist tradition anyway, especially in its European incarnation. There was Gramsci, Sartre, and the Frankfurt School to discuss, and later Habermas's and Giddens's reconstructions of historical materialism. Each of these authors had drawn on traditions outside Marxism to strengthen their analysis (see Held 1980, Habermas 1979, Giddens 1981).

Marxism and change in society

A second impulse came from the changing course of history itself. In the 20th century, Marxism has strained to account for far-reaching changes in the nature of society. The list is almost infinite. To start with, there are the continuing twists and turns of capital itself, including new regimes of capital accumulation based first upon mass consumption and latterly upon segmentation of the mass market and the flexible accumulation that goes with it, the growth of service industries, and the spread of a new international division of labour. Far-reaching social changes have also occurred and especially the rise of a service class of managers and professionals and the greater participation of women in the labour force. That is not all. Then there has been the rise of a large and comprehensive state apparatus with extensive disciplinary, welfare, and socialization functions. There has been the growth of socialist societies (Forbes & Thrift 1987, Thrift & Forbes 1986) with social dynamics which are often only marginally based upon capital accumulation and owe much more to the growth of bureaucracies. All round the world there are developing countries that have generated important social forces opposed to capitalism; the growth of a radical Islam is a case in point. The list of changes goes on and on.

The Marxist tradition has not found it easy to accommodate all of these changes. In order to survive it has been necessary to revise extant theory (Harvey 1982) and to broaden it to include the 'non-economic' factors of state and civil society.

The political impulse

A third impulse was political. By the end of the 1970s, the forces of the new right were asserting themselves in many countries. Certain of the demands of the new right, especially a radical individualism, clearly appealed to large sections of the population. The reaction of many on the left was to study such developments in detail with a view to formulating effective counter-strategies. But this was not all. New forms of politics started to come into existence in the 1970s and into the 1980s which were not based on the old axes of support such as class, but cut across them. The ecological movement, or what O'Riordan (1981), rather more generally calls 'environmentalism', has taken off worldwide; it directly challenges a number of the 19th-century ideas of industrial progress which have cast their shadow over the 20th century with often disastrous consequences for the environment. Similarly, the feminist movement has laid down a challenge to old ways of thinking , laying bare the subordination of women of all classes in patriarchal structures. None of these developments could be ignored for they were constitutive of new ways of thinking which were both political and personal, that is they involved not just planning out programmes to change social structures but also a deep commitment to changing 'oneself'.

The realist approach and attitudes to theory

A fourth impulse was theoretical. By the end of the 1970s some human geographers were beginning to catch the scepticism about the power of theory that had already infected the other social sciences, especially through the work of more extreme writers such as Foucault and Rorty. This is not to say that theory was to be consigned to the rubbish bin. Rather it was that there were limits to the applications of theory and its ability to illuminate geographic practice (Thrift 1979). Depending upon where these limits were placed, it was possible to argue for a thoroughgoing relativism or a thoroughgoing rationalism. But most commentators seemed happiest with the compromise formulations of realism.

In its present incarnation, realism has been associated primarily with the names of Bhaskar (1975, 1979, 1986) and Harré (1987). Its main routes of entry into human geography were through the work of Keat & Urry (1981) and Sayer (1984). Realism is a philosophy of science based on the use of abstraction as a means of identifying the causal powers of particular social structures, powers which are released only under specific conditions. In many ways, realism consists of a state-of-the art philosophy of science allowing for structural explanation but incorporating the scepticism about the powers of theory characteristic of late 20th century philosophy (Baynes *et al.* 1987). Certainly, realism has become the major approach to science in human geography, with special attractions for those pursuing the political-economy approach. But, as Sayer makes clear, it is not automatically radical:

> The changes are clearest in radical and Marxist research in geography . . . . However, this association of realism is not a necessary one: some radical work has been done using a nomothetic deductive method . . . and acceptance of realist philosphy does not entail acceptance of a radical theory of society – the latter must be justified by other means (Sayer 1985, p. 161).

Perhaps realism's greatest impact has been in promoting the thoughtful conduct of empirical research. The realist approach makes for a level-headed appraisal of what is possible. Thus Sayer points to what a viable 'regional geography' might ordinarily consist of:

> The best we can normally manage is an incomplete picture consisting of a combination of descriptive generalisations at an aggregate level (e.g. on changes in population and standards of living), some abstract theory concerning the nature of basic structures and mechanisms (e.g. concerning modes of production) and a handful of case studies involving intensive research showing how in a few, probably not very representative, cases these structures and mechanisms combine to produce concrete events. (Sayer 1985, p. 172).

This empirical connection was important. By the mid-1970s it had become crucial for those sponsoring a political-economy approach to demonstrate that they could do good empirical or concrete research, both in order to find out more about what was happening 'on the ground' and to demonstrate their skills in this area to colleagues sceptical about the work of abstract theorists. This meant that more careful attention had to be paid to how abstract theory could be applied in particular contingent situations. This, in turn, lead to more careful formulations of theory by realists aimed at eliciting the causal powers of particular social relations in a whole range of contingent situations. Thus, more attention was paid to such matters as the distinction between 'internal' and 'external' relations and the relative merits of 'intensive' and 'extensive' research (Sayer 1984). However, the problems of the relations between theory and empirical work have hardly been solved. For example, it is all very well to talk of levels of abstraction forming a neat hierarchy all the way from abstract theoretical proposals to empirical complexities, but it is difficult to find a mechanism which unambiguously allows the researcher to detect which theoretical objects occupy which levels (see Urry 1985).

Such problems begin to explain why rather more room was also given to the **hermeneutic** dimension (to the theory of the interpretation and clarification of meaning by those promoting political-economy perspectives. The hermeneutic tradition finally lapped upon the shores of human geography. Hermeneutics had been introduced into geography in the 1970s in a number of guises, including phenomenology and existentialism (see Pickles 1986). Its most important function was to underline the necessity for taking the act of interpretation seriously, whether as an awareness of theory as representative of a set of interpretive acts or as a set of procedures for explaining the interpretations which people give to their world (Jackson & Smith 1984). The results of its adoption by those involved in political-economy perspectives have been twofold. First, the contextual dimension of theory – its rootedness in particular times and places – is taken more seriously. Theories are themselves historical and geographical entities. Second, there is much greater awareness of the validity of a qualitative geography, made up of methods for getting at peoples' interpretations of their worlds in as rational and ordered a way as possible (Eyles & Smith 1988). For example, ethnography has gained a new respectability (Geertz 1973).

The importance of space as a constituent of the social

The last impulse was in many ways the most significant. It was a growing realization that space was rather more important in the scheme of societies than was envisaged at the farthest swing of the structural Marxist pendulum. Space is not just a reflection of the social but *a constitutive element of what the social is*; Massey (1984, p. 4) summarized these concerns well in *Spatial divisions of labour* where she pointed out that: 'It is not just that the spatial is socially constructed: the social is spatially constructed too.' She went on:

> The full meaning of the term 'spatial' includes a whole range of aspects of the social world. It includes distance, and differences in the measurement, connotations and appreciation of distance. It includes movement. It includes geographical differentiation, the notion of place and specificity and of differences between places. And it includes the symbolism and meaning which in different societies, and in different parts of given societies, attach to all of these things (Massey 1984, p. 5).

The renewed emphasis on the importance of space connects back to the realist project sketched above. In this project, space clearly makes a difference to whether the causal powers of particular social relations are activated, and the forms which these social relations can take. In other words, important social relations are necessary. For example, for the wage–labour relation to exist it is necessary for both capitalists and labourers to exist. But the existence and expression of such social relations in particular places relies upon the web of contingencies that is woven by the spatial fabric of society. The picture is immeasurably complicated by the fact that social relations, in their diverse locally contingent forms, continually constitute that spatial fabric. In Pred's (1985) terms, the social becomes the spatial, the spatial becomes the social'.

These five different impulses came together in the so-called 'structure–agency' debate. There were two main problems with this debate as it took place in geography. First of all, as is now hopefully apparent, its participants had quite different impulses motivating their participation in it. The opportunities for confusion were, therefore, legion. Second, especially in its initial stages, it was easier to point to what was unsatisfactory about extant theory and research than to cite examples of theory and research that met the standards being prescribed; critique piled upon critique in a wasteful duplication of effort. Five years later these two problems have become much less prominent; the different impulses have been negotiated and have even merged to produce new lines of thought, while a body of theory and research has been built up which can act as a template for further endeavours.

Responses

What, then, were the main foci of the structure–agency debate? Three foci were particularly important. First, there was a general concern with the individual. Parts of Marx (see Geras 1983) and Marxist writers such as Sartre show sensitivity to the question of the individual but, in practice, many commentators concede that the balance of Marxist theory has tipped away from the individual towards social structure. Of course, simply stating the need for more concern for the individual would, by itself, be an act of empty rhetoric. There was a need actively to expand this concern.

The chief efforts came from a group of human geographers who were interested in Hägerstrand's time-geography (Thrift & Pred 1981, Pred 1981). This interest rapidly transmuted into a concern with Giddens's structuration theory, the development of which has been partly influenced by human geography (see Giddens 1979, 1981, 1984, 1985, Thrift 1983). Structuration theory had a major influence on the political-economy approach in human geography for three main reasons. First of all, in Giddens's earlier works, it offered geographers a way out of the problem of structure and agency, precisely by concentrating on the importance of geography. Giddens's (1979, 1981, 1984) conceptions of locale and time–space distanciation (the stretching of societies over space) were meant to show how social structures were 'instantiated' in a particular geography, so that at any time social structure did not have to exist everywhere in order to have influence. Second, structuration theory emphasized the importance of hermeneutics at all scale levels, from the areas of day-to-day communicative interaction between individuals to the structures of communication (signification), power (domination), and sanction (legitimation) underpinning society as a whole. Third, especially in Giddens's later work, structuration theory offered a coherent and sympathetic critique of historical materialism, based in part on its lack of spatiality and in part on its lack of attention to matters of signification and legitimation.

Structuration theory has been attacked by some geographers for its schematic form, and for its glossing over of some major problems (Gregson 1986, 1987). But it remains one of the few examples of an advance made in social theory with explicit connections to human geography and with, in contrast to comparable schemes such as that of Habermas, an appreciation of geography as socially constitutive as well as socially constituted (Pred 1987).

A second general focus of research on structure and agency was on the reproduction of social structure. Quite clearly, with the impetus provided by structuration theory, the debate on structure and agency could not stay at the level of the individual and individual agency. It had to move towards analysis of social structure, and especially of how institutions come into being which are aimed at enforcing a particular order and a particular vision of social reality. There is no doubt that the reproduction of capitalism involves a good deal of crude coercion aimed at keeping workers disciplined. Nor is it to deny that many of the institutions of state and civil society have, as one of their functions, transmitting the multiple disciplines of capitalism: to produce, to reproduce, to consume, and so on. Rather is it to suggest that the ways in which capitalism is reproduced within these institutions are less direct than was once thought, leaving a number of social relations relatively untouched, and providing all kinds of sites from which it is possible to generate opposition and change. More than this, the processes by which the reproduction of capitalism is assured are not just negative ones of constraint but also processes in which people become positively involved. They are based on consent as well as coercion (Gramsci 1971).

Many strategies have been constructed by researchers in the social sciences to deal with these indeterminate elements of capitalist reproduction. In human geography three main strategies have been followed. The first of these is theoretical and still quite abstract. It is to use a theoretical system influenced by realism which explicitly invests social objects other than the capitalist economic system with causal powers, and especially the state. For example, Urry (1981) invested three 'spheres' – the state, certain entities within civil society (e.g. the family), and capital – with causal powers. The outcomes in any society – of the multiple determinations flowing into and out of these three spheres – will be complex, with different societies producing different resonances in capitalist social relations and different degrees to which capital is able to penetrate the state and civil society. Lovering (1987), in similar vein, provides an analysis of the way in which the state can direct the course of capital through the defence industry. Foord & Gregson (1986) provide patriarchy (the structures by which men oppress women) with a causal existence which is independent of capital although intertwined with it in various ways.

A second strategy is less abstract. It involves detailed study of how capitalist social relations overlap in societies as value systems and as symbols. This kind of work has focused in general upon the mechanics of cultural production and has been of two main types. First, there is a considerable amount of work involved with the varying modes of reproduction of the meanings attached to landscape (see Cosgrove 1985, Cosgrove & Daniels 1988). Second, there is all manner of work on communications media and the way they are used to promote capitalist and/or establishment values, with especial reference to use and manipulation of ideas of places (Burgess & Gold 1985). More recently, this work has burgeoned into consideration of how commodities are sold through the conscious use of symbolic systems which both draw on and reproduce particular lifestyles. Part of this process consists of the setting-up of places within which consumption and lifestyle can come together, reinforcing one another (Thrift & Williams 1987).

A third strategy is to study subcultures and ideologies which conflict with the dominant ones. These cultures are distinguished by their resistance to all or, more likely, a part of the capitalist system and the state. There are, of course, some oppositional cultures so strongly stigmatized that a muted but continuous opposition is their only choice if the integrity of the group is to survive. Such continually harassed groups as travellers are a case in point (Sibley 1981). However, most cultures choose a mixture of conflict and compromise. The classic cases of resistance can be found amongst 19th century working-class communities where the battle-lines between labour and capital were tightly drawn, amongst the 'Manchesters, Mulhouses, and Lowells' (Harvey 1985, p. 9). These communities have continued on in to the 20th century with the lines of battle often being drawn even more starkly, when a distinctive ethnic or religious composition strengthened community ties, rather than weakened them. However, these studies are not the only possibility. In the 20th century there has been an expansion in the number of urban movements, many of which are of middle-class, not working-class, composition (Castells 1983). The range of these movements is now very great indeed, and geographers have been studying all of them. There is the ecological movement, black, feminist and gay movements, the forces of nationalism, and so on. Each and every one of these oppositional cultures has a distinctive geography which is a vital part of their ability to survive and contest dominant orders.

These three research strategies have come together in certain literatures, and especially in that investigating the gentrification of urban neighbourhoods (Smith & Williams 1986, Rose 1988). But literature such as this also points towards one vexed question that cuts across all three research strategies – the question of class. Class has been, and continues to be, a focal point of the structure–agency debate since Thompson's interventions on the nature of class galvanized all manner of writers' pens into actions (Thompson 1963, 1978). What seems certain is that the Marxian depiction of class was too 'thin', concentrating too much on class struggle at work, important though this undoubtedly is. The social and cultural dimensions of class were neglected, even though they provide important forces dictating the intensity and direction of struggle at work, as well as being domains of class conflict in their own right. This bias has now been corrected and the full range of the permutations of conflict between capital and labour is now being revealed by a coalition of social historians, sociologists, and geographers. This is not to say all the problems have been solved; far from it. Many questions remain only partially conceptualized (Thrift & Williams 1987). In particular, space can now be seen to be a crucial determinant of class formation, but its exact role in particular situations requires much further work of both theoretical and empirical elaboration. The organization of space clearly alters the ability of classes to coalesce and pursue a class politics, rather than remain as separate islands of community (Harvey 1985).

The mention of space leads on to the third major focus of research on structure and agency: the place of space in the relations between human geography and social structure. Two particular areas of research have been developed here. First of all, there has been an interest in how structures are tied together in space by transport and communications innovations, from the invention of writing through the burst of new media of communication in the 19th and early 20th centuries (the train, the telegraph, the telephone, and so on) to the new instruments of mass communication and processing that dominate our worldview now: radio, television, video, even the computer (Gregory 1987). As a result, social structure has moved from reliance on face-to-face communication to reliance on indirect communication, from 'social' to 'system' integration (Giddens 1984). The notion of time space distanciation captures the uneven spatial dimensions of this integration. These changes have been crucial to the constitution of society in all kinds of ways. Economically, they have allowed multinational corporations and international finance to exist. Socially, they have allowed the state to spread its influence into all the corners of everyday life. Culturally, they have produced 'imagined communities', including nationalist and religious movements (Anderson 1983, Gellner 1985).

The second area of research, and one which has become very important indeed, has focused on the idea of locality. The idea of locality research sprang out of Massey's work in *Spatial divisions of labour* (1984) and a subsequent British research programme, sponsored by the Economic and Social Research Council, called the Changing Urban and Regional System Initiative, in which many of the chief proponents (and critics) of locality research were involved. At its simplest, locality research was an enquiry into the effects of international industrial restructuring on local areas, and especially into why different local areas produced different responses (Massey 1984). But the research soon ranged outside this initial area of enquiry, taking in issues of gender, class, and politics, as well as the consideration of flexible production and the rise of an economy based on the service industries (Murgatroyd *et al.* 1986, Cooke 1989). A complex theoretical debate soon began to rage about the degree to which localities could be defined and considered as independent actors with their own 'proactive' capabilities (Savage *et al.* 1987, Urry 1987). This debate has produced an enormous amount of heat, but it remains to be seen whether it will produce any light. However, the substantive pieces of locality research coming from Britain and the United States are clearly important.

In conclusion, what did the structure–agency debate achieve, in all its different guises? Three main things, perhaps. First, it focused attention on that old Marxian dictum, 'people make history but not in circumstances of their own choosing'. This was always a notoriously opaque statement, saying both everything and nothing. Now, however, it is possible to say more about almost every aspect of this statement. More is known about what people are, the social institutions they make, and the geographies within which they must make them. But second, none of this denies the power of the political-economy approach. What it does is extend and enrich the approach in all kinds of ways. Most particularly, against the background of the continuing effort to understand the shifting contours of capitalism, it makes a contribution to the social, cultural, and political knowledge necessary to withstand capitalism's depredations and understand capitalism's successes. Third and finally, the structure–agency debate underlined the fact that capitalism is not just a phenomenon of economic geography. It is also at one and the same time a social, cultural, and political geography which is equally made and disputed in each of these other realms.

The late 1980s: postmodernism and purity

By the late 1980s, a new issue had arisen within the political-economy approach. It can be summarized under the heading of **postmodernism** although this is a term which is currently used to excess (Punter 1988).

Postmodernism is a confusing term because it represents a combination of different ideas. It is, perhaps, most often seen as concerned with issues of method. As method, it is critical of the idea of totality that is typical of structural Marxism. Instead, it takes its cue from so-called post-structuralist theory, especially the work of those such as Derrida, Lacan, Kristeva, and Foucault (and, ironically, Althusser), which, although it is very different in a number of ways, shares common assumptions about the matters of language, meaning, and subjectivity (Dews 1987, Weedon 1987). In particular, this body of work assumes the following: that meaning is produced in language, not reflected by it; that meaning is not fixed but is constantly on the move (and so the focus of fierce political struggle); and that subjectivity does not imply a conscious, unified, and rational human subject but instead a kaleidoscope of different discursive practices. In turn, the kind of method needed to get at these conceptions will need to be very supple, able to capture a multiplicity of different meanings without reducing them to the simplicity of a single structure. Derrida's deconstruction, Foucault's genealogy, Lyotard's paralogism, the postmodern ethnography of anthropoligists such as Clifford (1988), the discourse analysis of various social psychologists – all these are attempts to produce a method that can capture history as a set of overlapping and interlocking fields of communication and judgement (discursive fields).

Postmodernism has also been used to describe the culture of a new phase of capitalism. Such commentators as Dear (1986), Jameson (1984), Davis (1985), and Harvey (1987a) have built on a variety of sources from the 'situationist' analysis of the consumer spectacle, through the power of financial capital, to the rise of 'flexible' methods of accumulation, to produce an analysis of a new phase of capitalist culture based upon a constant, self-conscious play with meaning and leading to the increased usage in everyday life of historical eclecticism, pastiche, and spectacle. There may be a dispute between those commentators about the point in time at which modern culture gave way to postmodern culture, even about the defining characteristics of postmodern culture, but all share a desire to link its advent to recent changes in the capitalist mode of production, in one way or another.

In contrast to these postmodern excursions, the end of the 1980s also saw the signs of a possible resurgence of a 'traditional' Marxist approach. Some commentators clearly felt that things had gone too far in the direction of eclecticism and that the Marxist core of the political-economy approach was under threat. It was time for the experimenters to return to the Marxist fold. Thus, locality research was subject to a sustained critique for its lack of grounding in grand theory and its apparently empiricist bent (Harvey 1987, Smith 1987a). Similarly, postmodern methodological approaches, although not the epochal developments, were lambasted by Harvey and others, who argued for a return to a more solid Marxism (Harvey 1987).

Other commentators have constructed theoretical halfway houses between the radical uncertainties of postmodernism and the radical certainty of the fully fledged structural approach. For example, some writers have commended the works of the French Regulationist School, whose members include Aglietta, Boyer, and Lipietz, which has developed an approach that holds on to notions which look suspiciously like base and superstructure, suitably altered for less rigid times (de Vroey 1984). Another approach has been to argue for a 'post-enlightment Marxism' in which it is possible to place the analysis of 'civil society on an equal footing with political economy in the theorisation of capital and the explanation of history and geography, while not insisting on subjecting them all to a dialectical totalisation' (Storper 1987, pp. 425–6). This approach would be close to the reconstitutions of historical materialism of such writers as Giddens and Habermas. Thus, in the 1980s, the political-economy approach in geography has continued in a state of flux. it continues to show signs of a healthy self-criticism. Hopefully, it continues to develop and grow.

Conclusions

The political-economy perspective in geography is barely 20 years old. Yet already we are able to chart its several periods of development. Radical ideas grew slowly and late because the discipline was conservative and because geographers had little experience in understanding and debating social theory. At first, therefore, radical geography was merely doing socially relevant work, with Marxism learned the painful way, through reading and interpreting the original, classical works. In the 1970s, structuralism in geography existed more as the reconstructive notions of eager critics than as a distinct and sophisticated school of thought. However, some time around the late 1970s and early 1980s the pace of change increased, the interaction between geography and social theory intensified, and, not coincidentally, fragmentation appeared in what still remained a relatively coherent perspective. Ideas of structuration theory, realism, and locality, towards which many geographers turned, were usually imported from points of origin outside the discipline, but political-economic geographers quickly began giving them new twists, applying them differently, then adding new dimensions. The quality of theorietical discourse improved as space and environment became burning issues of the day. In many ways the conflicts between different positions which typified the 1980s were a necesary part of improving the intellectual product in an era when people were beginning to listen.

Finally, at the end of the 1980s, we found some geographers pushing on through the postmodernist frontier, while others considered it more fruitful to improve on what had already been discovered. Of course, it is not the case that everyone joined each wave of interest, being carried along with the wave until it broke under criticism, then jumping to the next upswell of concern. Each new interest has left a residue of knowledge in all and made committed adherents out of some. This should be the case. For we are not talking here of knowledge as adornment but as interventionary ability. In the end, this is the original contribution of the political-economy approach. Knowledge for its own sake is unconvincing. Knowledge to make the world a better place becomes the only acceptable purpose.

References

Althusser, L. 1969. *For Marx* (Translated by Ben Brewster). London: Penguin.
Althusser, L. & E. Balibar. 1970. *Reading Capital*. London: New Left Books.
Ambrose, P. 1976. British land use planning: a radical critique. *Antipode* **8**, 2–14.
Anderson, B. 1973. *Imagined communities*. London: Verso.
Anderson, J. 1973. Ideology in geography: an introduction. *Antipode* **5**, 1–6.

Anderson, J. 1978. Ideology and environment (special issue). *Antipode* **10**, 2.
Anderson, P. 1980. *Arguments within English Marxism.* London: Verso.
Anderson, P. 1983. *In the tracks of historical materialism.* London: Verso.
Baynes, K., J. Bohman, & T. McCarthy (eds) 1987. *After philosophy. End or transformation?* Cambridge, Mass.: MIT Press.
Benton, T. 1984. *The rise and fall of structural Marxism: Althusser and his influence.* New York: St Martin's Press.
Berry, B. J. & A. Pred 1965. *Central place studies: a bibliography of theory and applications* (revised ed). Philadelphia: Regional Science Institute.
Bhaskar, R. 1975. *A realist theory of science*, 2nd edn. Brighton: Harvester.
Bhaskar, R. 1979. *The possibility of naturalism.* Brighton: Harvester.
Bhaskar, R. 1986. *Scientific realism and human emancipation.* London: Verso.
Billinge, M. 1977. In search of negativism: phenomenology and historical geography. *Journal of Historical Geography.* **3**, 55–67.
Blaut, J.M. 1974. The ghetto as an internal neo-colony. *Antipode* **6**, 37–41.
Boddy, M. 1976. Political economy of housing: mortgage-financed owner-occupation in Britain. *Antipode* **8**, 15–24.
Breitbart, M. 1975. Impressions of an anarchist landscape. *Antipode* **7**, 44–9.
Breitbart, M. (ed.) 1978–9. Anarchism and environment (special issues). *Antipode* **10**, 3 & **11**, 1.
Buch-Hansen, M. & B. Nielsen. 1977. Marxist geography and the concept of territorial structure. *Antipode* **9**, 1–12.
Burgess, J. & J. Gold. 1985. *Geography, the media and popular culture.* London: Croom Helm.
Burgess, R. 1978. The concept of nature in geography and marxism. *Antipode* **10**, 1–11.
Burton, I. 1963. The quantitative revolution and theoretical geography. *Canadian Geographer* **7**, 151–62.
Callinicos, A. 1976. *Althusser's Marxism.* London: Pluto Press.
Callinicos, A. 1985. *Marxism and philosophy.* Oxford: Oxford University Press.
Carney, J. 1976. Capital accumulation and uneven development in Europe: notes on migrant labour. *Antipode* **8**, 30–8.
Castells, M. 1977. *The urban question: a Marxist approach* (translated by Alan Sheridan). Cambridge Mass.: MIT Press.
Castells. M. 1983. *The city and the grassroots.* London: Edward Arnold.
Chorley, R. J. & P. Haggett (eds) 1967. *Models in geography.* London: Methuen.
Chouinard, V. & R. Fincher 1983. A critique of 'Structural marxism and human geography'. *Annals of the Association of American Geographers.* **73**, 137–46.
Clifford, J.C. 1988. *The predicament of culture. Twentieth century ethography, literature and art.* Cambridge, Mass.: Harvard University Press.
Cooke, P. 1986. The changing urban and regional system in the UK. *Regional Studies* **20**, 243–51.
Cooke, P. 1987. Clinical inference and geographical theory. *Antipode* **19**, 69–78.
Cooke, P. (ed.) 1989. *Localities.* London: Unwin Hyman.
Corbridge, S. 1986. *Capitalist world development: a critique of radical development geography.* London: Macmillan.
Cosgrove, D. 1985. *Social formation and symbolic landscape.* Totowa, NJ: Barnes & Noble.
Cosgrove, D & S. Daniels (eds) 1988. *The inconography of landscape.* Cambridge: Cambridge University Press.
Cosgrove, D. & P. Jackson 1987. New directions in cultural geography. *Area* **19**, 95–101.
Crompton, R, & M. Mann (eds) 1986. *Gender and stratification.* Cambridge: Polity Press.
Davis, M. 1985. Urban renaissance and the spirit of post-modernism. *New Left Review* **151**, 106–13.

Dear, M.J. 1986. Postmodernism and planning. *Environment and Planning D, Society and Space.* **4**, 367–84.
Dear, M. J. & J. V. Wolch 1987. *Landscapes of despair.* Cambridge: Polity Press.
de Vroey, M. 1984. A regulation approach of contemporary crisis. *Capital and Class* **23**, 45–66.
Dews, P. 1987. *Logics of disintegration: poststructuralist thought and the claims of critical theory.* London: Verso.
Duncan, J & D. Ley 1982. Structural marxism and human geography: a critical assessment: *Annals of the Association of American Geographers.* **72**, 30–59.
Elliott, G. 1987. *Althusser. The detour of theory.* London: Verso.
Eyles, J & D. Smith (eds) 1988. Quantitative methods in human geography. Cambridge: Polity Press.
Foord, J. & N. Gregson 1986. Patriarchy: towards a reconceptualisation. *Antipode* **18**, 186–211.
Forbes, D. K. & N. J. Thrift (eds) 1987. *The socialist Third World.* Oxford: Basil Blackwell.
Galois, B. 1976. Ideology and the idea of nature: the case of Peter Kropotkin. *Antipode* **8**, 1–16.
Geertz, C. 1973. *The interpretation of cultures.* New York: Basic Books.
Gellner, E. 1985. *Nations and nationalism.* Oxford: Basil Blackwell.
Geras, N. 1983. *Marx and human nature.* London: Verso.
Giddens, A. 1979. *Central problems in social theory.* London: Macmillan.
Giddens, A. 1981. *A contemporary critique of historical materialism.* London: Macmillan.
Giddens, A. 1984. *The constitution of society.* Cambridge: Polity Press.
Giddens, A. 1985. *The nation state and violence.* Cambridge: Polity Press.
Godelier, M. 1978. Infrastructures, societies, and history. *Current Anthropology* **19**, 4.
Gramsci, A. 1971. *The prison notebooks.* London: Lawrence & Wishart.
Gregory, D. 1978. *Ideology, science and human geography.* London: Hutchinson.
Gregory, D. 1987. The friction of distance? Information circulation and the mails in early nineteeth-century England. *Journal of Historical Geography* **13**, 130–54.
Gregory, D. & J. Urry 1985. *Social relations and spatial structures.* New York: St Martin's Press.
Gregson, N. 1986. On duality and dualism: the case of structuration and time-geography. *Process in Human Geography.* **10**, 184–205.
Gregson, N. 1987. Structuration theory: some thoughts on the possibilities for empirical research: *Environment and Planning D, Society and Space.* **5**, 73–92.
Habermas, J. 1979. *Communication and the evolution of society,* London: Heinemann.
Harré, R. 1987. *Varieties of realism.* Oxford: Basil Blackwell.
Hartshorne, R. 1939. *The nature of geography.* Lancaster, PA.: Association of American Geographers.
Hartshorne, R. 1955. 'Exceptionalism in geography' re-examined. *Annals of the Association of American Geographers* **43**, September, 205–44.
Hartshorne, R. 1959. *Perspective on the nature of geography.* Chicago: Rand McNally.
Harvey, D. 1969. *Explanation in geography.* London: Edward Arnold.
Harvey, D. 1973. *Social justice and the city.* Baltimore: Johns Hopkins University Press.
Harvey, D. 1975. The geography of capitalist accumulation: a reconstruction of the Marxian theory. *Antipode* **7**, 9–21.
Harvey, D 1982. *The limits to capital.* Chicago: University of Chicago Press.
Harvey D. 1983. Owen Lattimore – a memoire. *Antipode* **15**, 3–11.
Harvey, D. 1985. *The urbanisation of consciousness.* Oxford: Basil Blackwell.
Harvey, D. 1987a. Flexible accumulation through urbanisation: reflections on post-modernism in the American city. *Antipode.* **19**, 260–86.

Harvey, D. 1987b. Three myths in search of a reality in urban studies. *Environment and Planning D, Society and Space* **5**, 367–76.
Harvey, D. & N. Smith 1984. Geography: from capitals to capital. In *The left academy: Marxist scholarship on American Campuses*, vol. 2, B. Ollman, & E. Vernoff. (eds), 99–121. New York: Praeger.
Hayford, A. 1974. The geography of women: an historical introduction. *Antipode* **6**, 1–19.
Held, D. 1980. *Introduction to critical theory*. London: Hutchinson.
Hindess, B. & P. Q. Hirst 1975. *Pre-capitalist modes of production*. London: Routledge & Kegan Paul.
Horvath, R. 1971. The 'Detroit geographical expedition and institute' experience. *Antipode* **3**, 73–85.
Horvath, R. J. & K. Gibson 1984. Marx's method of abstraction. *Antipode* **16**, 23–36.
Hudson, B. 1977. The new geography and the new imperialism: 1870–1918. *Antipode* **9**, 12–19.
Jackson, P. & S. J. Smith 1984. *Exploring social geography*. London: Allen & Unwin.
Jameson, F. 1984. Postmodernism, or the cultural logic of late capitalism. *New Left Review* **146**, 53–92.
Johnston, R. J. 1986. *On human geography*. London: Edward Arnold.
Keat, R. & J. Urry 1981. *Social theory as science*. London: Routledge & Kegan Paul.
Kropotkin, P. 1902. *Mutual aid: a factor of evolution*. London: Heinemann.
Lancaster Regionalism Group 1986. *Localities, class and gender*. London: Pion.
Lattimore, O. 1950. *Ordeal by slander*. Boston: Little, Brown.
Lattimore, O. & E. Lattimore 1944. *The Making of modern China: a short history*. New York: Franklin Watts.
Lefebvre, H. 1972. *La Pensée marxiste et la ville*. Paris: Gallimard.
Ley, D. & M. Samuels (eds) 1978. *Humanistic geography*. London: Croom Helm.
Lovering, J. 1987. Militarism, capitalism, and the nation–state: towards a realist synthesis. *Environment and Planning D, Society and Space* **5**, 283–302.
MacKenzie, S., J. Foord & M. Breitbart (eds) 1984. Women and the environment, (special issue). *Antipode* **16**, 3.
Marx, K. 1970. *A contribution to the critique of political economy*. Moscow: Progress Publishers.
Marx, K. 1973. *Grundrisse*. Harmondsworth: Penguin
Marx, K. 1976. *Capital*, Vol. 1. Harmondsworth: Penguin.
Marx, K. & F. Engels 1975. *On Religion*. Moscow: Progress Publishers.
Massey, D. 1973. Towards a critique of industrial location theory. *Antipode* 5, 33–9
Massey, D. 1976. Class, racism and busing in Boston. *Antipode* **8**, 37–49.
Massey, D. 1984a. *Spatial divisions of labour: social structures and the geography of production*. London: Methuen.
Massey, D. 1984b. Geography matters. In *Geography Matters! A Reader*, D. Massey & J. Allen (eds), 1–11. Cambridge: Cambridge University Press.
Massey, D. & R. Meegan 1982. *The anatomy of job loss*. London: Methuen.
Massey, D. & R. Meegan (eds) 1986. *Politics and method: contrasting studies in industrial geography*. New York: Methuen.
Meillasoux, C. 1981. *Maidens, meal and money: Capitalism and the domestic community*. Cambridge: Cambridge University Press.
Newman, R. P. 1983. Lattimore and his enemies. *Antipode* **15**, 12–26.
Ollman, B. 1971. *Alienation: Marx's conception of man in capitalist society*. New York: Cambridge University Press.
Peet, R. 1977a. The development of radical geography in the United States. *Progress in Human Geography* **1**, 64–87.
Peet, R. 1977b. *Radical geography: Alternative Veiwpoints on contemporary social issues*. Chicago: Maaroufa Press.

Peet, R. 1978–9. The geography of human liberation. *Antipode* **10**, nos. 3 and 11, 1, 126–34.
Peet, R. 1985a. The destruction of regional cultures. In *A world in crisis: geographical perspectives*, R. J. Johnston and P. J. Taylor (eds), 150–72. Oxford: Basil Blackwell.
Peet, R. 1985b. Radical geography in the United States: a personal history. *Antipode* **17** 1–8.
Peet, R. 1985c. The social origins of environmental determinism. *Annals of the Association of American Geographers* **75**, 309–33.
Peet, R. (ed.) 1987. *International capitalism and industrial restructuring: a critical analysis*. Boston: Allen & Unwin.
Piaget, J. 1970. *Structuralism*. New York: Basic Books.
Pickles. J. 1986. *Phenomenology, science and geography: spatiality and the human sciences*. Cambridge: Cambridge University Press.
Poulantzas, N. 1975. *Classes in contemporary capitalism*. London: New Left Books.
Poulantzas, N. 1978. *State, power, socialism*. London: New Left Books.
Pred, A. 1981. Social reproduction and the time-geography of everyday life. *Geografiska Annaler*, Series B, **63**, 5–22.
Pred, A. 1985. The social becomes the spatial; the spatial becomes the social. Enclosure, social change and the becoming of places in the Swedish province of Skane. In *Social relations and spatial structures*, D. Gregory and J. Urry (eds), 337–365. London: Macmillan.
Pred, A. 1987. *Place, practice, structure*. Cambridge: Polity Press.
Punter, J. 1988. Post-modernism. *Planning Practice and Research*. **4**, 22–8.
Rees, J. (ed). 1986. *Technology, regions and policy*. Totowa, NJ: Rowman & Littlefield.
Rey, P. 1973. *Les Alliances des classes*. Paris: Maspero.
Rose, D: 1988. Homeownership, subsistence and historical change: the mining district of West Cornwall in the late nineteenth century. In *Class and space*, N. Thrift & P. Williams (eds), 108–53. London; Routledge & Kegan Paul.
Savage, M., J. Barlow, S. Duncan & P. Sanders 1987. 'Locality Research': the Sussex programme on economic restructuring, social change and the locality. *Quarterly Journal of Social Affairs* **1**, 27–51.
Sayer, A. 1984. *Method and social science*. London: Hutchinson.
Sayer, A. 1985. Realism, in geography. In *The future of geography*. R. J. Johnston (ed.), 159–73. London: Methuen.
Schaefer, F.K. 1953. Exceptionalism in geography: a methodological examination. *Annals of the Association of American Geographers* **43**, 226–49.
Schimidt, A. 1971. *The concept of nature in Marx*. London: New Left Books.
Scott, A. J. & M. Storper (eds) 1986. *Production, work, territory: the geographical anatomy of industrial capitalism*. Boston: Allen & Unwin.
Sibley, D. 1981. *Outsiders in an urban society*. Oxford: Basil Blackwell.
Slater, D. 1973. Geography and underdevelopment – Part I. *Antipode* **5**, 21–32.
Slater, D. 1975. The poverty of modern geographical inquiry. *Pacific Viewpoint* **16**, 159–76
Slater, D. 1977. Geography and underdevelopment – Part II. *Antipode* **9**, 1–31
Smith, N. 1984. *Uneven development: nature, capital and the production of space*. Oxford: Basil Blackwell.
Smith, N. 1987. Dangers of the empirical turn: the CURS initiative. *Antipode* **19**, 59–68.
Smith, N. & P. O'Keefe 1980. Geography, Marx and the concept of nature. *Antipode* **12**, 30–9.
Smith, N. & P. Williams (eds) 1986. *Gentrification of the city*. London: Allen & Unwin.
Sopher, D.E. 1967. *Geography of religions*. Englewood Cliffs, NJ: Prentice Hall.
Stea, D. & B. Wisner (eds) 1984. The Fourth World (special issue). *Antipode* **16**, 2.
Stoddart, D. 1986. *On geography*. Oxford: Basil Blackwell.

Stone, M. 1975. The housing crisis: mortgage lending and class struggle. *Antipode* **7**, 22–37.

Storper, M. 1987. The post-Englightenment challenge to Marxist urban studies. *Environment and Planning D, Society and Space* **5**, 418–27.

Storper, M. & R. Walker 1988. *The capitalist imperative*. Oxford: Basil Blackwell.

Tathem, G. 1933. Environmentalism and possibilism. In *Geography in the twentieth century*, G. Taylor (ed.), 128–62. New York.

Terray, E. 1972. *Marxism and 'primitive societies.'* New York: Monthly Review Press.

Thompson, E. P. 1963. *The making of the English working class*. London: Weidenfeld & Nicolson.

Thompson, E. P. 1978. *The poverty of theory and other essays*. London: Merlin Press.

Thrift, N. J. 1979. On the limits to knowledge in social theory: towards a theory of practice. Camberra: Australian National University, Department of Human Geography, Seminar Paper.

Thrift, N. J. 1983. On the determination of social action in space and time. *Environment and Planning D, Society and Space*. **1**, 23–57.

Thrift, N. J. & D. K. Forbes 1986. *The price of war: urbanisation in Vietnam 1954–1965*, London: Allen & Unwin.

Thrift, N. J. & A. Pred. 1981. Time-geography: a new beginning. *Progress in Human Geography*. **5**, 277–86.

Thrift, N. J. & P. Williams (eds) 1987. *Class and space*. London: Routledge & Kegan Paul.

Timpanaro, S. 1975. *On materialism*. London: New Left Books.

Urry, J. 1981. *The anatomy of capitalist societies. The economy, civil society and the state*. London: Macmillan.

Urry, J. 1985. Social relations, space and time. In *Social relations and spatial structures*, D. Gregory & J. Urry (eds), 20–48. London: Macmillan.

Urry, J. 1987. Society, space and locality. *Environment and Planning D, Society and Space*. **5**, 435–44.

Vogeler, I. (ed.) 1975. Rural America (special issue). *Antipode* **7**, 3.

Walker, R. (ed.) 1979. Human–environment relations (special issue). *Antipode* **11**, 2.

Walker, R. A. & D. A. Greenberg 1982a. Post-industrial and political reform in the city: a critique. *Antipode* **14**, 17–32.

Walker, R. A. & D. A. Greenberg 1982b. A guide for the Ley reader of marxist criticism. *Antipode* **14**, 38–43.

Weedon, C. 1987. Feminist practice and poststructuralist theory. Oxford: Basil Blackwell.

Wittfogel, K. 1957. *Oriental despotism: a comparative study of total power*. New Haven,: Yale University Press.

Wittfogel, K. 1985. Geopolitics, geographical materialism and marxism. *Antipode* **17**, 21–72. Geopolitik, geograhischer materialisumus und marxismus. Translation from G. O. Ulmen 1929. *Unter dem banner des marxismus* 3, 1, 4 and 5.

Wolpe, H. (ed.) 1980. *The articulation of modes of production*. London: Routledge & Kegan Paul.

Wolpert, J. & E. Wolpert 1974. From asylum to ghetto. *Antipode* **6**, 63–76.

Wright Mills, C. 1959. *The sociological imagination*. New York: Oxford University Press.

2 *Mathematical models in human geography: 20 years on*

Martin Clarke & Alan Wilson

Whatever happened to mathematical models?

In a book which is related to the publication of Haggett's and Chorley's *Models in geography* 20 years ago, but which is mostly not about quantitative models, it is useful to ask the question posed in the title of this section. The quantitative revolution in geography is usually formally dated from Ian Burton's 1963 paper; radical geography can perhaps be similarly dated from the 1973 publication of David Harvey's *Social justice and the city*. Fashions can change rapidly! One consequence of such change is that only a relatively small core of modellers have continued to work with the appropriate levels of technical expertise on the major research problems. Relatively few have attempted to engage in anything but knockabout debates on the relationship between modelling and radical geography. In this chapter we argue that it is important to understand what has happened to mathematical modelling, and that it does have a substantial contribution to make in the long term. Indeed, it can be argued that modelling (which also provides much of the conceptual basis of information systems) and what might be called *critical* (rather than radical) human geography form the two main strands of the subject for the forseeable future. Because of the differences in expertise between the two populations of practitioners, it is likely that much of this development will be separate. However, there is no intrinsic need for this subdisciplinary apartheid, and one of the arguments of this chapter is for greater mutual understanding as a basis for possibile future collaboration. It is useful to note in this context that some social theorists are arguing that analytical modelling and mathematics should have a rôle in contemporary studies which goes beyond the old arguments about positivism. What all sides have in common is a recognition that, even if the basis is heuristic rather than scientific, they have a part to play in handling complexity. For example, Turner (1987) argues that 'analytical models provide an important supplement to abstract propositions because they map the complex causal connections – direct and indirect effects, feedback loops, reciprocal effects'; and to quote from the same volume 'mathematical models have an essential place in our efforts to untangle the complexities of social realities' (Wilson in Giddens & Turner 1987).

However, it must first be appreciated that all the elements of quantitative geography must not and ought not to be lumped under one heading. It is not our purpose here to discuss essentially inductive, that is statistical, methodology; we restrict ourselves to *mathematical* modelling on the basis that this has a more direct contribution to make to the evolution of geographical theory in the long term. This distinction, between the statistical and the mathematical, has usually not been well understood. Another area of weakness has been that there appeared to be little explicit connection between what mathematical modelling had to offer to geographical theory and what might be called the classical contributions of such authors as von Thünen, Weber, Burgess and Hoyt, Christaller and Lösch – and these authors have provided the basis of much geographical textbook writing both before and after the advent of radical geography. (Perhaps their works constitute a neoclassical geography?) This was partly because the classical modellers had ventured into areas where the 1960s modellers had not the expertise to tread; and, more simply, because the effort of understanding what each perspective contributed to the other was not made. It could be argued that the contributions of modelling in the 1960s and early 1970s, exciting though they were at the time and useful though they remain in many ways, did not in fact address the central problems of geographical theory. However, this position has now changed and a brief articulation of the new contribution is a major purpose of our discussion.

A further complication then arose in the quantitative geography–radical geography debates: the arguments were conducted in rather simplistic terms (based on the outmodedness of positivism, and the perceived corollary that the positivist label could be used to dismiss anything to do with mathematical modelling) without the issues raised above being fully understood. In other words, debates were presented as arguments between incompatible paradigms with neither of the new paradigms being very closely related to the classical or neoclassical ones. No wonder that geography as a discipline seemed to be in a fragmented state.

It is now useful to try to improve upon this: to understand the development of mathematical modelling; its historical connections to classical theory; the levels of expertise that have now been achieved; and the possibilities for using it in the future in the light of the radical critique. We aim to show that modelling has a contribution in relation to geographical theory in general, but also that expertise is available for a wide range of applications which throw light on a variety of problems. We begin, therefore, by briefly outlining the state of the art and the history of modelling; next we look at the relationship of modelling to geographical theory; then we outline some illustrative problems in applied human geography; and, finally, we discuss the role of modelling, as we see it, in the future of geography.

What have modellers achieved?

There is a rich variety of approaches to modelling in ways which are relevant to geography – perhaps best distinguished in the first instance by a variety of disciplinary backgrounds. For example, this ranges from the 'new urban economics' school (cf. Richardson 1976) to ecological approaches (Dendrinos & Mullaly 1985). Here, we illustrate the argument with models based on spatial interaction concepts – initially rooted in entropy maximizing methods (Wilson 1970), but with a recently extended range of application through the use of methods of dynamical systems theory (Wilson 1981). This restriction both corresponds to our expertise but also, none the less, serves to illustrate most of the general points which have to be made: the kinds of development described within this subparadigm have also been achieved, or have to be achieved, in any of the alternative approaches. In other words, the gist of the argument would be preserved if it was rewritten as though from the viewpoint of another modelling perspective.

In the 1960s, a broad ranging family of spatial interaction models was defined and applied (cf. Wilson (1974) for a broad review). It was also recognized at an early stage that many such models, particularly the so-called singly-constrained varieties, also functioned in an important respect as *location* models. In the case of retailing flows, for example, the models could be used to calculate total revenue attracted to the shopping centre, as a sum of flows, and this of course is an important locational variable. However, such modelling exercises could only be carried out if a number of important geographical variables were taken as given: in particular, the spatial distribution of physical structures of, as in this example, shopping centres. In other words, no attempt was being made to model the main geographical structural variables – and this is where modellers had failed to tackle one of the problems of the classical theorists – in this case, Christaller and central place theory.

The situation was rectified in the late 1970s and this has led to dramatic advances in modelling technique as well as in the understanding of the contribution of modelling to geographical theory in a wider sense. The argument was first set out in Harris & Wilson (1978) in relation to the singly-constrained spatial interaction model and in particular to retailing, but it was realized from the outset that it had a much wider application. The modelling advance involved the addition of an hypothesis to spatial interaction models which specified whether particular centres at particular locations would grow or decline. It was then possible to model, in a dynamic context, not only the spatial flows within a geographical system but also the evolution over time of the underlying physical and economic structures. This method can be applied to any geographical location system which depends on spatial interaction as an underlying basis. These include agriculture (relating crop production to markets), industrial location (related to flows from input sources or to markets), residential location and housing (in relation to the journey to work and services), retailing, and a whole variety of services. In principle, subsystem models can be combined into whole system models and then comprehensive models such as those of Christaller (from an earlier generation) or Lowry (from a later generation) can be rewritten. This programme of rewriting has now been carried through and can be used to illustrate the application to the main areas of geographical theory:

(a) agriculture (Wilson & Birkin 1987);
(b) industrial location (Birkin & Wilson 1986a, b);
(c) residential location (Clarke & Wilson 1983);
(d) retailing (Harris & Wilson 1978, Wilson & Clarke 1979, Clarke & Wilson 1986, Wilson 1988a);
(e) health services as an example of a different kind of service (Clarke & Wilson 1984, 1985);
(f) comprehensive modelling (Birkin *et al.* 1984).

There are two kinds of achievement from these advances: first, there is a contribution of general insights from modelling to the development of geographical theory, and we take these up next; the second arises when the model developments can be fully operationalized and data are available for testing, and the models can then be applied in planning contexts. We take this up afterwards.

Modelling and geographical theory

We take the argument forward in two steps: first, the relationship of modelling in its current form to classical theory; and, second, the relationship of modelling to radical geography.

It is possible in each of the fields of classical theory which have been mentioned to take the classical problem and to reproduce it in the new modelling framework. But then, it is possible to use the model to progress beyond the restrictions of the traditional approach and to tackle more complicated problems. Indeed, what emerges is that the classical theorists were limited by technique. It did not help that they were fixated on a continuous space representation: in the modelling era, discrete zones were more natural for computer data bases and turned out to have intrinsic advantages in mathematical terms. We consider what can now be achieved in each of the major fields, summarizing in broad terms the arguments presented in more detail in the references listed above, at the end of the preceding section.

In the agricultural case, it is possible to reproduce von Thünen's rings for the example of a single market centre and uniform plain. However, with the new model there is no problem in having as many market centres as is appropriate and building in variable fertility on the plain (and also coping with the distorting effects of transport networks and so on). It is difficult to make more than theoretical progress in this field (though with hypothetical numerical examples) because real-world data are not systematically available.

A similar argument applies to the industrial location case: Weber's Triangle can be reproduced (with the different situations of the single firm in relation to the vertices), but the model can be extended to handle the competitive relationships of a set of firms. However, new complications also have to be built into the model. As soon as many firms are included, it is recognized that they should be classified into a number of centres, and the model also has to represent the input–output relationships between industrial sectors as well as the spatial relationships between all firms which are consistent with them. This, needless to say, is a very complex task. However, it can be accomplished, but again only using numerical examples rther than real data. In the residential location and housing case, the residential location part of the model has been available since the 1960s (though only developed in empirical applications relatively slowly because of the complexity of the problem). The new insights now allow housing to be added and modelled. What is achieved in this case is a rich generalization of Burgess's rings and Hoyt's sectoral patterns.

The retail case, building on the work of authors like Reilly, is interesting because spatial interaction models were used together with the equivalent of discrete zone systems. However, the models used were essentially unconstrained (and must have produced rather silly predictions for flows) and so the main use was to demarcate market areas between shopping centres – essentially a continuous space use of a discrete zone model. What is clear from modern spatial interaction models (and from all relevant empirical data) is that market areas do overlap substantially and it is better to focus on the flows directly and to model these rather than to worry about boundaries or market-area demarcations.

Interestingly enough, there is little or no classical work on public services. So the applications of models in fields such as health services analysis represent a new gain. This reflects the historical importance of the different sectors.

The subsystems can be combined and comprehensive models developed which can then be considered to replace central place theory (Wilson 1978, Birkin *et al.* 1984). That the state of the art is now highly developed can be seen from a number of reviews which have been compiled in the last few years. Examples are Weidlich & Haag (1983), Wilson & Bennett (1985), Bertuglia *et al.* (1987), Bertuglia *et al.* (1989), Nijkamp (1986), and Dendrinos & Mallaly (1985).

The new dynamic modelling methods also offer a different kind of insight for geographical theory, and this is the sense in which the argument applies to any modelling style. It turns on the existence of nonlinearities (from externalities, scale economies, or whatever) in these models. Analysis then shows that while in some sense the models represent general laws, there are in any application a large number of possible equilibrium states and modes of development. A particular one chosen, say in a particular city or region, will depend on the particular behaviour of local agents (or historical accidents). In other words, the modelling insights integrate the two sides of the uniqueness–generality debate (which still manifests itself in various forms in a variety of paradigms). Analysis also has a bearing on the agency–structure problem. It enables real world complexities to be understood and illustrates that it is impossible to forecast the future in a deterministic way. However, it does provide detailed accounts of the past, and is therefore of great importance in the context of historical geography; and it provides insights, but not precise forecasts, for the future in terms of the modes of possible development in different circumstances.

A further property of nonlinear models is that their structures are subject to instantaneous (or, in practice, rapid) change at certain critical parameter values. This is a phenomenon now very widely recognized in many situations in many other disciplines. An interesting research task in human geography in the future will be to identify empirically, and to model, rapid structural change of this type. These observations are relevant to the application of modelling ideas in the theory developed under the application of radical geography as well as to the alternatives.

How can we relate contemporary mathematical modelling to the radical critique? The first point to make is that it is important to distinguish alternative hypotheses or theories from issues of technique for representing those theories in models. Once this is achieved, then any disagreement can be shifted to where it ought to be: between theories rather than in terms of the validity of certain kinds of technique. It does not follow, as has sometimes been naïvely argued, that any piece of work involving mathematics is positivist. In practice, many models are based on the assumptions of neoclassical economics and are therefore subject to the criticisms which can be brought to bear on that perspective. However, most of the basis of the models used to illustrate the argument in this chapter are not so dependent: the entropy maximizing base does not depend on such economic assumptions – it is more reasonably seen as a combination of accounting and statistical averaging notions. Indeed, it can be argued that any good piece of geographical analysis should be underpinned by the appropriate accounts. In the case of the study of economic structure in a region, for example, this has been done both by neoclassical and by Marxist researchers. The first perspective leads to the input–output model, the second, through the work of Sraffa, leads to an alternative. But they both have, in principle, the same underlying set of accounts. The future of modelling could well be seen in this way: what contribution can it make to operationalize hypotheses?

Some might argue that there are deeper structuralist questions involved, that modelling inevitably engages with surface phenomena and as such fails to offer adequate in-depth explanation. This was to an extent true of the modelling techniques of the 1960s (though much of the information generated was useful in a variety of practical situations). This is much less true (at least in terms of potential) of the modelling methods of today. On the whole, whatever kind of theory can be clearly articulated can also be modelled. However, when we attempt to carry through this argument in relation to some examples of radical geography then the complexity of some of the issues raised becomes apparent, and this raises a new generation of modelling problems (cf. Wilson 1988b, on the potential for configurational analysis in this kind of situation). None the less, progress is being made and the works of Webber (1987), Webber & Tonkin (1987), and Sheppard (1987) all provide important examples of how modelling skills can be deployed in critical or political-economic approaches.

A contemporary view of applied mathematical modelling

We hope that we have illustrated the actual and potential contribution of model-based methods to various aspects of location theory. We now move on to examine how mathematical modelling can be used in an applied problem-solving context. Before doing so, it is worth making three general points. First, a distinction can be made between the use of models as frameworks for understanding and the use of models in some prescriptive way. It is the latter role that is most often attributed to modellers, perhaps because of the relationship in the 1960s between modelling and planning and the attempts to make planning model-based. What has emerged is that models have an important contribution to make in the understanding of how systems operate and in particular of their dynamics. Without this understanding, of course, prescription becomes a dangerous and difficult task. The second point, made earlier but worth repeating, is that through the work on dynamic modelling it has become clear that the conventional use of modelling in urban and regional planning – conditional forecasting – has to be replaced by a more qualitative approach where models are used to identify the possible range of developments rather than to specify fairly precisely the exact form of change. A final preliminary comment related to this is that while urban or town planning faces a real difficulty from the fact that there is only a small element of control, in other public sector systems, notably health and education, a much greater degree of control is possible. For example, in health care the size and location of new hospitals, the level of service provision, the setting of priorities, and so on are all within the power of health authorities and managers; this provides a much more promising opportunity for the use of model-based methods in planning (Clarke & Wilson 1986).

Given these comments, what differentiates applied modelling in the late 1980s from that in the 1960s? Is it possible to be confident about the contribution modellers can make and if so why? A list of points emerges in attempting to answer these questions:

More experience There is now a considerable amount of experience in applying models in practical contexts. This relates both to the technical aspects of model application, such as calibration and validation, as well as to the more strategic issues, such as model design and policy representation. This has resulted in a reduction of the naïve, simplistic applications that, often rightly, drew most criticism and an increase in more sophisticated, but more realistic studies. There has also been a recognition that model-based analysis is but part of a wider process of management and planning rather than the central feature of planning. It may still be the *critical* phase, however.

Better methods Although we focused on just one methodological approach, that of spatial interaction (see pp. 31–5), it was shown that developments in that method, for example, through the introduction of dynamics, had significantly improved the range of applicability. The same has been true in other areas such as optimization and also new methods, such as Q-analysis and microsimulation, which have been developed. The modellers' kit-bag of methods has, therefore, been improved and extended. This results in the availability of more appropriate methods for particular applications.

Better information In the late 1960s and early 1970s almost every paper written on applied modelling concluded that the full potential of a particular approach would be achieved only when better information and faster and bigger computers became available. Computing power is dealt with below. Information systems have improved, not necessarily in terms of the quality of data collected, but in the ways in which such data can be accessed and manipulated. For example, the 1981 Census is freely available on-line to academics through SASPAC (Small Area Statistics Package). There is a worrying trend, however, which Goddard & Openshaw (1987) term 'the commodification of information', whereby information becomes a valuable and traded commodity, collected and supplied by private organizations. With the present government's adherence to marketforce principles, this could reduce the quality and amount of data traditionally located within the public realm.

Better computers Increasing computer power in itself does little to improve the application of model-based methods. It does, however, remove certain types of barriers and creates opportunities. Perhaps surprisingly, it is not the increased power of computers that has heralded a change in mathematical modelling but the advent of the microcomputer, most notably the IBM PC and its clones. This modest computer has two distinct advantages over its mainframe brothers. First, it allows a model system developed on a PC in, say, Leeds, to be transferred with ease to any other compatible PC elsewhere in the world, this was simply not practical with programs developed on mainframes. Second, PCs have superb colour graphical facilities for displaying information and results that can vastly improve the quality of presentation – an important aspect of popularizing and selling modelling to both the initiated and the unconvinced, and a point to which we return later. The availablity of a new generation of PCs based on the Intel 386 chip means that computing will never present constraints for modellers, only opportunities.

Better packaging and presentation of outputs There was a time when modellers were instantly recognizable on a university campus. They trudged back and forth from the geography department to the computer centre, returning with vasts swathes of computer output, most of which was immediately dispatched to the bin. While the odd modeller might still engage in this practice, he or she is an endangered species. What the end user of a modelling system typically requires, whether this is a public sector planner or a marketing director of a private firm, is a succinct, informative, and well presented analysis of what is likely to happen if. In response to this a number of developments have occurred. One which we have already mentioned is computer graphics, where, to paraphrase an old saying, a colour map is worth a thousand lines of computer output. A second development has been in interactive computing, where the changes to the system are input at the terminal, the model run, and the results presented at the screen, to be selected, say, from a menu; if another run of the system is required this can be performed immediately. In a system called HIPS (Health Information and Planning Systems) which has been developed as a planning tool for health authorities (Clarke & Wilson 1985), a number of variants of the strategic plan could be examined in an afternoon using the interactive system. Finally, to allow the outputs of models to be interpreted meaningfully we have seen the development of Performance Indicators (Clarke & Wilson 1984), which can be seen as the outcome of transformations on either data or model outputs that relate stock or activity variables to consistent denominators.

More interest While the academic community has expended much energy in discussing the intricacies and merits of model-based methods, a number of people engaged in market research and management consultancy recognized the potential contribution of geographical models to problem solving in the public and private sector. Spurred on by the commercial success of small-area profiling systems such as ACORN (A Classification Of Residential Neighbourhoods) and an evident demand for locational analysis, these types of companies have been undertaking what we would recognize as applied human geography for a number of years. In another paper (Clarke & Wilson 1987) we have developed an explanation for why this has happened and the potential for the future. Suffice it to say that many organizations in both the public and private sector take location analysis – from locating a new supermarket, marketing a product, to allocating public funds – very seriously and wish to employ useful and appropriate methodologies.

The above six points do not, in themselves, either defend the appropriateness of mathematical models or suggest a transformation in the role of models in human geography. What it is hoped that they do illustrate is the maturity of the discipline and a concern with both usefulness and understanding combined with an interest from outside geography in their application. It is now possible to articulate a much longer list of model application areas which extends considerably beyond the realm of urban planning. We mentioned health and education earlier, but interesting applications exist in retailing, financial services, utilities (e.g. the water industry), leisure, and so on. Where progress is still most difficult is in economic and industrial analysis, although progress is being made. Evidence suggests that model-based analysis within human geography can retain a vital role within the discipline.

Conclusion

In this chapter we have attempted to describe the position of mathematical models within the discipline of human geography 20 years after the publication of *Models in geography*. We hope that we have illustrated that modelling does not exist in a technical vacuum (as many would like to think), but has strong links with traditional geographical location theory on the one hand and important contemporary applications on the other. In terms of application, model-based analysis in the 1960s and 1970s was strongly associated with urban planning. This association has weakened and the new relationships outlined above have emerged. There is every indication that these relationships are much stronger than those of the past and are, therefore, likely to be more enduring.

Given our bullishness about the prospects for model-based geography how can we rekindle this enthusiasm amongst colleagues and, perhaps most importantly, our students? The most promising way forward appears to rest in the development of an applicable human geography based on case studies and examples (Clarke & Wilson 1987) in which modelling plays a central role but one which is also based upon firm theoretical foundations.

References

Bertuglia, C. S., G. Leonardi, S. Occelli, G. A. Rabino, R. Tadei, & A. G. Wilson (eds) 1987. *Urban systems: contemporary approaches to modelling*. London: Croom Helm.

Bertuglia, C. S., G. Leonardi, & A. G. Wilson (eds) 1989. *Urban dynamics: towards an integrated approach*. London: Routledge.

Birkin, M., M. Clarke, & A. G. Wilson 1984. Interacting fields: comprehensive models for the dynamical analysis of urban spatial structure. Paper presented to the 80th Annual AAG meeting, Washington DC, April 1984. Also Working Paper 385, University of Leeds: School of Geography.

Birkin, M. & A. G. Wilson 1986a. Industrial location models I: a review and an integrating framework. *Environment and Planning A* **18**, 175–205.

Birkin, M. & A. G. Wilson, 1986b. Industrial location theory II: Weber, Palander, Hotelling and extensions in a new framework. *Environment and Planning A* **18**, 293–306.

Burton, I. 1963. The quantitative revolution and theoretical geography. *Canadian Geographer* **7**, 151–62.
Clarke, G. P. & A. G. Wilson 1985. Performance indicators within a model-based approach to urban planning. Working Paper 446. University of Leeds: School of Geography.
Clarke, M. & A. G. Wilson 1983. Exploring the dynamics of urban housing structure: a 56 parameter residential location and housing model. Working Paper 363. University of Leeds: School of Geography.
Clarke, M. & A. G. Wilson 1984. Models for health care planning: the case of the Piedmonte region. Working Paper 38. Turin, Italy: IRES.
Clarke, M. & A. G. Wilson 1985. Developments in planning models for health care policy analysis in the UK. In *Progress in medical geography* **10**, 427–51.
Clarke, M. & A. G. Wilson 1986. The dynamics of urban spatial structure: the progress of a research programme. *Transactions of the Institute of British Geographers* **10**, 427–51.
Clarke, M. & A. G. Wilson 1987. Towards an applicable human geography: some developments and observations. *Environment and Planning A* **19**, 1525–42.
Dendrinos, D. S. & H. Mullaly 1985. *Urban evolution: studies in the mathematical ecology of cities*. Oxford: Oxford University Press.
Giddens, A. & J. H. Turner (eds) 1987. *Social theory today*. Oxford: Polity Press.
Goddard, J. & S. Openshaw. Some implications for the commodification of information and the emerging information economy for applied geographical analysis in the United Kingdom. *Environment and Planning A* **19**, 1423–40.
Harris, B. & A. G. Wilson 1978. Equilibrium values and dynamics of attractiveness terms in production constrained spatial-interaction models. *Environment and Planning A* **10**, 371–88.
Harvey, D. 1973. *Social justice and the city*. London: Edward Arnold.
Nijkamp, P. (ed.) 1986. *Handbook of regional and urban economics* Vol. 1: *Regional economics*. Amsterdam: North Holland.
Richardson, H. 1976. *The new urban economics and alternatives*. London: Pion.
Sheppard, E. 1987. A Marxian model of the geography of production and transportation in urban and regional systems. In *Urban dynamics: towards an integrated approach*, C. S. Bertuglia, G. Leonardi & A. G. Wilson 189–250. London: Routledge.
Turner, J. H. 1987. Analytical theorizing. In *Social theory today*, A. Giddens & J. H. Turner (eds), 156–94. Oxford: Polity Press.
Webber, M. J. 1987. Quantitative measurement of some Marxist categories. *Environment and Planning A* **19**, 1303–21.
Webber, M. J. & Tonkin 1987. Technical changes and the rate of profit in the Canadian food industry. *Environment and Planning A* **19**, 1579–96
Weidlich, W. & G. Haag 1983. *Concepts and models of quantitative sociology: the dynamics of interacting populations*. Berlin: Springer.
Wilson, A. G. 1970. *Entropy in urban and regional geography*. London: Pion.
Wilson, A. G. 1974. *Urban and regional models in geography and planning*. Chichester: Wiley.
Wilson, A. G. 1978. Spatial interaction and settlement structure: towards an explicit central place theory. In *Spatial interaction theory and planning models*, L. Lundqvist, F. Snickars & J. W. Weibull (eds), 137–56. Amsterdam: North Holland.
Wilson, A. G. 1981. *Catastrophe theory and bifurcation: applications to urban and regional systems*. London: Croom Helm.
Wilson, A. G. 1988a. Store and shopping centre location and size: a review of British research and practice. In *Store choice, store location and market analysis*, N. Wrigley (ed.), 160–86. London: Routledge.
Wilson, A. G. 1988b. Configurational analysis and urban and regional theory. *Sistemi Urbani*.
Wilson, A. G. & R. J. Bennett 1985. *Mathematical models in human geography and planning*. Chichester: Wiley.

Wilson, A. G. & M. Birkin 1987. Dynamic models of agricultural location in a spatial interaction framework. *Geographical Analysis* **19**, 31–56.
Wilson, A. G. & M. Clarke 1979. Some illustrations of catastrophe theory applied to urban retailing structures. In *Developments in urban and regional analysis*, M. Breheny (ed.). London: Pion.
Wilson, T. P. 1987. Sociology and the mathematical method. In *Social theory today*, A. Giddens & J. H. Turner (eds), 383–404. Oxford: Polity Press.

Part II
NEW MODELS OF THE CITY

Introduction

Nigel Thrift

Modern towns and cities often seems to have little in common but their diversity. There are the mega-cities of the world, vast assemblages of people and concrete with their pockets of extreme wealth and extreme misery. There are still a few single-industry, working-class towns in poignant contrast, and there are opulent stockbroker-belt towns and villages, rural representatives of a new international financial system. There are desert retirement communities. There are heritage towns and cities pandering to the tourist dollar. There are cities which are centres of government and the military. The list goes on and on. How can we meld all these towns and cities into a single, coherent vision?

Urban processes

Towns and cities can perhaps best be viewed as the loci of a number of processes interacting with one another in all manner of ways (Logan & Molotch 1987). First of all, and most fundamentally, the city can be viewed as a focus of production. As Leitner and Sheppard, and Pratt point out in their chapters, the mode of commodity production is an essential first step in understanding the nature of towns and cities. It provides the *raison d'être* for many of the activities that take place in a city. But production is a difficult business for geographers; each moment of production has its own set of geographies which interact with other geographies in a complex synthesis. There is the point of commodity exchange. That will include a complex geography of transport and distribution. It will also include the buying in of labour power which must be done within a labourmarket geography which is continually in flux (Clark 1981). Then there is the point of production. That will include, most especially, the labour process, which can be a complex geography in its own right. It will consist of detailed functions like design, planning, production, and marketing which can be spatially differentiated. Production of parts of a commodity can also take place at many places in the world before the product is finally made up. In sum, cities are spatial divisions of labour, embedded in even larger spatial divisions of labour. They are, and they are part of, a mosaic of uneven development and redevelopment, the result of a constant dialogue between the location of production and the location of labourforces.

Currently, there appear to be monumental shifts occurring in the structure of commodity production, in the urban systems of many countries. For example, new flexible labour processes and technologies seem to be leading to clusters of light industry in certain cities of the world, with interesting social consequences

(Scott 1988, Storper & Scott 1989). At the same time, the growth of high-level producer services has brought a reconcentration of office activity into the downtown areas of some larger cities. As Leitner and Sheppard point out, such developments are conditioned by the supply of property by commercial institutions, mediated by the state. Whether these and other new developments add up to a new regime of flexible or disorganized accumulation is still – for all the talk of a *fin de siècle* shift in the nature of capitalism – a moot point. The effects of these new developments on cities and city systems are even harder to foresee, simply because modern towns and cities are not just loci of production.

Thus we come to other ways in which the city is a locus of processes. Second, the city is a locus of money, finance and credit, of the manipulation of money by banking and commercial capital (Thrift 1987). The city has too rarely been seen in this way by geographers, yet it might be argued that the time has never been more ripe for such a vision to take hold. For example, it might be argued that much of the growth of producer services in cities has had very little to do with production *per se*, and rather more to do with the growth of banking and finance capital. Harvey (1985a, p. 190) argues that financial markets can 'acquire a certain autonomy *vis à vis* production' and currently that autonomy seems very great indeed. Many cities have become 'centres of mercantilist endeavour' (Harvey 1985a, p. 217) in which banking and finance have an identifiably separate spatial division of labour with important social consequences (Friedman & Wolff 1982) such as social polarization.

A third and related way in which the city can be viewed is as a locus of consumption. Hence the state can be seen as an important progenitor of the city through its rôle as a provider of public goods and services, of collective consumption. There is a very large literature now on the geography of collective consumption (see Pinch 1985), following the original work of Castells (1977), on the state as provider of schools, housing, health care facilities, and the like, and as guarantor of the reproduction of labour power. However this work is now beginning to show its age. It is work borne of the late 1960s and early 1970s when state provision and corporatist solutions were at their height in many Western societies. A comparable body of work on private sector consumption, of schools, housing, health care as well as entertainment, leisure, and the like has still to be constructed. Certain objects of consumption like housing are very well covered (Ball 1985, Ball *et al.* 1988), but other elements like schooling and health care are not, whilst the explosion of luxury consumption in many countries has hardly been touched upon. Yet, as Harvey (1985a) pointed out, the mode of consumption and the spatial division of consumption surely deserve more attention:

> Production is typically separated from consumption under capitalism by market exchange. This has enormous implications for urbanisation and urban structure. Work spaces and times separate out from consumption spaces and times in ways unknown in artisan or peasant culture. The moment of consumption, like that of production, stands to be further fragmented. Vacation, leisure, and entertainment places separate from spaces of daily reproduction, and even the latter fragment into the lunch counter near the office, the kitchen, the neighbourhood drugstore or bar. The spatial division of consumption is as important to the urban process as the spatial division of

> labour – the qualities of New York, Paris and Rome as well as the internal organisation of these and other cities could not be understood without consideration of such phenomena. This is however, a theme that remains underexplored in Marxian theory, in part because of the tendency to focus exclusively on production because it is the hegemonic moment in the circulation of capital (Harvey 1985a, p. 189).

The mode of production and the mode of consumption are more and more closely linked together by the increasing tendency for towns and cities to be designed to sell commodities. From the invention of the department store (Benjamin 1973) to the consumer spectacles of postmodern cities, this tendency seems to be becoming stronger, more intrusive, and more internalized.

Fourth, the city is a locus of the reproduction of subjects. Harvey (1982) counts the synthesis of the varied and varying work on the reproduction of labour power as perhaps the most urgent task facing Marxian theory, and it is difficult to demur:

> It is . . . a task that must be undertaken in the clear knowledge that the reproduction of labour power throughout the lived life of the working classes is a quite different dimension to the analysis of the capitalist mode of production. It is not a mere addendum to what we already know, it constitutes a fundamentally different point of departure to that upon which the theory of *Capital* is based. The starting point is not the commodity, but a simple event – the birth of a working class child. The subsequent processes of socialisation and instruction, of learning and being disciplined may transform that human being into someone who also has a certain capacity to labour and who is willing to sell that capacity as a commodity. Such processes deserve the closest possible study (Harvey 1982, p. 447).

The socialization of human beings, from cradle to grave, is the result of a complex interdigitation of influences which cannot easily be reduced to a theory. But what is clear is that any such theory will have to take into account the competing or integrating influences of the economy, state, and civil society (Urry 1981). Thus it will involve disciplinary processes and processes of resistance at work and in the labourmarket. It will also involve the state's rôle in schooling, law, and other means of control – all those practices that are part of Foucault's biopolitics, his demography of power. It will include, as well, the various dimensions of civil society and the way that they structure subjects in households and elsewhere, most especially gender, ethnicity, and religion.

The chief importance of such a tripolar format is to point to the way that diverse social institutions can *interact* to form active and creative subjects by producing systems of social positioning which are simultaneously objective conditions of existence and categories of perception, what Bourdieu (1977, 1984, 1987) calls 'habitus'. Thus Willis (1978) in his famous work in a Midlands town, Jenkins (1982, 1983) in his work in Northern Ireland, and Macleod (1987), have all shown how working-class school children are piloted into manual working-class jobs by categories of perception which quite realistically reflect their life chances, but which also assure that their acts of resistance to the prevailing order will only reinforce these categories.

Leading out of and back into the process of reproduction of subjects are all manner of important topics. In her chapter Mackenzie points to the importance of gender as a distinctive subfield of work on the political economy of urban processes. In some of its manifestations this work links to work in geography on the regulation of the subject by the state, much influenced by the interest in the work of Weber and Foucault on processes of rationalization. In such a scheme, the city appears as an enormous disciplinary technology for the production of human beings and their distribution into various categories – worker or capitalist, man or woman, for example – especially through the stereotyping of groups as deviant (the technology of normalization). The city is a carceral machine 'for distributing individuals, ordering them along a graded scale in any of a number of institutional settings' (Dreyfus & Rabinow 1982, p. 142).

Last, the city can be seen as a locus of meaning, as a means of representation, as a source of understanding, as an anchor for culture (in its widest sense) (Williams 1973). It is striking the degree to which the structure of the city is used in literature as a metaphor for social life (Timms & Kelley 1985). In the opening to *Bleak House* Dickens (1853, p. 1) provides a famous example:

> London. Michaelmas Term lately over, and the Lord Chancellor sitting in London's Inn Hall. Implacable November weather. As much mud in the streets as if the waters had been newly retired from the face of the earth, and it would not be wonderful to meet a Megalosaurus, forty feet long or so, waddling like an elephantine lizard up Holborn Hill. Smoke lowering down from chimney pots, making a soft black drizzle, with flakes of soot in it as big as full-grown snowflakes – gone into mourning, one might imagine, for the death of the sun. Boys, undistinguishable in mire. Horses, scarcely better – splashed to their very blinkers. Foot passengers, jostling one another's umbrellas, in a general infection of ill-temper, and losing their foothold at street corners, where tens of thousands of other foot passengers have been slipping and sliding since the day broke (if this day broke), adding new deposits to the crust upon crust of mud, sticking at those points tenaciously to the pavement, and accumulating at compound interest.

Here Dickens conjures up London as a gloomy purgatory of mud and soot and money, creating a lasting interpretation which would colour its inhabitants' perception in time to come. Yet Dickens provided another interpretation as well, of London transformed by locomotive power into a quintessentially modern metropolis, exciting, threatening, even unstoppable (Schwarzbach 1978, Welsh 1971). Here mud and soot have turned into money:

> As to the neighbourhood which has hesitated to acknowledge the railroad in its straggling days, that had grown wise and penitent, as any Christian might in such a case, and now boasted of its powerful and prosperous relation. There were railway patterns in drapers' shops, and railway journals in the windows of its newsmen. There were railway hotels, office-houses, lodging houses, boarding houses; railway plans, maps, views, wrappers, bottles, sandwich-boxes and timetables; railway hackney coach and cabstands; railway omnibuses, railway streets and buildings, railway hangers on and

> parasites, and flatterers out of all calculation. There was even railway time observed in clocks, as if the sun itself had given in. Among the vanquished was the master chimney sweeper . . . he now lived in a stuccoed house three storeys high and gave himself out, with golden flourishes upon a varnished board, as contractor for the cleansing of railway chimneys by machinery.
>
> To and from the heart of the great change, all day and night, throbbing currents rippled and returned, incessantly, like its life blood. Crowds of people, departing and arriving scores of times in every four and twenty hours produced a fermentation in the place that was always in action. The very houses seemed disposed to pack up and take trips. Wonderful Members of Parliament, who, little more than twenty years before, had made themselves merry with the wild railroad theories of engineers, and given them the liveliest rubs in cross-examination, went down into the north with their watches in their hands, and sent on messages before by the electronic telegraph, to say that they were coming. Night and day the conquering engines mumbled at their distant work, or, advancing smoothly to the journey's end, and gliding like tame dragons into the allotted coves grooved out to the inch for their reception, stood bubbling and trembling there, making the walls quake, as if they were dilating with the secret knowledge of great powers yet unsuspected in them, and strong purposes not yet achieved (C. Dickens 1848, *Dombey and Son*, Ch. 15).

Thus, the city is used as a cultural resource, as a way of creating and expressing meaning, as a social formation.

Cities are not just conjured up in the texts of novelists, poets, even of government functionaries (Darnton 1984). They exist as symbols and images, never more so than in modern life-style advertisements, where a single image can conjure up a whole series of reciprocal meanings and feelings. Even their architecture is built to provide meaning, from the glacial imagery of the corporate towers to the traditional veneer of neo-vernacular housing estates.

These meanings are continually contested in myriad ways (Clark 1985, Harvey 1985b). They have a political resonance. A classic example is Haussmann's boulevards built, in part, to impose a new cultural order on the old Paris (Clark 1985, Herbert 1988). But cities are not only the spaces of representation of dominant groups. Events can be so ordered as to allow subordinate groups to mount a symbolic take-over of urban spaces and use them for their own purposes to shout defiance at the dominant cultural order. The example of various Afro-Caribbean carnivals in London is a case in point of two opposing interpretations (Jackson 1988a). On the one side, such carnivals could be represented by the media as problems of social control, even as threats to civilization (Burgess 1985). On the other side, they could be represented as symbolic rejection of the normal order of things.

Cities in action

So far, towns and cities have been presented as the loci of a set of five interrelated processes. But it is also crucial to capture towns and cities in such a way that the active, social nature of their creation is stressed (Gottdiener 1985).

That represents a move from a rather clinical compositional analysis to something more contextual and less cold-blooded. But this is not an easy shift to make, involving, as it does, the move from social science to the humanities, from analysis to interpretation, from universals to local circumstances, from structure to agency, that is at the heart of the new regional geography (Thrift 1983, Gregory 1989, Pudup 1988, Sayer 1989). The problem is how to go beyond the conventions of representation but not so far that the conventions can no longer be seen. Harvey's (1985b) firmly compositional account of Second Empire Paris is almost Tolstoyan in its range and intensity, but ultimately

> the attempt to force all the material into a Marxist framework does not succeed. His own data on Paris shows the limitations of an orthodox perspective. He is forced to add other trends, concepts and theories in an ad hoc fashion . . . and these are not integrated within his overall perspective (Urry 1988, p. 68).

For Urry, Harvey is captive to the conventions. On the other hand less conventional, more contextual approaches to representation like postmodern ethnography have their problems too:

> Whereas self consciousness about method is an indispensable characteristic of any serious ethnography, the current vogue for introspection and stylistic experiment leads all too easily into empty posturing and sophistry. In using our analyses of other societies to reflect our own experience, there is a danger that the empirical content of our descriptions will evaporate altogether, leaving only a never-never land of refractive images for endless but groundless speculation (Jackson 1988b, p. 232).

Perhaps all that can be done for the moment is to turn to active social movements and the way that they inscribe the city with their presence, and the city inscribes them. In terms of the political-economy approach, the most obvious strategy is to turn to class, as Pratt does in her chapter. In capitalism, class is a basic social relation both integrating and dividing society. That relation can and should be studied in compositional ways (see Wright 1985). However, in recent years more attention has been paid in the geography of towns and cities to the study of class formation, to how people crystallize this relation in different forms according to the different contexts they find themselves in: people make history but not in circumstances of their own choosing (Thrift & Williams 1987). Such an historical–geographical analysis will need to take in elements of each of the five registers outlined above in one way or another in order to reach a judgement on the ways in which people have been able to experience class, the degree to which class intermingles with other social relations like gender, and the extent to which class emerges as a primary force in political struggle.

The geography of class in and of the city has been rather infrequently examined in such a comprehensive fashion as this in geography, although there are many instances of work in social history and anthropology that prove that such a project is possible and, indeed, desirable. One partial exception in geography is provided by the example of the gentrification literature which, viewed as a corporate project, has covered each register in specific contexts. Here there has been work on the economic aspects of gentrification, especially in terms of Smith's (1987) rent gap theory and work on labourmarket geography. There has been some work on the connections between gentrification and the rise of producer services (Hamnett 1984). There has been considerable work now on the relations between consumption, reproduction, and culture, in the shape of work on the rise of the service class and its links to a vernacular postmodern consumption culture that dignifies its perceptions and tastes (Jager 1986). A total approach in the context of a city is probably best exemplified by the work of Zukin (1988) on New York.

Another way in which the active, social nature of the creation of cities has been captured is found in the literature on the rise of urban social movements (Lowe 1986). This work, again founded by Castells, has devoted considerable time to considering the distinctively urban character of new social movements (such as those centred around gender, housing struggles, and ecology). Castells (1983) has shown what complex phenomena these new movements are, arising partly from local senses of grievance and partly from displacements arising from a certain kind of compositional class politics:

> The philosophical rationalism of the political left and the one dimensional culture of the labour movement meant that the social movements of industrial capitalism tended to ignore sub-cultures, gender specificity, ethnic groups, religious beliefs, natural identities and personal experiences. All human diversity was generally considered a remnant of the past, and class struggle and human progress would help to supercede it until a universal fraternity was arrived at that would provide, paradoxically, the ideal stage for both bourgeois enlightenment and proletarian marxism. Between times, people continued to speak their languages, pray to their saints, celebrate their traditions, enjoy their bodies and refuse just to be labour or consumers (Castells 1983, p. 230).

Castells finds that these movements tend to delineate themselves as urban, citizen or city-related and that they tend to materialize around three central goals, namely: (a) greater shares of collective consumption; (b) matters of cultural identity based in specific pieces of territory; and (c) a greater degree of political self-management (especially involving attempts to win a degree of independence from the local arm of government). Such movements tend to use a language of local community rather than a language of class or race or gender, so that they can neutralize certain issues (although these issues can return to haunt them). According to Castells, it is only when all three of these goals combine in the practice of an urban social movement that social change is likely to occur. The separation of any one goal from the others reduces the potential of the social movement, and recasts it in the rôle of an interest group, in which narrowly partisan form it can be easily institutionalized. The theory of urban social movements is both comprehensive and also underlines their fringe nature. It

> correctly emphasises that they are not ready made agents for structural change but rather 'symptoms of resistance to domination'. They have their

> roots in a radical sense of powerlessness and though their resistances may have important effects on cities and societies, they are best understood as defensive organisations which are unlikely to be able to make the transition to more stable forms of politics. This lends these movements certain strengths as well as the obvious weaknesses. The utopian strands in their ideology, which demand the immediate satisfaction of needs, require totalising, historically feasible plans for economic production, communication and government. The movements are unlikely to be able to apply these without losing the very qualities which make them dynamic and distinct. Their orientation towards local governments and political institutions, on the immediate conditions in which exploration and domination are experienced, is a result of the simple fact that those whose grievances give the movements momentum have no other choice. They lack any credible sense of democracy other than the grass roots variety practised in their own organisations. As Castells puts it, 'When they find themselves unable to control the world they simply shrink the world to the size of their own community'. The 'politricks' of the system is replaced by an authentic immediate politics (Gilroy 1987, pp. 231–2).

A further gloss on the work on new social movements is provided by Cox's (1984) work on turf politics which notes the rise of locality-based movements cemented in the suburban environment. Cox argues that the nature of these movements is based in a commodification of the neighbourhood which enables people to identify their local areas as objects of desire from which others can be excluded.

Contemporary urban change

Currently, three major (related) changes are sweeping across the cities of the western world. One of these changes consists of the impacts of the internationalization of capital (in its industrial, financial, and commercial forms) (Thrift 1986). The most noticeable impacts of this change are to be found at the top end of the global urban hierarchy, in the so-called 'world cities' (Friedmann & Wolff 1982, Dogan & Kasarda 1988), places like New York (Sassen 1988), London (Thrift *et al.* 1987), Los Angeles (Soja *et al.* 1983), and Tokyo. There, the impacts of internationalization are the result of two related causal processes. First, there are impacts generated by the international expansion of banking and commercial capital (Thrift 1987), such as the expansion of a service class of professionals and managers; in their turn these developments lead, for example, to large amounts of office building, gentrification, and the building of new neo-vernacular housing. Like the architecture, the culture is decidedly post-modern, consisting of a dose of commodity aesthetics (Haug 1986, 1987), flavoured with pastiche (Jameson 1984).

Second, there are the impacts wrought by the international expansion of mainly industrial capital, resulting in a large working-class labourforce. These labourforces provide a pool of labour for sweatshops, subcontractors of subcontractors, cheap clerical labour, labour to work in retailers, labour to work in public services (e.g. nurses), and so on. Often such labour consists of 'new helots' brought from the Third World (Cohen 1987, Sassen 1988). The culture is defensively ethnic, often enclaved. Clearly, circumstances will vary with local urban context. Thus, New York and Los Angeles have a larger, more recent immigrant population than Tokyo, which has only its Korean enclave and expatriates working in the financial sector.

The second major change has had dramatic effects throughout the urban hierarchy. It results from the expansion of the tertiary sector of the economy. Employment in services (a term which hides a multitude of different determinations) has continued to grow and has important impacts, for example in terms of the feminization of the labourforce and the growth of a service class of professionals and managers (Urry 1987, Lash & Urry 1987). The growth of service industries (and the service class) has changed the nature of the urban hierarchy in a number of ways increasing the importance of those towns and cities with an economic base rich in producer services, government functions, specialist retailing, tourism, and so on (see Noyelle & Stanback 1984, Thrift & Leyshon 1987). As Harvey (1985a) points out, it has also increased the level of interurban competition in all kinds of ways, such as competition for cultural investments, command functions, and government revenues.

The third major change consists of the continuing spread of cities outwards from their cores. (The influence of demography is still something of a lacuna in work on the political economy of cities, but see Clark (1987).) The age of Gottman's megalopolis has been reached, almost without it being noticed. What constitutes a town or a city nowadays, when a mosaic of specialized and carefully designed urban and rural areas sprawls over vast tracts of space, is an interesting question (Lefebvre 1989, Thrift 1987b).

Cities outside the Western world

Last but by no means least there are all the towns and cities outside the Western world – in the socialist world, and in capitalist and socialist developing countries. There is still too little work on these cities from a political-economy perspective in geography, as Chatterjee notes in her chapter on Third World cities (but see Armstrong & McGee 1985).

One of the chief areas of work has been on urbanization processes, both in socialist developing counties (Drakakis-Smith 1987, Forbes & Thrift 1987, Kirkby 1985, Massey 1987, Thrift & Forbes 1986) and in developing countries within the capitalist orbit (e.g. Storper 1984). Other important areas of work have included work on squatter housing (e.g. Burgess 1981), on urban social movements (Slater 1986), and on gender (e.g. Mommsen & Peake 1988). But the fact remains that most work on new urban models has persisted in equating the urban with the occident. The relative paucity of work elsewhere must be counted as a wrong to be righted in future research.

References

Armstrong, W. & T. McGee 1985. *Theatres of accumulation. Studies in Asian and Latin American urbanisation*. Andover: Methuen.

Ball, M. 1985. *Housing policy and economic power. The political economy of owner occupation.* London: Methuen.

Ball, M., M. Harloe & M. Martens 1988. *Housing and social change in Europe and the USA.* London: Routledge.

Benajamin, W. 1973. *Charles Baudelaire. A lyric poet in the age of high capitalism.* London: Verso.

Bourdieu, P. 1977. *Outline of a theory of practice.* Cambridge: Cambridge University Press.

Bourdieu, P. 1984. *Distinction. A social critique of the judgement of taste.* London: Routledge & Kegan Paul.

Bourdieu, P. 1987. What makes a social class? On the theoretical and practical existence of groups. *Berkeley Journal of Sociology* **3**, 1–20.

Burgess, H. 1985. News from nowhere: the press, the riots and the myth of the inner city. In *Geography, the media and popular culture*, J. Burgess & J. Gold (eds). Beckenham, Kent: Croom Helm.

Burgess, R. 1981. Ideology and urban redevelopment in Latin America. In *Geography and the urban environment*, vol. 4, D. T. Herbert & R. J. Johnston (eds). Chichester: Wiley.

Castells, M. 1977. *The urban question.* London: Edward Arnold.

Castells, M. 1983. *The city and the grassroots.* London: Edward Arnold.

Clark, G. L. 1981. The employment relation and the spatial division of labour. *Annals of the Association of American Geographers* **71**, 412–29.

Clark, G. L. 1985. *Judges and the cities. Interpreting local autonomy.* Chicago: Chicago University Press.

Clark, T. J. 1985. *The painting of modern life. Paris in the art of Manet and his followers.* London: Thames & Hudson.

Clark, W. A. V. 1987. Urban restructuring from a demographic perspective. *Economic Geography* **63**, 103–25.

Cohen, R. 1987. *The new helots.* Farnborough: Gower.

Cox, K. 1984. Social change, turf politics and concepts of turf politics. In *Public service provision and urban development*, A. Kirby, P. Knox & S. Pinch (eds). London: Croom Helm.

Darnton, R. 1984. A bourgeois puts his world in order: the city as a text. In *The great cat massacre and other episodes in French cultural history*, R. Darnton. 105–40. London: Allen Lane.

Dickens, C. 1848. *Dombey and Son*, 1970 edn. Harmondsworth: Penguin.

Dickens, C. 1853. *Bleak House*, 1972 edn. Harmondsworth: Penguin.

Dogan, M. & J. Kasarda (eds) 1988. *The metropolis era* (2 vols). Newbury Park: Sage Publications.

Drakakis-Smith, D. W. (ed.) 1987. Socialist development in the Third World. *Geography* **72**, 333–63.

Dreyfus, H. L. & P. Rabinow 1982. *Michel Foucault. Beyond structuralism and hermeneutics.* Brighton: Harvester.

Forbes, D. K. & N. J. Thrift (eds) 1987. *The socialist Third World.* Oxford: Basil Blackwell.

Friedmann, J. & G. Wolff 1982. World city formation: an agenda for research and action. *International Journal of Urban and Regional Research* **6**, 309–44.

Gilroy, P. 1987. *There ain't no black in the Union Jack.* London: Hutchinson.

Gottdiener, M. 1985. *The social production of urban space.* Austin: University of Texas Press.

Gregory, D. 1989. Areal differentiation and post-modern human geography. In *New horizons in human geography*, D. Gregory & R. Walford (eds). London: Macmillan.

Hamnett, C. 1984. Gentrification and residential location theory: a review and assessment. In *Geography and the urban environment*, D. Herbert & R. J. Johnston (eds), vol. 6, 283–319. New York: Wiley.

Harvey, D. 1982. *The limits to capital*. Oxford: Basil Blackwell.
Harvey, D. 1985a. *The urbanisation of capital*. Oxford: Basil Blackwell.
Harvey, D. 1985b. *Consciousness and the urban experience*. Oxford: Basil Blackwell.
Haug, W. F. 1986. *Critique of commodity aesthetics. Appearance, sexuality and advertising in capitalist society*. Cambridge: Polity Press.
Haug, W. F. 1987. *Commodity aesthetics, ideology and culture*. New York: International General.
Herbert, R. L. 1988. *Impressionism. Art, leisure and Parisian society*. New Haven, Conn.: Yale University Press.
Jackson, P. 1988a. Street life. The politics of Carnival. *Environment and Planning D, Society and Space* **6**, 213–27.
Jackson, P. 1988b. Review of Marcus and Fischer, Anthropology and cultural critique. *Environment and Planning D, Society and Space* **6**, 231–2.
Jager, M. 1986. Class definition and the aesthetics of gentrification. Victoriana in Melbourne. In *Gentrification of the city*, N. Smith & P. Williams (eds), 78–91. Hemel Hempstead: Allen & Unwin.
Jameson, F. 1984. The cultural logic of late capital. *New Left Review*, **146**, 53–92.
Jenkins, R. 1982. *High town rules. Growing up in a Belfast housing estate*. Leicester: Leicester Youth Bureau.
Jenkins, R. 1983. *Lads, citizens and ordinary kids. Working-class youth lifestyles in Belfast*. London: Routledge & Kegan Paul.
Kirkby, R. 1985. *Urbanisation in China*. London: Croom Helm.
Lash, S. & J. Urry 1987. *The end of organised capitalism*. Cambridge: Polity Press.
Lefebvre, H. 1989. *The production of space*. Oxford: Basil Blackwell.
Logan, J. R. & H. W. Molotch 1987. *Urban fortunes. The political economy of place*. Berkeley: University of California Press.
Lowe, S. 1986. *Urban social movements. The city after Castells*. London: Macmillan.
Macleod, J. 1987. *Ain't no making it. Levelled aspirations in a low income neighbourhood*. Boulder, Colorado: Westview Press.
Massey, D. 1987. *Nicaragua: some urban and regional issues*. Milton Keynes: Open University Press.
Mills, C. A. 1988. Life on the upslope: the postmodern landscape of gentrification. *Environment and Planning D, Society and Space* **6**, 169–89.
Mohan, J. 1988. Spatial aspects of health care employment in Britain 1: Aggregate trends. 2: Current policy initiatives. *Environment and Planning A* **20**, 7–24, 203–18.
Mommsen, J. H. & L. Peake (eds) 1988. *Women in the Third World*. London: Longman.
Noyelle, T. J. & T. M. Stanback 1984. *The economic transformation of American cities*. Totowa, NJ: Rowman & Allanheld.
Pinch, S. 1985. *Cities and services. The geography of collective consumption*. London: Routledge & Kegan Paul.
Pinch, S. 1989. The spatial division of services. *Environment and Planning A* **21** (forthcoming).
Pudup, M. 1988. Arguments within regional geography. *Progress in Human Geography* **12**, 369–90.
Sassen, S. 1988. *The mobility of capital and labour. A study in international investment and labour flow*. Cambridge: Cambridge University Press.
Sayer, A. 1989. The new regional geography and problems of narrative. *Environment and Planning D, Society and Space* **6** (forthcoming).
Schwarzbach, F. S. 1978. *Dickens and the city*. London: Allen Lane.
Scott, A. J. 1988. *Metropolis*. Berkeley: University of California Press.
Slater, D. (ed.) 1986. *Social Movements*. Amsterdam: Centre for Latin-American Research and Documentation (CEDLA).
Smith, N. 1987. Of yuppies and housing: gentrification, social restructuring and the urban dream. *Environment and Planning D, Society and Space* **5**, 151–72.

Soja, E., R. Morales & G. Woolf 1983. Urban restructuring: an analysis of social and spatial change in Los Angeles. *Economic Geography* **59**, 195–230.
Storper, M. 1984. Who benefits from industrial decentralization? Social power in the labour market, income distribution and spatial policy in Brazil. *Regional Studies* **18**, 143–64.
Storper, M. & A. J. Scott 1989. The geographical foundations and social regulation of flexible production complexes. In *The power of geography*, J. Wolch & M. Dear (eds), 21–40. London and Winchester, Mass.: Unwin Hyman.
Thrift, N. J. 1983. On the determination of social action in space and time. *Environment and Planning D, Society and Space* **1**, 23–57.
Thrift, N. J. 1986. The geography of economic disorder. In *World in crisis*, R. J. Johnston & P. J. Taylor (eds), 12–67. Oxford: Basil Blackwell. (Reprinted in *Uneven redevelopment* (1988), D. Massey & J. Allen (eds), 6–46. London: Hodder & Stoughton.)
Thrift, N. J. 1987a. The fixers. The urban geography of international commercial capital. In *Global restructuring and territorial development*, J. Henderson & M. Castells (eds). London: Sage Publications.
Thrift, N. J. 1987b. Manufacturing rural geography. *Journal of Rural Studies* **3**, 77–81.
Thrift, N. J. & D. K. Forbes 1986. *The price of war. Urbanisation in Vietnam, 1954–1985.* Hemel Hempstead: Allen & Unwin.
Thrift, N. J. & A. Leyshon 1987. The gambling propensity: banks, developing country debt exposures and the new international financial system. *Geoforum* **19**, 55–69.
Thrift, N. J., A. Leyshon & P. W. Daniels 1987. Sexy greedy. The New international financial system, the City of London and the south east of England. Producer Services Working Paper 8.
Thrift, N. J. & P. Williams (eds) 1987. *Class and space.* London: Routledge & Kegan Paul.
Timms, E. & D. Kelley 1985. *Unreal city. Urban experience in modern European literature and art.* Manchester: Manchester University Press.
Urry, J. 1981. Localities, regions and social class. *International Journal of Urban and Regional Research* **5**, 455–74.
Urry, J. 1987. Some social and spatial aspects of services. *Environment and Planning D, Society and Space* **5**, 5–26.
Urry, J. 1988. Review of Harvey, Studies in the history and theory of capitalist urbanisation. *Antipode* **20**, 66–8.
Welsh, A. 1971. *The city of Dickens.* Oxford: Clarendon Press.
Williams, R. 1973. *The country and the city.* London: Paladin.
Willis, P. 1978. *Learning to labour.* Farnborough: Gower.
Wright, E. O. 1985. *Classes.* London: Verso.
Zukin, S. 1988. *Loft living* (revised edn). London: Radius.

3 *The city as locus of production*

Helga Leitner & Eric Sheppard

Introduction

Economic activities in cities, like those in capitalist society as a whole, may be broadly divided into two groups. The first contains activities primarily associated with the production of commodities for profit, including not only the production of capital goods (such as semi-finished products), wage goods (those consumed as final demand), and buildings, but also the various producer services which make the co-ordination of production possible. The second contains activities primarily associated with the consumption of commodities by households: housing, transportation, and other goods necessary for the social reproduction of households within the city, and the various institutions which influence and organize consumption behaviour. The topic of this chapter is the former: the location of commodity production.

The changing location of commodity production in cities has been examined in a number of ways. In order to organize and evaluate this body of literature we begin by identifying some of the principal considerations which an adequate theoretical account of this subject should address. These may be divided into two broad issues: the treatment of societal structure and individual behaviour, and the inclusion of both supply- and demand-side factors influencing location.

Many authors have pointed to a failure in economic geography, as in social science in general, to integrate analyses of the evolution of society with detailed explanations of individual actions (cf. the chapters by Smith and Schoenberger in vol. 1). The conventional literature briefly reviewed below as traditional approaches has tended to concentrate at the latter, individualist scale. In general, certain attributes of the commodities produced and of the methods of production used are taken to be the cause of the location of commodity production in cities. Thus neoclassical theorists assume that a set of production techniques, a given quantity of capital and labour inputs, and certain consumer preferences exist and then derive location as a rational response to these factors. Those using either a location factor approach or the Lowry model assume that the type of commodity being produced dictates either the method of production or the degree to which consumer demand depends on distance from the production facility, and then derive location patterns from these characteristics. One immediate effect of this strategy is that the resulting explanations are ahistorical; changes in production technologies, for example, can only be introduced as *deus ex machina*. More seriously, we shall argue that what is taken as exogenous in these explanations is in fact an outcome of the evolution of the broader socio-economic system; which in this case is the system of social and production relations characteristic of capitalism.

Transportation technologies, for example, are not invented by chance at a certain point in time thereby affecting intra-urban location. Rather, there is an industry providing transportation as a commodity. Innovations in the transportation industry depend in part on competition in that industry and in part on the need (expressed as demand) for new transportation products from other industries, both of which depend on broader social, economic and, geographic conditions in society (see the section on manufacturing decentralization). In reality, then, there is not some given backcloth of technologies, working conditions, and preferences against which individuals seek to maximize their economic wellbeing. Rather these contain within them both the structure and history of society, as it evolves and creates the particular mix of conditions of production which represents the situation faced by economic actors in some city and at a particular point in time.

If we neglect to examine the societal factors which operate at various spatial scales to structure local conditions, our analysis will by default concentrate on individual responses to these conditions, giving the impression that the location of production in cities can be understood as the result of freely taken economic decisions – an excessively voluntaristic conclusion. The political-economic approach avoids this bias by delving deeper. The production and social relations of the capitalist system structure the responses of individuals in far deeper ways than they imagine, and it is by digging below the appearances of the conditions of production to reveal the classes and conflicts they hide that a more penetrating analysis of intra-urban production is possible. We review the political-economic approaches below (see our discussions of production, social relations and location and then of the supply side) and see that their first, and sometimes only concern is to examine how the location of commodity production in cities has its origins in the process of production and thereby in the political and social forces immanent in that production process. We shall also see how the global reach of both capitalism as a mode of production, and the individual corporations engaged in commodity production, implies an analysis which must also involve examining how the situation of a city within the international economy influences the conditions faced by local producers.

It is important to note that this conceptualization should not be interpreted as implying that individual actions taken in the city are unimportant because they are structurally determined by the mode of production. These actions must also be studied because the mode of production cannot fully determine behaviour: the ingenuity of human response to even the most draconian constraints can provide an unpredictability and variety that we should attempt to understand and not suppress in our analysis. The eventual influence of behaviour on societal structures also cannot be ignored, particularly if our analysis is to be historical.

The second dimension to an adequate explanation is ensuring that all important factors influencing the changing intra-urban location of commodity production are included. We divide these factors into two groups, which we label demand and supply (Walker 1981). On the demand side, the production requirements for producing a certain commodity influence the locations capitalists seek. In the section on production, social relations, and location we survey research which attempts to explain how capitalists' decisions about the kind of production process, and the location at which to employ it, are structured by the processes through which the conditions of production evolve. This understanding is, however, insufficient. In the city, the supply of sites for commodity production depends on the availability of land and buildings, which are either suitable for the activity or can be converted into suitable facilities at reasonable cost. The cost of land and buildings is often conceptualized as land rent, and we consider attempts to explain the production and social relations influencing land rents below in our discussion of Marxian rent theory and the concept of the rent gap. The production of the built environment, however, is itself commodity production, where for our purposes the commodity being produced is commercial property. Therefore, we also examine the details of this production process; the private and state induced actions through which commercial property is produced (under our discussion of the economics of commercial property development and then of state intervention in the property develoment process).

Traditional approaches

Traditional approaches to modelling the geography of economic production within metropolitan areas can be broadly divided into three categories: the neoclassical approach, the location factor and morphological approaches, and the Lowry model. Space allows for only the briefest summary of these approaches.

The micro-economic neoclassical approach comes closest to formulating a single coherent theory. This conceptualizes the intra-urban location of economic activities as the result of a bidding process for land regulated by the urban land market, very much in the manner pioneered by von Thünen (1826) for agriculture (Muth 1962, Alonso 1964). Alonso assumed that a city is made up of a large number of individual households and entrepreneurs all wishing to purchase land somewhere in the city. Individuals have a price they are willing to pay for any location in the city (the bid rent), which is the highest at the city centre. The bid rent offered by entrepreneurs equals the profit per acre that can be made at a location, after the costs of transporting the product to a market in the city centre have been subtracted.

The land market operates by simply allocating each piece of land to the individual who can offer the highest bid rent at that location. The result is a series of concentric land use zones centred on the city centre. Typically, it is assumed that office activities profit most from being downtown, but have the most to lose from being at non-central locations due to their need to be at the centre of information networks (Goddard 1975). Thus, compared to other activities their bid rents are highest near the CBD but fall off most rapidly with distance. Alonso also hypothesized that if rich households had a stronger preference for open space than poor households, they would be more tolerant of long commutes and thus have a lower and flatter bid rent curve; a hypothesis that, although never tested, is nevertheless often simply assumed to explain the suburbanization of middle- and upper-income households in America. Casual empiricism in the American context has tended to support this pattern. As a result this has turned out to be a tremendously popular approach in North

America. It in turn has generated the school of New Urban Economics (NUE), attempting to generalize Alonso's model into a complete neoclassical model of the spatial structure of the city (Richardson 1977).

The main addition that NUE brought was an explicit consideration of commodity production. Typically, each sector is assumed to have a given technology, which determines the profit maximizing mix of capital, labour, and land for any combination of prices for these factors. The bid rent offered by producers of a particular commodity at a given distance from the city centre is equal to the marginal productivity of land multiplied by the net revenue at the point of production. Land is allocated to the highest bidder, assuming perfect competition and that markets are cleared (cf. Mills & Hamilton 1983). Extensions and improvements within this approach, adding any number of rococo twists, abound in the urban economics literature. Some of the more interesting additions include: several income groups (Yinger & Dantzinger 1978); racial prejudice (Rose-Ackerman 1977); multi-centred cities (Papageorgiou & Casetti 1971), and the aging of buildings (Brueckner 1980).

In neoclassical economics, production is rooted in the supply of, and demand for, factors of production (land, labour, capital, and technical knowledge). The production process itself is reduced to a mathematical production function, which dictates how much of each factor should be used given the prices at which they are available. So much effort goes into conceptualizing markets of exchange (as can be seen in standard economics texts that always start with the laws of supply and demand), that even production is virtually reduced to the exchange of production factors. Production itself is simply a technical issue (Pasinetti 1981, Ch. 1).

The technology assumed in production functions dictates that the productivity of any production factor decreases if the relative physical abundance of that factor increases. Under perfect competition, the price of each production factor equals its marginal productivity; consequently, the price of a factor falls as its abundance increases. This explains, for example, why high-density development occurs downtown. Land is scarce and expensive there and should be economized on.

With its emphasis on exchange, the neoclassical model completely neglects the question of how factors of production are themselves produced. It turns out that if the production of capital goods and the reproduction of the labourforce are incorporated into the picture then the internal logic of the neoclassical model falls apart. Indeed, the space-economy no longer operates harmoniously at all, but is riven with unemployment, capital over- and underutilization, and social conflict (Sheppard & Barnes 1986). Thus the problems of the neoclassical approach to the intra-urban geography of production do not really stem from its often criticized depiction of an over-simplified radially symmetric city, but lie much deeper – in its model of capitalism.

The **location factor** approach is based on a large number of case studies that identify location factors pulling certain kinds of production activities to particular locations. Examples include high fashion textiles and business services congregating in the city centre due to external and agglomeration economies (cf. Hoover & Vernon 1959, Daniels 1975), and manufacturing activities which relocate to the suburbs given their land requirements and the location of their markets (cf. Yeates & Garner 1980). In such analyses, the authors identify immediate factors to which entrepreneurs respond, and then attribute their importance in a particular situation either to the nature of the commodities being produced or sold, or to the production method employed.

The **morphological approach** is a variant of the location factor approach. In this case particular location patterns are identified, and the existence of these patterns is explained by reference to location factors. This has been most widely used in discussion of the location of commercial (retail) activities in cities, drawing on the work of Berry *et al.* (1963; see also Yeates & Garner 1980, Lichtenberger 1986) to identify ribbon developments, centres (including shopping centres), and 'special districts'.

From the many empirical studies performed, a long list of location factors for industry and services emerges, factors whose importance varies from case to case (cf. Scott 1982a). Each case study might be seen as a particular application of some grand location theory consisting of a complete list of all possible location factors. Can such a list, however, really be called a location theory? A theory is supposed to be a concise and penetrating explanation, but such a list does not account for the relations between the various factors. Furthermore, explanations of the relative importance of factors often do not go beyond saying that they were identified as such by entrepreneurs, or that they reflect characteristics of the commodity or of the methods of production used.

The Lowry model begins with a division of production activities into two groups: basic and non-basic. Basic activities are those which sell their product mostly outside the city. Non-basic activities are those for which the market is predominantly local, and are often identified as services. Once the location of basic industry is known, workers and their families are assumed to cluster around these workplaces, allowing for transportation costs and the residential capacity of different areas. Non-basic retail activities are then located close to these workers. These non-basic activities hire more workers, creating a secondary local demand for housing and retailing, and so on until all multiplier effects of the initial basic industries have been exhausted (Lowry 1964).

The Lowry model has been widely used in Britain and Sweden to develop operational models of land use patterns (Wilson *et al.* 1977, 1981, Khakee 1985). Its developments and shortcomings have been detailed by Webber (1984). There is no theory of the location of basic activities. Accessibility to the consumer is assumed to be the paramount determinant of the location of non-basic activities, and this in turn is governed by the need for workers to be close to their jobs. There is no attempt to investigate why this should be so; it is simply taken as self-evident that we are all concerned with how long it takes to get to work or to travel to shop.

Within the confines set for itself, the research briefly summarized in this section is often competent and useful. None of these approaches, however, explicitly attempts to ground the explanation in any broader picture of society. There are underlying reasons why a particular production technology is used at a particular place and time (Lipietz 1984), why particular retail services have become important (cf. Miller 1983), or even why accessibility has come to play an important rôle in everyday lives (Rose 1981, Walker 1981). Furthermore, approaches which do not question the origins of the location factors which we observe to be important can have a distinctly status quo bias. 'What is central to the reproduction of capitalist societies as capitalist . . . is the normal routine of "going to work" . . . [yet it is through] the taken-for-granted character of [such] social interaction [that] capitalist societies . . . are routinely, though contingently, reproduced.' (Hudson & Sadler 1986, p. 185; see also Sheppard 1980.)

Production, social relations, and location

Recent years have seen increased attention paid to developing an explanatory framework for the location of production in the city which attempts to ground locational behaviour more adequately in a theorization of the conditions of production. Studies of 1870s Chicago demonstrated the historical specificity of location factors. Fales & Moses (1972) showed that high intra-urban transportation costs in the 1870s drew heavy industry close to central rail and water terminals for interurban transportation in a manner consistent with Alfred Weber's theory of industrial location, turning downtown neighbourhoods into working-class neighbourhoods (Moses & Williamson 1967). These papers attempted to apply what has been dubbed the theory of production approach (Moses 1958, Norcliffe 1984). This approach does not explain the location of production by the location factors observed to be important for a particular industry but rather takes as its starting point the current technological conditions which represent the method of production used by firms in a particular industry. From these technological conditions it is then possible in principle to deduce which location factors are important.

While these studies showed how changes in production technologies influence the location of industry, they did not delve into the question of how an understanding of the evolution of society can in turn account for the changes in technologies, products, and production processes that then determine which location factors are important. The researchers whose work we examine take the level of analysis one step further to do exactly this. The predominant theme in this literature is the attempt to understand location choice. It is, then, location theory with a political-economic foundation; attempting to explain the demand for different locations within the city (as measured by the profitability of those locations) within a broader theory of the evolution of capitalism. Interestingly, and despite the anti-empiricist bias of political-economic approaches, this literature can be divided chronologically. Attempts to explain the decentralization of production in cities were written at a time when there was much concern about the movement of industry to non-metropolitan areas. Subsequent work emphasizing the spatial clustering of these activities is occurring in a period when old city centres are experiencing a resurgence of growth.

Manufacturing decentralization

Two phases in the decentralization of manufacturing within American cities have been scrutinized by political-economic theorists: the movement at the turn of the century away from the kinds of urban core locations described by Fales and Moses, and the postwar exodus of industry from central cities into suburban and non-metropolitan areas. Conventional explanations of the former phase stress the rôle of new transportation technologies (notably the truck), and new, capital intensive, space consuming production methods (Borchert 1967, Logan 1966). A re-examination of these theses by Gordon (1978) and Walker (1978) shows, however, that decentralization was underway before the truck became available. Gordon also notes that new technologies implemented at suburban locations (he uses the example of open hearth furnaces for steel production in Gary, Indiana on the eastern fringe of Chicago) were sometimes less space consuming than old technologies. Thus changing technologies are better seen as a part of societal developments that included the decentralization of industry, not as the exogenous cause of decentralization.

In order to gain insight into the societal conditions influencing decentralization of manufacturing from the downtown at the turn of the century, it is necessary to understand the conditions which led to choice of the downtown in the first place. Webber (1982) identifies both economic and geographical conditions which accounted for the development of many industries at downtown locations. Economically, manufacturing production in the 19th century was more competitive than is the case today, with firms acting as price takers not price makers. This meant that the way to realize profits was to reduce production costs. In addition, as manufacturing industry became the core of capital accumulation, it both inherited and created a spatial economy in which the advantages of low transportation costs, high labour accessibility, and good access to markets could be found together at city centre locations. Some cities attracted manufacturing activities. These tended to be the larger ones growing around key transport nodes such as ports, chosen because of their locational advantages and also because of lower social resistance to the discipline of factory life (Gordon 1978). New industrial cities (such as Pittsburgh in the US or Birmingham in the UK) also developed, growing up around the locations of manufacturing plants, because in that era the location of labour was determined by capital rather than the reverse (Webber 1982). In either event, the advantages of downtown locations attracted manufacturing industry which then reinforced the advantages of these locations in a cycle of cumulative causation.

Yet the very spatial and economic structure thus created, when interpreted in terms of the evolving relations of capitalism, sowed the seeds for its own dissolution. First, the concentration of workers in downtown neighbourhoods intensified the class struggle between workers and their employers in the inner city. Gordon (1978) cites evidence to show capitalists' awareness of growing conflicts, which could be avoided by decentralizing production. Walker (1978) notes it became increasingly undesirable for factory owners themselves to live in the proximity of the working class, providing a further impetus to decentralization. Second, the very dynamics of competitive capitalism drove individual entrepreneurs to try and gain advantage over their competitors by adopting more efficient, capital intensive production methods; taking over the markets and plants of weaker competitors; and continually searching for new markets and products. These economic forces also encouraged decentralization: many of the new production methods required large plots of land, and the search for new markets meant that firms were less likely to regard residents of the city where they located as their primary market and thus saw city centre locations as less important (Webber 1982).

Furthermore, factors associated with class struggle and capital accumulation interacted. Thus labour troubles in the inner city would no doubt stimulate employers to a more rapid search for locational and technological strategies to reduce costs, and enthusiastically to adopt transportation and production technologies which allowed them to substitute both capital for labour and urban fringe for central city locations. Furthermore, transportation was not simply a technology, but was an industry like any other. The producers of transportation were under similar pressures to adopt more efficient production methods, both to save labour costs and to sell a better (i.e. faster and cheaper) product than their competitors. The exact nature of these interactions and the relative strength of the two sets of factors is uncertain, but the general result was the decentralization of industry since both sets had a centrifugal effect given the spatial structure of the 19th-century city.

A similar approach that roots locational change in the evolution of capitalist society has been applied to the relocation of economic activities in American cities during the last 15 years. Initial concern was with the suburbanization of industry; a response to empirical analyses of the 1970s showing a loss of manufacturing from central cities and from large metropolitan areas in general, particularly those of the American manufacturing belt. Scott (1982b, p. 191) described this as a 'pervasive tendency to vacate the cores of large metropolitan areas and to spread progressively outward.' Muller (1981) and others documented similar trends for office space and, in the context of the long established suburbanization process in general, the vision emerged of central cities progressively losing all their locational advantages by comparison to suburban and non-metropolitan locations. Even Sunbelt cities, seen as the destinations of activities abandoning the manufacturing belt, were shown to exhibit similar trends (cf. Rees 1978). Two kinds of explanations were initially proposed for this trend: urban agglomeration diseconomies, and the product cycle.

Larger and older cities, it is argued, exhibit **urban agglomeration diseconomies** due to congestion, high rents, and an outdated built environment. The technologies of production and intra-urban transportation in the postwar era enabled industry to avoid these by moving to the suburbs. This suggestion, that there is some inherent economic attribute associated with the size and age of cities which affects their attractiveness for population and industry, has been heavily criticized. Theoretically it makes little sense to treat cities primarily as isolated systems whose economic growth depends on their internal characteristics (Richardson 1973). Empirically this simply did not explain why larger older SMSAs such as Boston and Cleveland were decentralizing more slowly than newer smaller SMSAs like Miami or Atlanta, or why Los Angeles was experiencing a centralization of manufacturing when Dallas–Fort Worth showed a relative decentralization (Scott 1982b, Table 2; data for 1972–7). From our perspective, while inner-city locations do have high rents, congestion, and out-dated built environment and so forth, representing the conditions of production to which capitalists were responding, once again we need to understand how these have resulted from the evolving production process under capitalism.

According to the **product cycle** thesis, the commodities produced by entrepreneurs go through a life cycle from experimental products produced in a non-routine way, to standardized and established products that are mass-produced by routine methods (see Schoenberger, Ch. 5, vol. 1). In the early stages locations in the metropolitan core, the seedbeds of entrepreneurial invention, are attractive (although this is disputed: Struyk & James 1975, Leone & Struyk 1976). In later stages, however, decentralized locations using less skilled labour are more profitable. Thus the maturing of product cycles brings about the decentralization of production (Norton & Rees 1979). The product cycle approach, like the location factor literature discussed above, assumes something inherent in all commodities means they go through a product cycle. Thus the cause of decentralization lies in the nature of products, not the nature of the social system under which they are produced (Storper 1985). A more recent reformulation of this idea by Markusen (1985), in terms of the cycles of profitability of economic sectors, goes some way towards showing how such cycles are not a feature of products but a feature of corporate capitalist competition. There is still, however, an essentialist tone to the argument. Accepting the empirical existence of these cycles in some cases, it is still hard to understand why there should be a universal sequence of stages through which all industries must pass.

Scott (1980, 1982b) attempts a more thoroughgoing analysis. He sees the locational strategies of entrepreneurs as responses to the evolving relation between labour and capital. He argues that capitalist entrepreneurs were driven progressively to substitute capital for labour because this increases the productivity and profitability of production, thus enhancing capital accumulation. This then brought about a fourfold change in production methods: more efficient technologies using larger plants; greater standardization of production; the use of less skilled labour; and the geographical separation of blue- from white-collar work (Scott 1982b; pp. 191–2). On the one hand, he sees trends to larger facilities and integrated production causing the demise of older downtown industrial agglomerations of smaller plants, while on the other hand the suburbanization of industry is encouraged because suburban locations have higher wages and more land, both of which give a comparative advantage to capital intensive single-storey production facilities (Scott 1980). If office activities decentralized more slowly that was because of the difficulty of adopting mass production techniques in the sector. In the case of back office activities, however, which require a large, educated but not highly trained labourforce (due to the labour intensive but routinized nature of such activities as answering queries about customers' charge accounts), suburban locations have a comparative advantage in the form of a large, docile, and cheap labourforce of suburban housewives (Nelson 1986).

While Scott's approach goes some way towards a deeper analysis, it still leaves something to be desired. First, his use of trade theory to establish why suburbs have a comparative advantage for large-scale mechanized production is flawed. He uses a theory of comparative advantage whose predictions are inconsistent with the political-economic approach (Steedman 1979, Webber 1982, Sheppard & Barnes 1986). Second, in concentrating on individual capitalists adopting new technologies he tends to neglect discussion of the societal forces driving these changes. Thus Norcliffe (1984) can misinterpret Scott's work as falling into the 'theory of production' tradition. Both Webber (1982) and Gordon (1978) provide an analysis that attempts to overcome the latter shortcoming. Essentially, their proposition is derived from the argument that capitalism has undergone a transition from competitive to corporate capitalism. This is characterized by the concentration of production into fewer, larger corporations with: larger, national and global, marketing strategies; the monopoly power that enables them to influence prices; and the resources to develop and implement increasingly automated and geographically fractionated production systems. As a result, individual plants are not oriented as much to local suppliers and customers but are more likely to be buying inputs from, and shipping their product to, distant corporate affiliates. These dynamics fuel the technological changes described by Scott.

The clustering of production activities

The emphasis on decentralization has more recently been balanced by realization that industry is not simply diffusing across the landscape but coalescing in new clusters. Areas such as Silicon Valley are new centres for clusters of fast-growing production activities; old metropolitan areas such as Boston show unexpected new life as they make the socially costly but economically beneficial transition to high technology (Harrison 1982); and CBDs no longer stagnate as the last activities to decentralize pick up and leave, but develop new clusters of skyscrapers and new economic momentum. In a sense the work to be described in this section had a tendency to treat decentralization as a universal phenomenon. It is now better recognized that all activities do not dance together to the tunes of centralization and decentralization. Rather, while some activities within a sector, or within a corporation, decentralize others cluster in space, as empirical studies in Los Angeles and Toronto have tentatively shown (Scott 1983b, 1984, Gad 1985).

It is certainly not novel to examine the clustering of economic activities in cities and the agglomeration economies which bring this about. Detailed studies of this date back to the 1920s, as Scott has reminded us. The novelty lies in the argument that current production conditions allow an ever more rapid redefinition of the activities which can be centralized and decentralized, and of the places where this occurs. If we define such 'territorial clusters' (Scott 1986) as the core of the urban or regional economy, there is no permanent core or periphery, and there is no steady diffusion of development from core to periphery. Rather, we can observe continual and unpredictable reorientations of the space economy (Scott & Storper 1986, Clark 1980).

In seeking an explanation for this situation, Scott (1983a, 1986) suggests that '*[t]he most essential* element . . . is . . . organizational and labour market processes . . . and . . . the dynamics of the social division of labor' (Scott 1986, p. 26; emphasis in original). He argues that geographical clustering of firms occurs as a result of vertical and horizontal disintegration; the replacement of large single enterprises by interdependent groups of smaller ones. Disintegration occurs when the following conditions hold: different components of the manufacturing process do not need to be integrated into one plant; the market for the product is uncertain and unstable; specialized producer services are more economically subcontracted out; and segmentation of the labourmarket makes it possible to subcontract out stages of the production process. Geographical clustering results when the linkages are small-scale, uncertain, and difficult to negotiate. The existence of such clusters, furthermore, changes the conditions of production in such a way as to encourage further disintegration and clustering. Scott's contribution lies in integrating the rôles of subcontracting, global sourcing, and fractionated production (see the chapter by Schoenberger, vol. 1 for a more detailed description of these practices) into explanations of agglomeration and external economies, thus introducing the important dimension of linkage whose existence and content are, at least in part, controlled by the major producers who co-ordinate production (Holmes 1986, Cowling 1986). This brings the concept of external economies up to date with contemporary corporate practices (cf. Walker & Storper 1981, Storper & Walker 1983).

While Scott strongly hints that the location and production strategies of producers, and the conditions to which centralization and decentralization react, are grounded in the evolution of capitalism, he again does not make these connections satisfactorily. He pays more attention to the labourmarket and the capital–labour relation as a consequence of such changes than as a cause. We wish to understand, however, why it is that the new methods of production described by Scott have become important. Without attempting to sketch the evolution of the global economy, suffice to say that the historical concentration of industrial production in certain countries, and in particular within certain cities and regions of those countries has been associated with increasing wages and better working conditions for labour. The historical agglomeration of capital in the industrial belts of the advanced capitalist economies was matched by a concentration of resistance to poor working conditions as workers learned to press effectively for improvements. This was not entirely unfavourable for capital accumulation as increased wages also meant increased local markets for consumer goods. At the same time, higher wages could be maintained without eliminating profits by expanding investment in Third World countries for the low-cost production of raw materials due to very low wages. The result was a continuing and, at the international scale, increasing polarization of wages and of other labour costs between industrialized and non-industrialized regions (cf. Peet 1983, 1986). Storper & Walker (1983) have shown how these geographical conditions, together with the communications technologies developed to co-ordinate multiplant corporations, led to the development of new production processes that allowed components of a larger production process to be automated and located at those foreign and domestic sites in the global economy where labour costs were lower (Amin & Smith 1986).

An example of a clustering of production activities within the city is the concentration of producer services and corporate headquarters in the CBD of American cities. Despite the suburbanization of office activities, which some have taken as evidence that the last economic activities downtown are finally decentralizing, there has been a reconcentration downtown of producer services and corporate headquarters since the mid-1970s in many American cities. To explain why these activities are locating in increasing numbers in the CBD, we start with the restructuring of the national and global economy coming with the development of corporate capitalism. Production processes are increasingly placed in peripheral locations with less skilled and cheaper labour, whereas activities associated with providing the communications, finance, and technical assistance for co-ordinating these wide-flung production systems are located in particular urban centres both within the nations of western Europe and North America as well as in selected Third World metropolises. Thus the shift to an economy of white-collar activities in the United States, and the consequent huge demand for office space in urban areas, can be related to shifts in the spatial division of labour, part of which is a concentration of corporate administrative and producer service activities in global, national, and regional centres (Cohen 1981, Noyelle & Stanback 1983, Westaway 1974).

The production processes utilized in corporate administrative and producer service activities encourage vertical and horizontal disintegration of these activities. Productivity depends on the ability to combine large amounts of specialized knowledge in order to make strategic decisions within complex organizations, requiring the intense, non-routinized communication among diverse specialists dubbed by office location theorists 'face-to-face contacts'. These conditions suffice to explain clustering, whether in the CBD or in suburban office parks, according to Scott's schema. In this way the increased demand for downtown office space for certain production activities can be explained by changes in the organization of commodity production.

The spatial concentration of offices downtown is, however, not a result of demand alone. The economic and political mechanisms of the property market govern the supply of land and buildings, which in turn clearly influences the locations at which suitable office space is available. Thus Ball (1985) has argued that the requirements of the financial institutions involved in financing commercial property development (requirements that stem in part from the changing rôle of financial institutions in the global economy), clearly influence the location of property development. Furthermore, national and local-level intervention in the property market in the US, themselves triggered by problems resulting from the geographically uneven development of the space economy, enhanced the supply of commercial property in general and in downtown locations in particular.

The research summarized here is gradually developing the links between the evolution of society, the conditions of production, and the intra-urban behaviour of producers in a way that integrates the different levels of analysis necessary to understand the demand for production sites in the city. However, the organization of space has to be conceived not only in terms of the demand for locations, or their value to producers as a source of profits, but also in terms of supply. If a firm rents space in an office building or industrial park its location decision is clearly subject to where these facilities will be located. Even if the firm relocates to a new building which it also owns, the costs of the move will depend on the feasibility and cost of constructing the appropriate kind of building at that site. Therefore the sites chosen by producers depend also on the supply of the built environment, making an understanding of the real estate and land development process essential to a complete explanation of the intra-urban location of production, as Walker (1981) and Ball (1985) have emphasized.

A particularly spectacular example of this was the location of a massive new single-story General Motors Cadillac assembly plant close to downtown Detroit in the 1970s. In order to persuade GM not to follow the trend of placing such facilities at peripheral locations, the local municipal authorities used their legal power, over heated neighbourhood opposition, to raze the Poletown neighbourhood and create a site identical in design to any suburban factory – even to the extent of providing three times as much land as necessary for the production facility itself to meet GM's requirement that the parking lots for Cadillac workers should not be multi-story (for a fuller analysis see Jones & Bachelor, 1986). This is an example of how location decisions are influenced by the supply of real estate, in this cases effectuated by the local state. More typically, the connections are more subtle. There are, however, intimate relationships between real estate capitalists and industrial capitalists. Sometimes industrial capitalists lead the development process by their own site location decisions. At other times, property capital makes strategic land purchase decisions that shape development by attracting industrial corporations. It is these relationships that we seek to clarify in the next section.

The supply side: land rent and the property development process

The supply of property for production facilities may be characterized initially by its rent and physical characteristics. Even when the latter are not suitable, decisions to convert to other land uses depend on the cost of purchase and construction and the value of the new building. Thus it is logical to begin an analysis of the supply side with the value of property. There is a long tradition of Marxian land rent theory. This theory shows that rents do not depend on site and situation characteristics alone – these are intervening factors whose influence in turn depends on the broader conditions of production: the economic power of landlords, the existence of monopolies and oligopolies among producers, and general wage levels and profit rates. '[P]roduction relations structure the conditions under which competition over land takes place' (Ball 1985, p. 503). This theory was initially conceived for agricultural land, however, and there is considerable dispute about its applicability in an urban context, where the cost of buildings dominates the cost of land. Urban property is increasingly treated as just another form of investment. Accordingly, an alternative perspective is to examine the supply of property as the result of all of the factors influencing real estate investment. These include the way in which real estate investment is influenced by financial institutions and developers (discussed below under the economics of the commercial property development process), as well as the rôle of state intervention in manipulating the attractiveness of various real estate investment possibilities (discussed below under state intervention in property development).

Marxian rent theory and the concept of the rent gap

Under neoclassical assumptions, land rent is conceptualized as no more than the price of land, representing either its marginal productivity as a production factor, or its marginal utility as a consumer good. It then follows that the highest rent at any location is bid by that activity for which land exhibits its highest marginal productivity (or utility), and that rent maximization allocates land to its 'highest and best use'.

The political-economic approach challenges these assumptions. Ricardo and Marx argued that rent is first and foremost a result of production. It is through production that a surplus is made in the space economy, and it is from this profit that rents must be paid to owners of land (Richardson 1977, p. 25; Barnes 1984). Marxian rent theory therefore proceeds by grounding rent within the process of production, and thereby in the political and social forces immanent in that production process. Land rent depends on a series of factors, all of which influence the ability of landowners to demand payment for a particular location. Four sets of factors are recognized: the different attributes of locations (differential rent); the monopoly power of landlords as a group which affects their ability to demand rent in general from other groups (class monopoly rent); the monopoly power of producers within an economic sector which enables them to make above average profits wherever they locate (monopoly rent); and the ability of producers in a generally profitable economic sector to prevent others from investing in that sector (absolute rent).

Differential rent is a rent paid out of the extra profit accruing at a location because a certain commodity can be produced more cheaply there than at some other location. Locational differences in the cost of producing and marketing a product may be due to site and situation characteristics of a location (such as soil fertility or access to markets; Differential Rent I), or differences in the production method used (Differential Rent II). In either event, cost differences enable the producer at the cheaper location to make higher profits, which the landlord is free to claim as rent (Fine 1979, Harvey 1985, Ch. 4).

Marxian differential rent seems superficially very similar to von Thünen's theory of land rent; even though von Thünen pays little attention to different production methods. Appearances are, however, misleading. All forms of rent must be paid out from profits made in production. If geographical patterns of differential rent were unaffected by the general rate of profit in society, it would not be necessary to examine the profitability of production in order to understand differential rents. This, however, is not the case. Recent research has shown that spatial patterns of differential rent, and spatial patterns of land use, indeed depend on the general rate of profit (Sheppard & Barnes 1986). This in turn implies that the site and situation characteristics of a location, and the definition of what is a more or a less productive method of production, cannot be defined exogenously. Instead, and in contradistinction to von Thünen and the neoclassical approach, the way the attributes of a location affect the value of that location depend on the conditions of production: '[B]oth fertility and location are social appraisals' (Harvey 1985, p. 94).

Class-monopoly rent is defined as an increment to land rents due to the monopoly power that landlords as a class possess over the supply of land:

> This . . . arises because there exists a class of owners of 'resource units' – the land and the relatively permanent improvements incorporated in it – who are willing to release the units under their command only if they receive a positive return above some arbitrary level . . . . The realization of this rent depends on the ability of one class-interest group to exercise its power over another class-interest group and thereby assure for itself a certain minimum rate of return (Harvey 1974, pp. 240ff).

Some see class-monopoly rent as an ubiquitous increment to land rents due to the overall economic power that landowners possess in society. At times when land owners have more economic power, this increment will increase, in the same way that profit and wage levels reflect the relative power of workers and capitalists in society. Scott (1980) has introduced an analogous concept which he terms scarcity rent, defined as a general increase in rent levels due to a general physical shortage of land. This, however, does not recognize that scarcity can be socially created.

An alternative approach argues that the owners of particular land or property in the city can realize class-monopoly rents (Harvey 1974) as a geographically localized phenomenon. Strategies which create class-monopoly rent include: allowing property of a certain kind to deteriorate in order to create scarcity and eventually raise rents; influencing the city to limit the amount of land zoned for a particular use, with the same effect; and exploiting the desirability of certain locations to raise rents. Edel (1976) argues, however, that such actions are better described as extracting monopoly rent.

Monopoly rent is defined as the extra rent extracted at locations occupied by corporations with monopoly power. This power both allows corporations to make above average rates of profit, and makes it easier for them to pass any cost increases (including higher rents) on to consumers by raising prices to maintain their profit margin (Lauria 1985). For both of these reasons, the owners of land occupied by corporations with monopoly power can demand higher rents than the owners of land occupied by smaller firms, *ceteris paribus*.

Absolute rent is possible at locations occupied by an economic sector with an above average profit rate. Marx argued that any economic sector with a low organic composition of capital (i.e. a high ratio of direct labour relative to other inputs, measured in labour value terms) has the potential for above average profits since profits accrue from employing labour. Normally any such sector would attract investment until overproduction drives the profit rate down. He argued, however, that if producers in this sector were able to prevent others from investing in it, then the excess profits at all locations occupied by this sector would be appropriated by landlords as absolute rent.

In the urban context it is argued, such barriers to investment are used by landowners to realize absolute rent by limiting investment in real estate, thus delaying suburban development or urban renewal (Edel 1976). Ball (1985) notes, however, that the process described by Edel reflects monopoly power, not absolute rent. It is also suggested that housing costs are high because the construction sector has a low organic composition of capital and uses barriers to extract absolute rent (Lipietz 1984). Yet it is now recognized that a mathematically correct reworking of Marx's theory shows that a low organic composition of capital in a sector does not necessarily imply the possibility of higher profits (Ball 1985, Sheppard & Barnes 1986). The whole question of absolute rent is still hotly debated (Ball 1980, 1985, Fine 1979, 1982, Harvey 1974, 1982, Lauria 1985, Lamarche 1976, Bruegel 1975, Edel 1976, Scott 1976, Barnes 1984).

The various categories of Marxian rent theory are useful in showing explicitly how different social agents intervene in the provision, allocation, and cost of land, and in creating a classification of potential mechanisms for the appropriation of rent. It illustrates how rents depend on social relations, and the dynamic of capital accumulation. Yet there is considerable dispute about which of these categories, if any, are useful in an urban context, or whether they can even be conceptually distinguished from one another. Further, there is much doubt about whether the concept of absolute rent can survive the mistakes made by Marx in his analysis of the relation between labour values and prices.

Notwithstanding these problems, in trying to come to terms with inner-city renaissance, Smith (1979) has suggested that changes in land development and land use patterns can be usefully analyzed by means of a modified rent theory, dubbed the 'rent gap'. The rent gap is defined as the difference between the potential rent generated by a new land use, and the actual rents paid by the activity currently using a location. '*When, and only when*, this rent gap between actual and potential ground rent becomes sufficiently large, redevelopment and rehabilitation into new land uses becomes a profitable prospect, and capital begins to flow back into the inner city' (Smith 1982, p. 149; our emphasis). Crucial to his analysis is the recognition that the built environment of the past, constructed when the conditions of production were different, need not be profitable today. Because the built environment is a major and durable investment, it cannot be torn down as soon as conditions change. Specifically, he argues that the fixed nature of the inner-city built environment encouraged suburbanization, and that suburbanization reduced the value of downtown property more rapidly than could be explained simply by the physical aging of the property. As a result, a gap emerged between the current value of downtown sites and their potential value, and gentrification followed.

The concept of a rent gap is used beyond the Marxist literature. In neoclassical models the aging of the built environment leads to its replacement only when the rental stream expected from the new land use exceeds that expected from the current use by a margin sufficient to cover the costs of conversion or redevelopment (Brueckner 1980). There are two crucial differences between the two explanations, however. First, in the neoclassical approach the rate of depreciation is not related to the changing spatial structure of the city, but rather to the characteristics of the property in question. Second, in a neoclassical world any change in land use must be an improvement in efficiency.

When Smith (1979, p. 545) refers to 'ground rent capitalized under the present land use', this implies that ground rent can not be rent for the land alone, excluding improvements, because rents depend on the present land use. Calculating rents as 'capitalized under current land uses' implies that an investor takes into account the time horizon within which he wishes to recoup his investment, the resale value of the property, and possible tax benefits. If so, it is almost tautological that a rent gap must exist in order for land use change to occur, since otherwise there would be no profit to the entrepreneur (unless the state subsidizes the investment). It is not clear, however, why the existence of an adequate rent gap is sufficient to lead to land use change (as Smith himself acknowledges when he draws on an additional condition, the switching of capital into the built environment, which we shall discuss below).

There is general disagreement as to what should be included as land rent, and therefore the utility of rent theory for urban property. Some argue that all returns from investment in a site may be regarded as rent (Boddy 1981, Walker 1981, Smith 1982, Harvey 1985). Against this, Ball (1985) argues that *ground* rent is the subject of rent theory, and that ground rent

> if it is to have an identifiably separate meaning, has to be used only for the revenue payment to landowners . . . [R]ent . . . paid for the use of the building . . . is a payment to a building owner . . . Where the building owner owns the land outright no ground rent relation exists [because] land has no separate physical or social identity from the building (Ball 1985, p. 518).

Since ground rent, by this definition, is not an important component of building prices in urban areas, and since, he claims, other aspects of building price cannot be illuminated using the principles of rent theory, Ball concludes that current Marxian urban rent theorists cannot explain urban property values. Harvey (1985) also argues that the distinction between land rent and other forms of capital investment is disappearing, and must disappear if investments in the built environment are to be smoothly co-ordinated with other aspects of capital accumulation:

> [G]round rent, when capitalised at the going rate of interest, yields the land price . . . . The money laid out by the buyer of land is equivalent to an interest-bearing investment . . . . Title to land becomes a form of fictitious capital, in principle no different from stocks and shares, government bonds, etc. (Harvey 1985, p. 96).

Yet Harvey still clearly feels that rent theory is important to understanding the built environment. This debate may best be resolved once we understand how rent theory may be integrated with theories of real estate and other forms of investment.

The economics of the commercial property development process

The production of built structures, its timing and location reflect investment decisions by many different agents such as financial institutions, developers, land owners and brokers, as well as by various levels of the state. The immediate factors influencing these decisions are those determining the estimated rate of return on a new or converted building. It is not enough, however, to construct a list of factors and examine them individually, but rather they should be grounded in a broader analysis of the structural conditions driving property development.

The rate of return on a piece of property is measured as the annual net income expected from a project as a proportion of the cost of the initial investment. A higher rate of return means this investment can be recouped more quickly and that the developer can keep a higher profit margin. The rate of return then plays a rôle analogous to the rate of profit on commodity production, and just as is the case for the rate of profit depends on the various costs and revenues associated with property development and on the relationship between them. Revenues include rents or lease payments and future resale value of the property. Both are important because developers often expect to make more money from the future resale of a project than from rental income. The costs include money spent on project design, land purchase and assembly, demolition and construction or conversion, obtaining finance, and maintenance (Hamnett *et al.* 1982). In addition, both costs and revenues are influenced by actions of the state apparatus at its various geographical scales, by means of tax regulations, direct and indirect development subsidies, and urban development planning, discussed in detail below under state intervention in the property development process.

Investment in property is a large-scale fixed capital investment in a permanent structure, and only yields its value to the investor over a period of years. Expectations and speculations about future revenues from the property then greatly influence the supply of new or converted built structures. These expectations in turn are based on analyses of present demand, and guesses about future demand, for that kind of project at that location; demands which, as we have seen, are a part of the evolving geography of production in the global economy.

The costs of land purchase and assembly, and of conversion of old property, vary significantly within the city depending on how the city has developed. The historical geography of the city provides a pre-existing stock of buildings of varying quality and age into which any new developments must be inserted. The locational pattern of buildings and vacant land, the economic value of these buildings, and the degree to which they have been maintained, all profoundly influence both the cost of providing a site for a new development and whether it should involve rehabilitation or demolition of current buildings. Much of the current stock was built to serve the economic needs of earlier generations, may now be outmoded, and constitute a barrier to current production needs (Walker 1981). The economic value of this outmoded stock will then be depressed, making it cheaper for developers to purchase. For example, the growing size of producer service corporations and their requirement for offices that can accommodate sophisticated communications technology have encouraged the construction of huge office complexes containing 'smart office' technology. As a result, the economic value and rents of existing office space have fallen, less is invested in maintenance thus advancing the date at which it becomes profitable to tear down and replace these buildings.

Yet, even when a rent gap exists, because revenues exceed costs, development will not occur unless the necessary finance capital is made available. Developers usually do not have the funds to finance many projects, nor do they wish to risk sinking the assets they do have into just a few buildings. Therefore it is necessary to examine the increasingly complex structures of capital provision for property development – one of the most important determinants of when, where, and how real estate is developed.

Money is made available in many forms from a variety of sources: mortgages and loans; various degrees of equity ownership; or purchase of the property, which then may be leased back to the developer on a 'sale-leaseback' arrangement (see National Council for Urban Economic Development 1980, Boddy 1981, Downs 1985). Which arrangement is offered, if any, depends on general financial conditions, the lenders' evaluation of the potential of the project (which includes its size, type, and location; its likely rate of return; and the reputation of the developer; Knecht 1985, p. 76), and the institutional nature of the lender.

General financial conditions refer to the availability of money for real estate investment, which influences the general dynamics of property development. Harvey (1978) and Walker (1981) have argued that the process of commodity production under capitalism implies that large amounts of capital will be made available at certain points in time for real estate investment. Harvey (1978) conceptualizes capital flows in the economy as divided into three circuits: a primary circuit representing investment in commodity production; a secondary circuit representing investment in fixed capital (notably the built environment) and in items that facilitate consumption; and a tertiary circuit representing investment in science and technology. In this view, the steady accumulation of capital through expanded commodity production (i.e. investment in the primary circuit) creates a realization crisis for capital investment due to an oversupply of commodities. This can, however, be resolved by switching investment into the secondary circuit, an important component of which is the built environment. Thus the momentum of capital investment in capitalist economies switches periodically between investment in commodity production and investment in the built environment.

Data on office space construction worldwide do show remarkable similarities in the timing of development booms, most recently from 1968–73, and again from 1978–84. Yet the quantity of money available in a nation also depends on the factors influencing international capital flows and thus on the evolving global economy. International differences in interest rates, shifts in the relative value of national currencies, and national fiscal policies have led to large capital flows between economies, influencing the size of office space booms in Australia, Britain, and the United States (Daly 1982, Badcock 1984, Leitner 1986b). Furthermore, Harvey's model does not easily explain why the more recent office space boom in the United States coincides with a resurgence in the profitability of investment in Harvey's primary circuit via the supply-side policies of the Reagan Administration. This raises the question of how distinct Harvey's circuits of capital really are, and also whether all investment in the built environment should be classified as part of the secondary circuit.

The general availability of investment funds, the relative profits of different investment opportunities, and the investor's interpretation of what constitutes a secure investment portfolio set the context in which individual decisions to finance particular property developments are made. Given this, a financial institution evaluates the potential of a particular project and decides what kind of financing to offer. Since real estate is regarded increasingly as just another form of capital investment, and since financial institutions are themselves increasingly involved in ownership of property (with about 90 per cent of commercial property sold in the United States being bought by life insurance companies and other financial institutions managing pension funds in 1981; Feagin 1983, p. 54), the evaluation of a project by financial institutions may be expected to coincide more and more with developers' evaluations. Boddy (1981, p. 282) argues that this structural shift to more direct investment by financial institutions in commercial property can reduce the stability of the financial system by exposing it to the effects of booms and slumps in property values.

The first consideration in such an evaluation is the expected rate of return. The location of the project within the city is an important component of this evaluation, since it affects both land and development costs as well as future revenues. When money is in short supply or other investment opportunities are more profitable a higher rate of return is required from property, which in practice means that only 'blue chip' projects at the best locations will be funded. Non-financial considerations also enter the evaluation of potential projects, however, such as the symbolism of power inherent in a property: '"I knew [the deal] was working', Cushman recalls, 'when Rutland and Morrison were on their knees looking at how much higher [their] building was than Security Pacific"' (Knecht 1985, p. 76; Cushman, a New York real estate broker, describing negotations for a Crocker Bank headquarters in Los Angeles).

Different kinds of financial institutions have particular goals and legal constraints which influence how and where they invest in property. In the United States, for example, commercial banks currently provide most short-term construction loans for industrial and commercial developments. By contrast, life insurance companies and pension funds have become the prominent source of long-term mortgages and equity financing for larger non-residential properties in Britain and the United States. The long-term nature of pension fund and life insurance liabilities leads these companies to seek secure long-term investments in large projects at prime locations (Urban Land Institute 1980, Ball 1985). By increasing the part of their portfolio representing equity investment in property, these companies can invest in real assets that historically have kept their value in the face of inflation (Boddy 1981, Urban Land Institute 1980).

Since the conditions of production driving investment decisions in commercial and industrial real estate are rooted in the structure of the national and international economy, it is not surprising to observe a convergence of the institutional structure of the property industry and that of the economy in general. Not only is the industry becoming dominated by large developers and increasingly by large financial institutions (the realty division of the Prudential Insurance Company, was the fifth largest American real estate developer in 1985 (Anders 1986)), but multinational corporations with formerly little interest in property are purchasing real estate companies. Foreign investment capital is becoming an increasingly important component of property investment, and the real estate companies are themselves becoming international in scope (Apgar 1985, Thrift 1986). This increased interdependence has also brought the rôle of the state as mediator of the provision of commercial property to the fore.

State intervention in the property development process

State intervention in the property market attempts to manipulate the timing, nature, and location of real estate development and investment, thus affecting the supply of property and influencing the possible locations for economic activity in the city. At the national level, state intervention has two aspects: influencing general levels of real estate investment by altering the relative financial benefits of different kinds of investment as a component of national sectoral economic management, and directing investment to the urban built environment as a part of national urban policies. The local state, by which we mean the state apparatus in a municipality (including regional authorities operating within the city (Heiman 1985)), is more concerned with intra-urban management. Its agents encourage investment at a location, in the kind of property which it is believed should occupy the site, by attempting to bridge any gap that may exist between the rate of return required by developers and/or investors, and that which this site is expected to generate. There are many ways to subsidize the rate of return. Investing in public improvements along with favourable zoning regulations in a particular district can increase developers' expectations about how rapidly new property built there will appreciate in value. Property tax relief can increase net rental income. Public sector financing, using a variety of tools such as second and third mortgages, equity investment, low interest loans, and loan guarantees, reduces development costs and the risk of financing for private investors.

We wish to understand the conditions under which different levels of the state intervene in the property market, in order to explain how the timing and location of intervention influences the supply of commercial property and thus the intra-urban location of economic activities. This requires a political-economic theory of state action in capitalist society (O'Connor 1973, Offe 1974, 1978, Habermas 1976, Carnoy 1984; see also the chapters by Johnston and Fincher in vol. 1).

Generally speaking, we can characterize the capitalist state as involved in mediating between two conflicting requirements: accumulation and legitimation. On the one hand, sustaining the economy requires supporting the accumulation of capital by ensuring that favourable conditions for private investment exist. This includes intervening in the market to correct imbalances and imperfections, ensuring that adequate labour is available at a reasonable cost, and subsidizing necessary capital investments that are too large or insufficiently profitable for private investors.

On the other hand, public legitimation for actions of the state must be secured by some combination of providing for the social needs (collective consumption) of the electorate, and promoting ideological conditions through which individuals take for granted the validity of state intervention and indeed the responsibility of the state in certain areas. Since 'the late capitalist welfare state bases its legitimacy on the postulate of a universal participation in consensus formation and on the unbiased opportunity for all classes to utilize the state's services and to benefit from its regulatory acts of intervention' (Offe 1978, p. 396), it is subject to various class and popular democratic struggles. State intervention, therefore, cannot necessarily serve the actual needs of individual capitalists, but must generally be responsive to political pressure from a variety of sources. These include different branches of capital and labour, components of the state apparatus and political grassroots groups. The form and content of reactions by the state to these pressures depend on the balance of power between these forces as well as on the structure and condition of the economy and society. This implies that there is a significant degree of historical specificity to state action; depending on the economic conditions and power relationships in society, the means used by the state to ensure stability of the socio-economic system as a whole differ in time and place.

Within the United States, the 1960s and early 1970s were periods when Federal policies supporting consumption were favourable for both accumulation and the welfare of the middle class. Walker (1981) shows that one important component of this policy was state support for suburbanization, encouraging the intra-urban decentralization of economic activities. One cost of promoting accumulation by stimulating consumption was increased wage levels, encouraging the relocation of investment in production outside the United States, as well as increasing the competitive advantage for indigenous producers from low-wage countries in the American market. Together these created a new form of accumulation crisis in the 1970s (O'Connor 1984). As a result, state intervention shifted from stimulating consumption to stimulating production; the 'Reagan revolution' (Harvey 1985). In the property market this was signalled by a shift in emphasis in Federal policies away from supporting housing provision and neighbourhood revitalization, in favour of constructing commercial and industrial property (Lauria 1986). Federal tax regulations were a principal mechanism. The Economic Recovery Tax Act of 1981 significantly reduced tax rates of real estate investment, dramatically accelerated the rate at which property could be depreciated (increasing tax deductions and reducing the time needed to recover invested capital), and introduced generous tax credits for property rehabilitation (see National Council for Urban Economic Development 1983, Walsh 1983). With the introduction of Urban Development Action Grants in 1977, providing public support for those property developments which would leverage a large amount of private investment, Federal urban policies also explicitly stimulated urban growth by directly subsidizing commercial and industrial real estate investment.

The hierarchical nature of the state apparatus implies that local state intervention is formulated taking into account political and economic influences operating at many different spatial levels (Clark & Dear 1984, pp. 135ff), which are then filtered through the particular rules and power arrangements of local political systems (Swanstrom 1985, p. 33). Consider, for example, the central cities of those older metropolitan areas within the American manufacturing belt progressively losing the competition for investment in the 1970s, as evidenced in a new out-migration of economic activities due to changing economic conditions. This eliminated jobs and created fiscal problems for the local state by reducing the tax base (enhanced by a reduction in the flow of Federal funds). The reorientation of Federal urban programmes and the new tax regulations, along with the deteriorating economic base and the fiscal strain in these central cities, both necessitated and provided an opportunity for agents of the local state to attempt to use their power in the landmarket as a lever to attract economic activities back to the downtown. As a result, in many of these cities the local state actively encouraged prospective investors, using a wide variety of planning, financial, and legal tools to subsidize the expected rate of return on real estate investments in the downtown.

Local state activities have become increasingly entrepreneurial, working closely with developers and using financial incentives ranging from land purchase subsidies and tax rebates through low-interest loans and loan guarantees to equity financing and even outright property ownership and leasing. These are combined with assistance through public improvements, land use control, and management as well as project development assistance (for details see the Urban Land Institute 1980, Walker & Heiman 1981). The more disadvantaged a city in the inter-urban competition for economic activities, the greater the concessions perceived as necessary to attract investment capital, even though there is much dispute about the effectiveness of these incentives in influencing the locational choice of individual investors (Pascarella & Raymond 1982, 1986, Beauregard & Holcomb 1983, Wolkoff 1985, Swanstrom 1985, Marlin 1986). Firms play one community off against another by demanding greater subsidies and concessions to reduce development costs and the risk of their investment (Kirby 1985). At the same time, the harder the local state tries to defray the risks for private real estate investors, the more it assumes responsibility for these risks itself.

Since such activities often require that the local state shift its rôle somewhat from a provider of social services to an ally of private capital, this can threaten its legitimacy in the eyes of the general public. One widely adopted strategy to minimize this problem has been a division of labour within the local state apparatus; separating accumulation-oriented activities, located in public or quasi-public development agencies that are not subject to popular democratic pressure, from legitimation oriented activities (Friedland *et al.* 1984). In addition, powerful pro-growth coalitions have emerged in many American cities 'in which certain dominant interests – of banking and finance capital, of property capital and construction interests (including laborers and their unions), of developers and ambitious agents of the state apparatus – typically call the tune' (Harvey 1985, p. 157). One important activity of these coalitions has been propagating the belief that all residents benefit from subsidizing, for example, downtown commercial property development; promoting broader support for attempts by the state to foster local capital accumulation (Molotch 1976, Mollenkopf 1983, Swanstrom 1985). Such boosterism has, however, been challenged at times by local social movements such as San Franciscans for Reasonable Growth (Feagin 1983, Hartman 1984, Clarke 1984, Fainstein & Fainstein 1985, Leitner 1986b).

Many observers have noted that the logic of both economic and political processes, with investors seeking to increase their rate of return and public officials seeking to improve the tax base and reap political benefits, biases the results of local state action towards the development industry, and reduces the likelihood that the local state's control over land use and public resources is used to redirect some of the profits of development towards the residents of the city (Peattie *et al.* 1985, Molotch & Logan 1985). This would imply that state intervention, while important, does not at this time fundamentally challenge the trend towards treating property increasingly as merely another branch of capital investment (fictitious capital). By the same token, an understanding of the supply of property and its influence on the location of production in the city requires an understanding of finance capital and its intersection with other forms of capital and the state.

Conclusion

A political-economic approach to the geography of the city implies that the point of departure for a comprehensive account should be with the logic of economic and political processes governing commodity production in the city. While we would disagree with those who argue that all aspects of the city can be deduced from an understanding of the intra-urban geography of commodity production (cf. Scott 1980, 1986), we find the opposite approach of starting with the realm of consumption to be inadequate (cf. Castells 1976). We do not pretend that such a comprehensive account of the city is within reach, nor do we attempt to develop a full-blown theory of the location of production. Our aim has been simply to identify progress to date, and indicate unsolved problems.

One issue that distinguishes the political-economic accounts is the necessity of grounding the immediate factors which are perceived to influence location within the broader structural, or structurating, societal context. While we are convinced that this is vital to an adequate understanding, we do not wish to take the extreme position that every study of the location of economic activity pays equal attention to all issues ranging from the evolution of the global economy to the fine details of locational behaviour. There must be some division of labour between those studying the broader issues in detail and those concerned with the specifics of individual cities. Yet it is necessary for those pursuing more local analyses to be aware of the broader influences on the situation they are examining, if they are to understand properly the causes of the changes they observe.

The second major point we have made is that the location of production does not just depend on the requirements and locational demands of producers. In the crowded and durable built environment of the city, it is at least as important to understand how the built environment is made available for production purposes, since this constrains the sites available and influences the profitability of such options as moving to 'greenfield' locations versus relocating within the city. Our review of the literature shows that both the location requirements of producers and the supply of commercial property have been analyzed, but that these two bodies of literature have rarely been brought together to explain the changing location of production activities and land use patterns.

Third, we have attempted to argue that a middle level of analysis is necessary in order to mediate between the influences of societal structure and individual behaviour in our explanation. We see this middle ground as an analysis of the strategies adopted by key actors which they perceive as best for achieving their goals. For producers this might be a profit maximizing method of production, for financiers an interest maximizing investment strategy, and for agents of the state the maximization of revenues over expenditure. In each case we would wish to ask: what societal forces influence the strategy adopted, and how does this affect the actions taken?

As we have tried to show at various points in this chapter, capitalist society is not a homogeneous landscape with all cities responding in lock-step to the beat of the mode of production. Indeed, if this were so, a geographical account would be superfluous. Rather the broad, regional, national, and international forces of capitalism combine with the history and situation of individual cities to create a unique combination of production conditions and responses. The result is a variety of situations differentiating even apparently similar cities, which can be bewildering. The success of a political-economic approach must be measured by its ability to marshal an understanding of the geographical and historical evolution of society in order to give insight into this variety. An adequate explanation is subject to continual reformulation as it faces up to the continuous contradiction between the necessity to generalize and abstract our explanations, and the necessity of accounting for the historical and geographical specificity of society.

References

Alonso, W. 1964. *Location and land use*. Cambridge, Mass.: Harvard University Press.

Amin, A. & I. Smith (1986). The internationalization of production and its implications for the UK. In *Technological change, industrial restructuring and regional development*, A. Amin & J. Goddard (eds), 41–76. London: Allen & Unwin.

Anders, G. 1986. Empire builders: insurance companies, pension funds become big office landlords. *Wall Street Journal*, 1 April, 1986.

Apgar, M. (1985). The changing realities of foreign investment. In *Development review and outlook 1984–5*. Washington, DC: The Urban Land Institute.

Badcock, B. 1984. *Unfairly structured cities*. Oxford: Basil Blackwell.

Ball, M. 1980. On Marx's theory of agricultural rent: a reply to Ben Fine. *Economy and Society* **9**, 304–9.

Ball, M. 1985. The urban rent question. *Environment and Planning A* **17**, 503–25.

Barnes, T. J. 1984. Theories of agricultural rent within a surplus approach. *International Regional Science Review* **9**, 125–41.

Beauregard, R. & B. Holcomb 1983. Enterprise zones: the non-manipulation of economic space. *Urban Geography* **4**, 223–43.

Beavon, K. S. O. 1980. *Central place theory*. London: Longman.

Berry, B. J. L., R. J. Tennant, B. J. Garner & J. W. Simmons 1963. *Commercial structure and commercial blight*. Research paper no. 85. Chicago: University of Chicago Press.

Boddy, M. 1981. The property sector in late capitalism: the case of Britain. In *Urbanization and urban planning in capitalist society*, M. Dear & A. J. Scott (eds), 267–86. London: Methuen.

Borchert, J. R. 1967. American metropolitan evolution. *Annals of the Association of American Geographers* **57**, 301–32.

Brueckner, J. K. 1980. Residential succession and land use dynamics in a vintage model of urban housing. *Regional Science and Urban Economics* **10**, 225–40.

Bruegel, I. 1975. The marxist theory of rent and the contemporary city: a critique of Harvey. In *Political economy and the housing question*, Political Economy of Housing Workshop. London: CSE Books, pp. 34–46.

Carnoy, M. 1984. *The state and political theory*. Princeton, NJ: Princeton University Press.

Castells, M. 1976. Theory and ideology in urban sociology. In *Urban sociology: critical essays*, C. Pickvance (ed.). London: Tavistock.

Clark, G. 1980. Capitalism and regional disparities. *Annals of the Association of American Geographers* **70**, 222–37.

Clark, G. & M. Dear 1984. *State apparatus: structures and language of legitimacy*. Boston, Mass.: Allen & Unwin.

Clarke, S. 1984. The local state and alternative economic development strategies: gaining public benefits from private investment. Discussion paper 16. University of Colorado, Boulder: Center for Public Policy Research.

Cohen, R. P. 1981. The new international division of labor, multinational corporations and urban hierarchy. In *Urbanization and urban planning in capitalist society*, M. Dear & A. J. Scott (eds), 287–319. London: Methuen.

Cooke, P. 1983. *Theories of planning and spatial development*. London: Hutchinson.

Cowling, K. 1986. The internationalization of production and de-industrialization. In *Technological change, industrial restructuring and regional development*, A. Amin & J. Goddard (eds), 77–99. London: Allen & Unwin.

Daly, M. T. 1982. *Sydney boom Sydney bust. The city and its property market 1850–1981*. Boston: Allen & Unwin.

Daniels, P. W. 1975. *Office location: an urban and regional study*. London: Bell.

Daniels, P. W. 1979. *Spatial patterns of office growth and location*. London: Wiley.

Daniels, P. W. 1982. An exploratory study of office location behavior in greater Seattle. *Urban Geography* **3**, 58–78.

Downs, A. 1985. *The revolution in real estate finance*. Washington, DC: The Brookings Institution.

Duncan, S. & M. Goodwin 1982. The local state and restructuring social relations: theory and practice. *International Journal of Urban and Regional Research* **6**, 157–86.

Edel, M. 1976. Marx's theory of rent: urban applications. In *Housing and class in Britain*, Political Economy of Housing Workshop, 7–23. London: CSE Books.

Fainstein, S. S. & N. I. Fainstein 1985. Economic restructuring and the rise of urban social movements. *Urban Affairs Quarterly* **21**, 187–206.

Fales, R. N. & L. N. Moses 1972. Land use theory and the spatial structure of the nineteenth century city. *Papers of the Regional Science Association* **28**, 49–80.

Feagin, J. R. 1983. *The urban real estate game: playing monopoly with real money*. Englewood Cliffs, NJ: Prentice-Hall.

Fine, B. 1979. On Marx's theory of agricultural rent. *Economy and Society* **8**, 241–78.

Fine, B. 1982. *Theories of the capitalist economy*. London: Edward Arnold.

Friedland, R., F. F. Piven & R. R. Alford 1984. Political conflict, urban structure, and the fiscal crisis. In *Marxism and the metropolis: new perspectives in urban political economy*, 2nd edn, W. Tabb & L. Sawers (eds), 273–97. Oxford: Oxford University Press.

Gad, G. 1985. Office location dynamics in Toronto: suburbanization and central district specialization. *Urban Geography* **6**, 331–51.

Goddard, J. B. 1975. *Office location in urban and regional development*. Oxford: Pergamon.

Gordon, D. 1978. Capitalist development and the history of American cities. In *Marxism and the metropolis: new perspectives in urban political economy*, W. Tabb & L. Sawers (eds), 25–63. Oxford: Oxford University Press.

Habermas, J. 1976. *Legitimation crisis*. London: Heinemann Educational Books.

Hamnett, C. 1982. Urban land use and the property development industry. In *Market processes II*, C. Hamnett, 3–54. Milton Keynes: Open University Press.

Harrison, B. 1982. The tendency towards instability and inequality underlying the 'revival' in New England. *Papers of the Regional Science Association* **50**, 41–65.

Hartman, C. 1984. *The transformation of San Francisco*. Totowa, NJ: Rowman & Littlefield.

Harvey, D. 1974. Class-monopoly rent, finance capital and the urban revolution. *Regional Studies* **8**, 239–55.

Harvey, D. 1978. The urban process under capitalism: a framework for analysis. *International Journal of Urban and Regional Research* **2**, 101–31.

Harvey, D. 1982. *The limits to capital*. Oxford: Basil Blackwell.

Harvey, D. 1985. *The urbanization of capital*. Oxford: Basil Blackwell.

Heiman, M. K. 1985. Public authorities as agents of the quiet revolution in land development. Paper presented at the Annual Meeting of the Association of American Geographers, Detroit.

Holmes, J. 1986. The organization and locational structure of production subcontracting. In *Production, work, territory*, A. J. Scott & M. Storper (eds), 80–103. London: Allen & Unwin.

Hoover, E. M. & R. Vernon 1959. *Anatomy of a metropolis*. Cambridge, Mass.: Harvard University Press.

Hudson, R. & D. Sadler 1986. Contesting works closures in Western Europe's old industrial regions: defending place or betraying class? In *Production, work, territory*, A. J. Scott & M. Storper (eds), 172–93. London: Allen & Unwin.

Jones, B. & L. W. Bachelor 1986. *The sustaining hand: community leadership and corporate power*. Kansas: University Press of Kansas.

Khakee, A. 1985. Urban models and municipal planning. *Urban Studies* **6**, 48–68.

Kirby, A. 1985.. Nine fallacies of local economic change. *Urban Affairs Quarterly* **21**, 207–20.

Knecht, G. B. 1985. Real estate's matchmaker. *New York Times Magazine* 27 October 1985.

Kutay, A. 1986. Effects of telecommunications technology on office location. *Urban Geography* **7**, 243–57.

Lamarche, F. 1976. Property development and the economic foundations of the urban question. In *Urban sociology: critical essays*, C. Pickvance (ed.), 85–118. London: Tavistock.

Lauria, M. 1985. The implications of marxian rent theory for community controlled development strategies. *Journal of Planning Education and Research* **4**, 16–24.

Lauria, M. 1986. The internal transformation of community controlled implementation organizations. *Administration and Society*.

Leitner, H. 1986a. Investitionen in den Downtowns US-amerikanischer Großstädte: Das Beispiel Minneapolis/St Paul. In *Verhandlungsband zum 45. Deutschen Geographentag Berlin*. Wiesbaden: Steiner.
Leitner, H. 1986b. Pro-growth coalitions, the local state and downtown development: the case of six cities. Paper presented at the Annual Meeting of the Association of American Geographers, Minneapolis.
Leone, R. A. & R. Struyk 1976. The incubator hypothesis: evidence from five SMSAs. *Urban Studies* **13**, 325–31.
Lichtenberger, E. 1986. *Stadtgeographie 1: Begriffe, Konzepte, Modelle, Prozesse*. Stuttgart: Teubner Studienbucher der Geographie.
Lipietz, A. 1985. A marxist approach to ground rent: the case of France. In *Land rent, housing and urban planning: a European perspective*, M. Ball, V. Bentivegna, F. Folin & M. Edwards (eds), 129–55. Beckenham: Croom Helm.
Logan, M. I. 1966. Locational behavior of manufacturing firms in urban areas. *Annals of the Association of American Geographers* **56**, 451–66.
Lowry, I. S. 1964. *A model of metropolis*. Santa Monica: Rand Corporation.
Markusen, A. R. 1985. *Profit cycles, oligopoly and regional development*. Cambridge, Mass.: MIT Press.
Marlin, M. R. 1986. Reevaluating the benefits and costs of industrial revenue bonds. *Urban Affairs Quarterly* **21**, 435–41.
Miller, R. 1983. The Hoover® in the garden. *Environment and Planning D, Society and Space* **1**, 73–88.
Mills, E. & B. W. Hamilton 1983. *Urban economics*, 3rd edn. London: Scott, Foresman.
Mollenkopf, J. 1983. *The contested city*. Princeton, NJ: Princeton University Press.
Molotch, H. 1976. The city as a growth machine. *American Journal of Sociology* **82**, 309–32.
Molotch, H. & J. R. Logan 1985. Urban dependencies: new forms of use and exchange in U.S. cities. *Urban Affairs Quarterly* **21**, 143–69.
Moses, L. 1958. Location and the theory of production. *Quarterly Review of Economics* **73**, 259–72.
Moses, L. N. & H. F. Williamson 1967. The location of economic activity in cities. *American Economic Review* **57**, 211–22.
Muller, P. 1981. *Contemporary suburban America*. Englewood Cliffs, NJ: Prentice-Hall.
Muth, R. 1962. The spatial structure of the housing market. *Regional Science Association, Papers and Proceedings* **7**, 207–20.
National Council for Urban Economic Development 1980. Financial instruments, financing incentives, sources of financing. Advanced Training Institute Notebook. Washington, DC: National Council for Urban Economic Development.
National Council for Urban Economic Development 1983. Development tax incentives. *Information Service*, no. 26. Washington, DC: National Council for Urban Economic Development.
Nelson, K. 1986. Labor demand, labor supply and the suburbanization of low-wage office jobs. In *Production, work, territory*, A. J. Scott & M. Storper (eds), 149–71. London: Allen & Unwin.
Norcliffe, G. B. 1984. Nonmetropolitan industrialization and the theory of production. *Urban Geography* **5**, 25–42.
Norton, R. D. & J. Rees 1979. The product cycle and the spatial decentralization of American manufacturing. *Regional Studies* **13**, 141–51.
Noyelle, T. & T. M. Stanback 1983. *The economic transformation of American cities*. Totowa, NJ: Rowman & Allenfeld.
O'Connor, J. 1973. *The fiscal crisis of the state*. New York: St Martin's Press.
O'Connor, J. 1984. *Accumulation crisis*. Oxford: Basil Blackwell.
Offe, C. 1975. The capitalist state and policy formation. In *Stress and contradiction in modern capitalism*, L. Lindberg (ed.). Lexington, Mass.: Lexington Books.

Offe, C. 1978. Political authority and class structures. In *Critical sociology*, P. Connerton (ed.), 388–421. London: Penguin.

Papageorgiou, Y. & E. Casetti (1971). Spatial equilibrium residential land values in a multicenter setting. *Journal of Regional Science* **13**, 385–89.

Pascarella, T. A. & R. D. Raymond 1982. Buying bonds for business – an evaluation of the industrial revenue bond program. *Urban Affairs Quarterly* **18**, 73–89.

Pascarella, T. A. & R. D. Raymond 1986. Rejoinder to Matthew R. Marlin's 'Reevaluating the benefits and costs of industrial revenue bonds'. *Urban Affairs Quarterly* **21**, 443–7.

Pasinetti, L. L. 1981. *Structural change and economic growth.* Cambridge: Cambridge University Press.

Peattie, L., S. Cornell & M. Rein 1985. Development planning as the only game in town. *Journal of Planning Education and Research* **5**, 17–25.

Peet, R. 1983. Relations of production and the relocation of United States manufacturing industry since 1960. *Economic Geography* **59**, 112–43.

Peet, R. 1986. Industrial devolution and the crisis of international capitalism. *Antipode* **18**, 78–95.

Rees, J. 1978. Manufacturing change, internal control and government spending in a growth region of the USA. In *Industrial change: international experience and public policy*, F. E. I. Hamilton (ed.), 155–74. London: Longman.

Richardson, H. W. 1973. *The economics of urban size.* Lexington, Mass.: Lexington Books.

Richardson, H. W. 1977. *The new urban economics.* London: Pion.

Rose, D. 1981. Accumulation versus reproduction in the inner city: *The Recurrent Crisis of London* revisited. In *Urbanization and urban planning in capitalist society*, M. Dear & A. J. Scott (eds), 339–82. London: Metheun.

Rose-Ackerman, S. 1977. The political economy of a racist housing market. *Journal of Urban Economics* **4**, 150–69.

Sayer, A. 1984. *Method in social science: a realist approach.* London: Hutchinson.

Scott, A. J. 1976. Land and land rent: an interpretative review of the French literature. *Progress in Geography*, vol. 9, 103–45. London: Edward Arnold.

Scott, A. J. 1980. *The urban land nexus and the state.* London: Pion.

Scott, A. J. 1982a. Location patterns and dynamics of industrial activity in the modern metropolis. *Urban Studies* **19**, 111–42.

Scott, A. J. 1982b. Production system dynamics and metropolitan development. *Annals of the Association of American Geographers* **72**, 185–200.

Scott, A. J. 1983a. Industrial organization and the logic of intra-metropolitan location I: theoretical considerations. *Economic Geography* **59**, 233–50.

Scott, A. J. 1983b. Industrial organization and the logic of intra-metropolitan location II: a case study of the printed circuits industry in the greater Los Angeles region. *Economic Geography* **59**, 343–67.

Scott, A. J. 1984. Industrial organization and the logic of intra-metropolitan location III: a case study of the women's dress industry in the greater Los Angeles region. *Economic Geography* **60**, 3–27.

Scott, A. J. 1986. Industrialization and urbanization: a geographical agenda. *Annals of the Association of American Geographers* **76**, 25–37.

Scott, A. J. & M. Storper 1986. The geographical anatomy of industrial capitalism. In *Production, work, territory*, A. J. Scott, and M. Storper (eds), 301–11. London: Allen & Unwin.

Sheppard, E. 1980. The ideology of spatial choice. *Papers of the Regional Science Association* **45**, 197–221.

Sheppard, E. & T. J. Barnes 1986. Instabilities in the geography of capitalist production, *Annals of the Association of American Geographers* **76**, 493–507.

Smith, N. 1979. Toward a theory of gentrification: back to the city movement by capital, not people. *Journal of the American Planning Association* **45**, 538–48.

Smith, N. 1982. Gentrification and uneven development. *Economic Geography* **58**, 139–55.

Steedman, I. 1979. *Trade amongst growing economies*. Cambridge: Cambridge University Press.

Storper, M. 1985. The product cycle: essentialism in economic geography. *Economic Geography* **61**, 260–82.

Storper, M. & R. A. Walker 1983. The theory of labour and the theory of location. *International Journal of Urban and Regional Research* **7**, 1–41.

Struyk, R. & F. J. James 1975. *Intrametropolitan industrial location*. Lexington, Mass.: Lexington Books.

Swanstrom, T. 1985. *The crisis of growth politics: Cleveland, Kucinich, and the challenge of urban populism*. Philadelphia, Penn.: Temple University Press.

Thrift, N. 1986. The internationalization of producer services and the integration of the Pacific Basin property market. In *Multinationals and the restructuring of the world economy*, M. J. Taylor & N. J. Thrift (eds), 142–92. Beckenham: Croom Helm.

Thünen, J. H. von 1826. *Der isolierte Staat*. Hamburg: Perthes.

Urban Land Institute, The 1980. *Downtown development handbook*, Community Builders Handbook Series. Washington, DC: Executive Group of the Urban Development/ Mixed-Use Council of The Urban Land Institute.

Walker, R. A. 1978. The transformation of urban structure in the nineteenth century and the beginnings of suburbanization. In *Urbanization and conflict in market societies*, K. Cox (ed.), 165–212. Chicago: Maaroufa Press.

Walker, R. A. 1981. A theory of suburbanization: capitalism and the construction of urban space in the United States. In *Urbanization and urban planning in capitalist society*, M. Dear & A. J. Scott (eds), 383–430. London: Methuen.

Walker, R. A. & M. Heiman 1981. Quiet revolution for whom? *Annals of the Association of American Geographers* **71**, 67–83.

Walker, R. A. & M. Storper 1981. Capital and industrial location. *Progress in Human Geography* **5**, 473–509.

Walsh, A. A. 1983. Real estate investment after the tax acts of 1981 and 1982. Development Component Series. Washington, DC: The Urban Land Institute.

Webber, M. J. 1982. Location of manufacturing activities in cities. *Urban Geography* **3**, 203–23.

Webber, M. J. 1984. *Explanation, prediction and planning: the Lowry model*. London: Pion.

Westaway, J. 1974. The spatial hierarchy of business organizations and its implications for the British urban system. *Regional Studies* **8**, 145–55.

Wilson, A. G., P. H. Rees & C. M. Leigh 1977. *Models of cities and regions: theoretical and empirical development*. London: Wiley.

Wilson, A. G., J. D. Coelho, S. M. MacGill & H. C. W. L. Williams 1981. *Optimization in locational and transport analysis*. Chichester: Wiley.

Wolkoff, M. J. 1985. Chasing a dream: the use of tax abatements to spur economic develoment. *Urban Studies* **22**, 305–15.

Yeates, M. & B. Garner 1980. *The north American city*. New York: Harper & Row.

Yinger, J. & S. Dantzinger 1978. An equilibrium model of urban population and the distribution of income. *Urban Studies* **15**, 201–14.

4 *Reproduction, class, and the spatial structure of the city*

Geraldine Pratt

In the 1970s and 1980s, urbanists working within the tradition of a Marxist political economy transformed our thinking about the spatial structure of the city. It was argued that urban form must be theorized in relation to the concept mode of production. This chapter reviews the major theories developed over the last 15 years to explain the relationship between the spatial structure of the capitalist city and the capitalist mode of production.[1] In particular, the chapter focuses on the way in which urban form reflects, and helps reproduce, a basic social relationship in capitalist society: that of class. In exploring the links between class, reproduction, and the spatial structure of the city, this chapter demonstrates the importance of extending critical social theory into the realms of urban social geography. Not only do social processes shape the city, urban spatial structures also shape capitalist social relations.

In the way of preparation, I first outline some key features of the capitalist mode of production as conceptualized within Marxian theory emphasizing the basis for class divisions in capitalist society. The discussion then reviews important applications and extensions of these concepts to urban space.

The capitalist 'Mode of Production'

'Mode of production' is a controversial analytical category, even within Marxist theory (see Edel (1981) for a brief review of this controversy and Harvey (1982) for a review of the various ways Marx defined the term). While the details of the debate need not concern us here, a few key features of the capitalist mode of production should be emphasized.

From the perspective of Marxist theory (Marx 1967), the class relation between capital and labour is the fundamental social cleavage in the capitalist mode of production. It is an inherently conflictual relationship. Capitalists own the means of production, control the labour process, and derive profits through the exploitation of wage labourers. Exploitation of wage labour refers to a situation in which workers produce surplus value by creating more in production than they receive in wages. Workers' wages and the amount they produce are determined by two separate processes: wage levels are set historically by subsistence defined for a particular point in time; productivity is determined by the nature of the technology and equipment, and by how hard, long, and fast the worker can be induced/forced to work. The rate of exploitation is under continual negotiation. Workers must sell their labour in order to survive. But they are involved in a constant struggle to reduce the rate of exploitation through increased wages and improved conditions of work.

From the capitalist's perspective, labour power is a commodity that must be purchased in the same way as machinery and raw materials. However, Marx argues, labour power differs from other commodities in that it has the special capacity to produce surplus value. Labour power also differs because workers, as the 'bearers' of labour power, are human beings. If human beings are to be effective workers they must be healthy and well rested. At a minimum, they require adequate housing and food to reproduce themselves materially. In addition, different capitalists require workers with varying skills. There is, therefore, a need (not necessarily met) for society to reproduce adequate types and amounts of labour power (Althusser 1971, Marx 1967). Capitalist social relations must also be constantly reproduced to maintain social harmony, and economic and political stability. Workers are conscious and intentional human beings. It is within their capacity, as a class, to resist exploitation. If capitalist society is to function smoothly, both workers and capitalists must accept the legitimacy of class relations and their respective rôles as sellers and buyers of labour power. The more general point is that class relations are dynamic; they are continually reproduced in changed variations of the same general type.

Applying this theory of capitalist class relations to the urban context, one begins to understand how class relations are reproduced on a daily basis. Class relations are not reproduced in the abstract; they are reproduced in concrete ways on the ground. In fact, the spatial organization of classes plays a significant rôle in the reproduction, and by extension, the maintenance of capitalist society.

Several theories developed over the last 15 years elucidate this relationship between spatial structure and social reproduction – these can now be outlined. It should be noted at the outset, however, that the manner of conceptualizing this relationship has shifted over these years, especially in terms of the type of explanation offered. One current critique of earlier explanations is that they are functionalist. Simply, this means that events are explained in terms of the functions they serve. For example, increased rates of homeownership are explained by referring to the function homeownership serves in stabilizing the social order. Critics of functionalist explanation wish to disentangle cause and function (or effect), arguing that the two may not be the same and, at the very least, the latter does not exhaust the former. This epistemological concern has been central to debates during the last decade; it will be raised continually throughout this review.

The discussion is organized into four main sections. Several theories linking the built environment and the reproduction of the capitalist class through capital accumulation are reviewed. Attention is then directed to the way the spatial structure of the city mediates the reproduction of labour and class conflict. In a final section, directions for future research are outlined.

Accumulation and reproduction of the capitalist class

Capitalists, individually and as a class, are distinguished by their ability to accumulate capital through the exploitation of labour. Capital accumulation is by no means automatic; competition among capitalists and attempts by the working class to reduce the rate of exploitation continually undermine the potential for accumulation. The conditions for capital accumulation must be continually re-established if capitalist class relations are to be maintained.

Numerous theories have been developed to explain the way in which the spatial organization and construction of the city both reflect and establish the potential for capital accumulation, thus mediating the reproduction of the capitalist class. The selection of theories reviewed here is structured by a distinction made by Harvey (1976) between factions within the capitalist class. In conceptualizing conflict around the built environment, Harvey distinguishes three fractions of capital on the basis of how each realizes surplus value: as rent; as interest and profit realized through the construction of the built environment; and by capital 'in general', which regards the built environment as an outlet for surplus capital and a support for capital accumulation. We consider how the spatial organization of the city is implicated in the reproduction of each class faction in so far as it creates their potential to realize surplus value.

Residential differentiation and the realization of class-monopoly rents

In his early work in the political-economy tradition, Harvey (1974, 1976) argues that housing and land markets necessarily create a subset of capitalists whose interests revolve around the appropriation of rent, directly as landlords or property companies, or indirectly as financial intermediaries. The capitalist city creates a specific configuration of interests and conflicts *within* the capitalist class, depending on how surplus value is realized. Of particular interest is Harvey's contention that the institutional and geographical structure of the urban housing market creates the conditions for the extraction of class-monopoly rents.

Harvey develops this position by extending Marxian rent theory to the urban context. He views rents as transfer payments to landowners resulting from their monopoly power over land and other urban resources:

> Class-monopoly rents arise because there exists a class of owners of 'resource units' – the land and the relatively permanent improvements incorporated in it – who are willing to release the units under their command only if they receive a positive return above some arbitrary level (Harvey 1974, p. 241)

The realization of class-monopoly rents is based on the fragmentation of urban space into 'a series of man-made islands on which class monopolies produce absolute scarcities' (Harvey 1974, p. 249). These 'absolute spaces', though mediated by speculator-landlords and speculator-developers, are primarily structured by financial and governmental institutions. Segmentation of the housing market is controlled by a network of private and public institutions which channel loans and investments to different submarkets. In Baltimore 13 submarkets were identified in 1970 (Harvey 1974, Harvey & Chatterjee 1974). For example, the inner city was distinguished by the almost total absence of institutional or governmental involvement. In this context private landlords extracted class-monopoly rents through controlled scarcity via property deterioration and abandonment. In contrast, affluent groups buying houses in the suburbs relied on savings and commercial banks for mortgage credit and here class-monopoly rents were extracted through the manipulation of zoning and status concerns.

The effect of this complex pattern of institutional involvement is the creation of 'absolute spaces', housing submarkets with relatively stable boundaries. The rigidity of the residential structure allows for the realization of class-monopoly rents. For example, inner-city households are unable to obtain access to other housing submarkets, such as owner-occupier and public housing. Effectively trapped within the inner city, they are vulnerable to the extraction of monopoly rents. Housing market segmentation provides the basis for extracting monopoly rents. The subset of capitalists who survive through the extraction of class-monopoly rents would not exist without the residential segmentation found in contemporary capitalist cities (and vice versa). This is one example, then, of the dialectical relation between spatial structure and social class.

Large builders and postwar suburbanization

Focusing on the fraction of capital appropriating profits through the construction of the built environment highlights the dynamic quality of the reproduction of class relations. The types of capitalists involved in the construction industry have changed dramatically in the last 40 years. All accumulate capital by constructing houses, but each does so in a different way. These variations affect patterns of urban growth.

In the United States, Checkoway (1980) documented the restructuring of the residential construction industry, from small craftsmen to large builders, in the postwar period. Large builders are qualitatively different from previous builders in their size, scale, and operating structure. This transformation reflects more than technological capacity. The federal government actively promoted it: 'Large builders more easily received credit advances and more easily negotiated with the FHA. Large operators and powerful economic institutions were among the principle beneficiaries of federal programs. Small operators were either excluded, penalized, or driven from the market' (Checkoway 1980, pp. 32–3). Large builders brought mass-production techniques to residential construction, and this had implications for the location of their projects. Mass production required large, inexpensive tracts of land. These were found typically in suburban areas beyond the city limits.

A similar transformation in Britain, from small to large builders, was noted by Ball (1983). He detected a further restructuring within the housebuilding industry since the mid-1970s: there is a larger presence of long-term investment capital, large conglomerates that entered the housebuilding industry as an investment outlet for surplus money capital.

The ramifications of the involvement of long-term investment capital in the housebuilding industry for patterns of residential differentiation are significant.[2] It may have the effect of heightening residential differentiation. Firms tend to obtain financing from within the conglomerate, and are relatively invulnerable to business cycles. They purchase land during downturns and bank it until house prices rise, at which point they develop the land. Because profits are obtained through manipulation of the land market, they do not expend resources to introduce technical efficiencies in the actual production of houses. They tend to assemble a diversified land bank and build small batches of houses geared to different markets on each site: output is then increased by spatial product differentiation. When a large site is developed, the arrangement may take the form of a consortium whose members differentiate their products to avoid competition.

These studies highlight the dynamic quality of class reproduction, indicating two rounds of restructuring within the housebuilding industry since World War II in terms of the type of capital involved. In each case, old modes of capital accumulation became obsolete. Small and later large capitalists with concentrated interests in house building are unable to compete with long-term development capital. Because housebuilding firms are a critical outlet for conglomerates with surplus money capital, they play a rôle in the reproduction of conglomerates by sustaining their potential for capital accumulation. Further, the way each type of housebuilder accumulates capital has real implications for the urban landscape, in terms of location and size of housing developments. Once again, there is a dialectical relationship between the built environment and social class.

Crises of overaccumulation and the construction of the built environment

Capitalists must accumulate capital to reproduce themselves as a class; yet Marxian theory indicates an inherent tendency for crises of capital accumulation. Two principal forms of class contradiction create such crisis tendencies: contradiction within the capitalist class and conflict between capitalists and workers. Harvey (1978, 1982) stresses the first contradiction and interprets economic instability as reflecting crises of overaccumulation. O'Connor (1984) focuses on the second contradiction and views the current economic crisis in the US as one of *under*accumulation. This theoretical controversy will not be addressed here. Harvey's analysis is outlined because, with Walker (1978, 1981), he has drawn detailed links between accumulation, crises, and urban growth.

Following Marx, Harvey notes a tendency for overaccumulation within the primary circuit of capital – that employed in production. Overaccumulation refers to a situation where more capital is produced than opportunities for its employment. The symptoms of this tendency include: a glut of commodities, falling rates of profit, surplus capital (which can take the form of idle productive capacity or uninvested money-capital), and surplus labour. The secondary circuit of capital flows into fixed capital investments (aids to the production process, such as office buildings and transport networks) and the consumption fund (aids to consumption, such as washing machines and houses). The tertiary circuit of capital comprises investment in science and technology, as well as a wide range of social expenditures for the reproduction of labour power (including health and education). Individual capitalists tend to overaccumulate in the primary circuit and underinvest in the secondary and tertiary circuits. At times of overaccumulation, however, capital flows may switch from the primary to the secondary circuit (depending on the existence of adequate money-market and state support) in an attempt to stave off crisis. When there are no productive outlets for capital in the primary circuit, capital may flow into property speculation, housebuilding, or other investments in the built environment. The construction of the built environment occurs in cycles of 15–25 years linked to rhythms of accumulation in the primary circuit. The timing of both suburban expansion (Walker 1978, 1981) and gentrification (Smith 1979, 1982,

1986) have recently been explained by cycles of capital accumulation rather than changing consumer preferences and demographic trends.

Harvey believes that investments in the built environment also eventually become unproductive: 'At such a time the exchange value being put into the built environment has to be written down, diminished, or even totally lost' (1978, p. 106). Though the built environment is 'devalued', it none the less provides support, or infrastructure, for the next cycle of capital accumulation. For example, many urban mass transit systems were bankrupt at the turn of the century, yet nevertheless remained to be useful in the next stage of accumulation.

The inevitability of the devaluation of the built environment is a matter of debate. Gottdiener (1985, p. 97) argues that Harvey provides little more than the assertion that such devaluation necessarily occurs, rather than clarifying the conditions under which devaluation is likely to exist. Harvey's work is also criticized for positing a functionalist theory of the state (e.g. Gottdiener (1985) who extends a critique of Poulantzas's theory of the state to Harvey). Harvey attributes to the state a significant rôle in mediating switches from the primary and secondary circuits of capital and, more generally, in reproducing capitalist society. This ascribes a high degree of rationality to the accumulation process and a unified relationship between capital and the state. A related issue is the degree to which Harvey envisages cycles of accumulation to occur undisturbed by class conflict (Saunders 1981). For example,

> the overall rhythms of accumulation remain broadly intact in spite of the variations in the intensity of working class struggle. . . . But if we think it through, this is not, after all, so extraordinary . . . if [capitalist] society has survived then it must have done so by imposing those laws of accumulation whereby it reproduces itself (Harvey 1978, p. 119).

There has been extended discussion in recent years over the adequacy of this type of reasoning; the debate is now associated with Giddens (1979). The concern is that it reifies social processes. By indicating that society (a thing) reproduces itself, such explanations stop short of specifying the complexity, the contingency, and the range of agents involved. A more complete explanation of accumulation cycles and crises most likely requires a fuller treatment of working-class action and government policy. Regardless of these reservations, Harvey's analysis forges important links between the built environment and the reproduction of the capitalist class.

Suburbanization and the maintenance of effective demand

A further connection between accumulation and urban structure is made by arguing that the tendency towards overaccumulation is averted through the increasing penetration of market relations into all parts of life (e.g. Aglietta 1979, Castells 1978, Ewen 1976, Harvey 1982, Hayden 1981, Mackenzie & Rose 1983, O'Connor 1973, 1984). Mass consumption plays an important rôle in maintaining effective demand for commodities as households purchase more and more goods and services previously either not needed or provided informally through non-market means. Residential suburbanization is extremely favourable to market expansion of this kind. The increased spatial separation between home and work through the 20th century necessitated the purchase of transportation; in the North American context it typically created the need for the purchase of at least one automobile by each household.

The expansion of homeownership and the single-family house also fosters a consumption life-style (Aglietta 1979). United States single-family housing developments are distinguishable by their absence of communal facilities (see Popenoe (1977) for a comparison of the US and Swedish suburb) necessitating the purchase by individual households of a considerable array of consumer durables, such as washing machines and recreational equipment. Two complementary changes in the design of North American suburban housing also promoted mass consumption: the space alloted to the domestic production of goods such as quilts, clothing, and home-canned vegetables was reduced (Belec *et al.* 1986, Wright 1981); and labour saving gadgets and appliances were introduced as important components of the marketing strategies for mass-produced single-family homes. Levitt houses, built in the 1940s and 1950s, and now taken as prototypes of North American suburban housing, tell the story:

> Levitt spent more money on consumer research than any builder of small houses in history. . . . Each house was small, detached, single family, Cape Cod in style. . . . To insure the sale, each house came complete with radiant heating, fireplace, electric range and refrigerator, washer, built-in television, and landscaped grounds (Checkoway 1980, p. 28).

It is increasingly recognized (e.g. Mackenzie & Rose 1983) that the form of suburbanization occurring in North America, though functional to the maintenance of the capitalist class, is not a simple reflection of capitalist needs. There is a meaningful distinction between cause and effect; the causes of suburbanization cannot be understood only in terms of its success in extending the demand for consumer goods (in other words, by its effects). At the very least, the pattern of suburbanization was the outcome of a contested struggle.

The National Association of Real Estate Boards and the National Association of Home Builders lobbied hard for the 'suburban solution': the homebuilders spent over $5 million in their efforts to shape the 1949 Housing Act. In fact, their tactics in promoting their suburban vision prompted a congressional investigation (Checkoway 1980). So too, the transport policy that led to the expansion of car ownership resulted from intense lobbying by a coalition of auto-oil-rubber-construction interests, dating from 1916 (O'Connor 1973, Yago 1978). Thus, while suburbanization effectively extended mass consumption, its development was not an automatic reflection of capitalist needs.

This review of the links between accumulation, reproduction of the capitalist class, and the built environment has been extremely selective; it merely provides an introduction to some important contributions, stressing that the reproduction of the capitalist class is no less of an issue than the reproduction of labour. The connections were drawn in relation to three fractions of capital identified by Harvey: landlords, capitalists who profit from the construction of the built environment, and capital in general. In each case, the built environment is an important vehicle for capital accumulation and, in this sense, for the reproduction of the capitalist class. Financial and governmental institutions create fragmented housing submarkets; residential segregation then provides a means of extracting class-monopoly rents. The physical appearance of capitalist cities reflects the dominant type of housebuilder and the way that each type accumulates capital. At the same time, suburbanization and the production of single-family homes has provided a significant outlet for surplus money capital, while simultaneously increasing the effective demand for consumer goods. The spatial structure of capitalist cities both reflects characteristics of the capitalist class and acts as a vehicle for its reproduction.

The spatial structure of the city and the reproduction of labour power

People cannot work if they lack the basic means for maintaining health, the requisite skills, or the means for getting to work. Reproduction of labour power refers to the processes by which workers produce and reproduce themselves in the quantity, quality, and location required by various types of employer. A number of the important spatial dimensions will be drawn in this section.

Residential differentiation and the reproduction of skills

Residential differentiation by class is a distinctive characteristic of industrial capitalist cities (Ward 1975, Zunz 1982). In the preceding section, two complementary explanations were provided for residential segregation: patterns of mortgage lending create absolute spaces and housebuilders diversify their product to enhance marketability.

A third explanation focuses on changes in the production process generating wage and status differentials among the working class, allowing for different levels or amounts of housing consumption. Technological innovations, a desire to impose social control and, in some cases, negotiations between capitalists and workers[3] have increased the separation between mental and manual work and continually deskilled both types of workers (Braverman 1974). These status and skill differentials are reinforced by varying income levels, incentives, security, and career paths. Theories of monopoly capitalism (Baran & Sweezy 1966, O'Connor 1973) indicate an increasing bifurcation of the workforce into monopoly and competitive sectors (Edwards 1979).[4] The monopoly or primary sector is comprised of well paid, unionized workers with considerable job security. The competitive or secondary sector is comprised largely of minorities and women who perform non-unionized, low-paid jobs with little security.

Distinctions like these within the working class determine the relative access of workers to different types of housing. Owner-occupation is generally only available to workers in the primary sector with higher incomes and job security who can obtain mortgage credit. Stability of employment may also affect a household's willingness to invest in homeownership: workers without job security may need the relative flexibility afforded by private rental accommodation. Alternatively, they may be tied to central locations near flexible modes of transportation. Level of income provides the basis for further residential differentiation.

A demand-side explanation of residential differentiation such as this complements the previous two: distinctions among workers in terms of income and job security provide the basis for discrimination by mortgage lenders, and for product differentiation by housebuilders. The important point is that residential differentiation not only reflects the increasingly skill differentiated and hierarchically co-ordinated workforce, it also serves to reproduce these differences.

Varying access to public services is one way in which residential differentiation reproduces the division of labour. The theme of comparative access to public services by residents living in different areas of the city received considerable attention from radical urban geographers in the early 1970s (e.g. Harvey 1973, 1975, Gale & Moore 1975, Neenan 1970, 1974). It was shown that patterns of public service provision tend to exacerbate, rather than lessen, existing income differences across the city, a situation reinforced by metropolitan fragmentation in the United States (Child Hill 1974, Hoch 1979, 1984, Markusen 1984). Higher-income residents living in suburban areas have better access to medical facilities and recreational opportunities. They suffer less from the negative externalities of living near noxious facilities, such as air and noise pollution. Their use of public utilities and urban infrastructure has been subsidized by inner-city residents (Neenan 1970). In this way, the effects of labourmarket segmentation have been extended and intensified by residential differentiation and metropolitan fragmentation.

Differential access to public services not only exacerbates variations within the existing division of labour, it also plays a rôle in reproducing them (Peet 1975). The uneven quality of education across the city reproduces skill differentials (Gray 1976). There is ample documentation (e.g. Gale & Moore 1975) that resources allocated to education in the United States vary directly with wealth within the school board jurisdiction. The effects of uneven resources are amplified by varying education priorities so that the children of higher-income parents are given more encouragement, both at home and at school, to obtain educational qualifications that prepare them for high-paying jobs (Bourdieu 1974, Bowles & Gintis 1977, Scott 1980). Beyond direct constraints within the educational system, neighbourhoods are social milieux within which children learn the value (or lack of value) of skills obtained through formal education (Herbert 1979).

This line of argument is not without its critics. Giddens argues the insufficiency of the functionalist claim: 'Education, in a capitalist society, has the function of allocating individuals to positions in the occupational division of labor' (1984, p. 296). In his view this statement merely establishes a problem that demands research and 'fudges over the differences between intended and unintended aspects of social reproduction' (Giddens 1984, p. 297). It implies a high degree of intentionality on the part of the state. Giddens cites Willis's ethnography (1977) of working-class children in a poor area of Birmingham, England as a prototype of a non-functionalist explanation of social reproduction. An important quality of Willis's study is that he demonstrates the complex interplay between the intentionality and unintended consequences in the reproduction of unskilled labour in Birmingham.

Willis studies a group of boys who reject the authority of the school and, in opposition, develop a joking culture, 'having a laff'. This rebellion against the authority and educational goals of the teachers is interpreted as an outcome of their 'partial penetration' of the limited life chances open to them. The unintended consequence of the rebellion is that they actively perpetuate the conditions limiting those life chances: they leave school at the first opportunity to take unskilled manual jobs. 'The working class lad is likely to feel that it is already too late when the treacherous nature of his previous confidence is discovered. The cultural celebration has lasted, it might seem, just long enough to deliver him through the closed factory doors' (Willis 1977, p. 107). While Willis's study is not explicitly *about* the effects of residential differentiation on unequal educational opportunities, it offers geographers considerable insight into the processes by which children living in working-class milieux reproduce themselves as unskilled labourers. The causal argument Willis develops is essential for an adequate understanding of the effects that residential differentiation have in reproducing skill variations.

Residential differentiation thus influences the reproduction of the division of labour through the educational process. When the primary source of funds for education is the local municipality, as it is in the United States, children living in poorer areas have fewer educational resources. Further, neighbourhoods are social milieux in which children come to understand their life chances and the likely value of formal education. They act on that understanding and make it true.

The ghetto and the dual labourmarket

Workers in competitive or secondary sector jobs are trapped in the inner-city rental housing market because they are unable to obtain mortgage credit for home purchase. Several outcomes of this entrapment have been reviewed: vulnerability to the extraction of class-monopoly rents and deteriorating housing conditions, as well as impoverished quality of life due to restricted access to public services.

Urban political economists extended the analysis of the ghetto to an understanding of why ghetto residents tend to hold jobs in the secondary sector of the economy, if they are employed at all. The educational process offers insight into the perpetuation of low skills. Attention has also been given to the rôle the spatial organization of jobs and transportation plays in the effective exclusion of ghetto residents from primary sector jobs.

The geographical restructuring of manufacturing jobs, from central city to suburb, and the spatial organization of transit systems and housing markets in American cities constrain inner-city residents' access to primary sector jobs. Inner-city residents find it difficult to reach these jobs. The suburban housing market has been closed to lower-income households, because suburban housing is for the most part owner-occupied and many suburban municipalities have actively prevented the construction of lower-cost housing through exclusionary zoning (e.g. Johnston 1984). Constrained to live in the inner city, lower-class residents have limited access to information regarding suburban job opportunities. The problem is further exacerbated by urban transportation systems not meeting the needs of persons trying to commute from central city to suburb. North American transportation policy has been biased towards the automobile. Individuals forced to rely on inadequate public transportation systems are thwarted by the reverse commuter problem – public transportation systems tend to be oriented to the CBD. In many North American cities, there are few crosstown and outward-bound buses serving peripheral workplace locations, further limiting the employment opportunities of inner-city residents (Deskins 1970, Elgie 1970, Gale & Moore 1975). In this sense, the spatial structure of the city works against inner-city residents finding employment in primary sector occupations located in the suburbs.

Blaut (1974, 1975, 1983) offers a different, though complementary, explanation for the tendency of North American black and hispanic ghetto residents to find employment in the secondary sector of the economy. He proposes that black and hispanic slums are qualitatively different from the slums of other immigrant groups: 'In Chicago we have some black ghettos of very respectable age which show no sign of turning into simple slums. The Latin *barrios* are growing larger, and poorer, and more purely Latin all the time' (Blaut 1974, p. 37). Blaut explains the persistence of black and hispanic ghettos through an extension of the theory of monopoly capitalism (Baran & Sweezy 1966, O'Connor 1973). A split between the monopoly (primary) and competitive (secondary) sector of the labourmarket is an essential feature of monopoly capitalism. It is a lever for extracting surplus value and a central element in legitimating monopoly capitalism. Workers in the organized monopoly sector are well paid at the expense of those in the competitive sector who are 'super-exploited'.

Blaut's (1974, 1975, 1983) contribution is to connect this process to residential segregation in the American city and imperialist expansion. 'In the case of Puerto Rico, about half of the labor force, and *nearly one-half of the total population*, have been translocated to the United States . . . part of its space has been internalized in the guts of the metropolitan world' (Blaut 1974, p. 40; emphasis in the original). A number of authors (Blaut 1974, Doherty 1973, Feldman 1977) cite the functionality of this arrangement for the capitalist class. Under- and unemployed immigrants form a labour reserve which depresses wages and acts as a scapegoat for the inadequacies of the capitalist system (i.e. unemployment or inconsistent job histories are blamed on a particular racial or ethnic group rather than the socio-economic system).

Blaut admits that his argument is 'impressionistic' and 'not entirely clear' (1985, p. 162). More detail is required to moderate the directly functionalist nature of the argument. We need to know more about the process of translocation; it seems extremely unlikely that ghetto residents are only drawn to US cities at times of economic expansion (when it meets the needs of American capital), as Feldman (1977) suggests. This complex migration process cannot be reduced to a single factor. More attention must also be given to processes through which the ghetto becomes a repository and trap for competitive sector workers and the unemployed. As well, we need to understand differential entrapment within the ghetto and competitive sector of the labourmarket. In other words, more contextual detail is required to understand the different immigrant histories of blacks, hispanics, and other groups. None the less, Blaut draws some important connections between the ghetto, the theory of labourmarket segmentation, and capitalist imperialism.

Urban spatial structure and the mediation of class conflict

The class relation between capitalists and labourers, a central point of conflict in capitalist society, involves a continuing struggle over the relative appropriation of surplus value. Through collective action, workers can resist exploitation by, for example, bargaining for shortened work hours and improved work conditions, and by retaining skill levels and control over the work process. Collectively, they can redefine their social needs to include improved standards of housing, education, health care, and recreation, thus necessitating either higher real wage levels or state intervention.

By most accounts, broadly organized, direct working-class struggle has declined in the postwar period in countries such as the United States, Canada, and Britain. In trying to understand this decline, a number of authors have examined changes in the spatial organization of home and work, as well as other transformations in workers' residential environments, and linked these in a causal fashion to the question of political consciousness and levels of political activism among the working class. Three changes have received attention: (a) increasing spatial separation between home and work; (b) increasing levels of residential segregation, between and within classes; and (c) expansion of homeownership and the suburban life-style to fractions within the working class.

Spatial separation between home and work

Increasing separation between home and work has been a distinctive spatial trend within the industrial capitalist city since the early 19th century. This represents a significant transformation from the pre-capitalist city, in which households of different classes not only lived in the same areas but often on different floors of the same house. In the pre-capitalist city, distinctions between trades were more important spatially than distinctions between classes (Katznelson 1979, Stedman Jones 1973–4).[5]

This situation changed in the early 19th century, as the upper classes abandoned the industrial quarters of the city centre for the suburbs. With the introduction of the factory system of production, there was a diminished need to supervise directly all aspects of the work process. Employers had sufficient discretionary time to commute between home and work. Industrialization also brought a number of negative externalities, such as noise and air pollution, making the inner city a less desirable place to live. Increased class conflict created fear among the upper classes and a desire for spatial distance from the working classes. Rising inner-city land values and property taxes increased the attractiveness of selling inner-city land and relocating the home in a suburban location. Finally, the pastoral ideal exerted a pull towards the suburbs (Walker 1978). As a result of these various processes, the upper classes moved their residences to suburban locations. In this context their homes became private spheres:

> For the private citizen, for the first time the living-space became distinguished from the place of work. The former constituted itself as the interior. The office was its complement. The private citizen who in the office took reality into account, required of the interior that it should support him in his

> illusions. This necessity was all the more pressing since he had no intention of adding social pre-occupations to his business ones. In the creation of his private environment he suppressed them both (Benjamin 1973, pp. 167–8).

At times of economic depression and social unrest, however, the upper classes were not afforded the luxury of indifference and the spatial separation between themselves and the working classes became a focus of concern.

By the late 19th and early 20th centuries, the spatial separation between home and work became a mass phenomenon for the skilled working class as well. As an indication of the extent of suburban migration, the residential population of the City of London declined from 75 000 in 1871 to 38 000 in 1891, while the day-time population increased from 170 000 in 1866 to 301 000 in 1891 (Stedman Jones 1973–4, p. 486). A number of scholars have questioned the influence of the spatial separation between home and work on the class consciousness of working-class suburbanites.

The claim that separation between home and work weakens working-class consciousness and organization is substantiated by two types of analysis. On the one hand, proximity is considered crucial to explaining the high levels of union organization and labour unrest in company towns with employer-provided housing (Feldman 1977). Studies of US industrial towns such as Lowell, Massachusetts (Dublin 1979) and Pullman, Illinois (Buder 1969) suggest that industrial unrest was heightened by both the spatial proximity of workers, which allowed them to organize with relative ease, and the self-evident unfairness of low wages and high prices in a situation where the employer set both.

Other studies provide more direct assessments of the influence of the spatial separation of work and home on working-class consciousness. Stedman Jones (1971, 1973–4) detects a long-term decline in trade unionism in London during the last quarter of the 19th century, accompanied by a remarkable transformation of working-class culture. He attributes both trends largely to the combined effects of increased leisure time and suburban migration. At mid-century there existed a work-centred working-class culture:

> Workers generally lived in the immediate vicinity of their work. Political discussion, drinking and conviviality took place either at the work place itself or at a local pub which generally served as a house of call and a center of union organization. . . . Homes were cramped and uncomfortable; where they were not the place of work, they were little more than places to sleep and eat in (Stedman Jones 1973–4, p. 468).

By the mid-1870s, working class culture became more home-centred. Even among those who continued to frequent the pub, there were important differences:

> If the man commuted to work . . . the regular pub visited would no longer be the trade pub near the work place, but the 'local'. At the 'local' they would mingle with men of different trades and occupations. Conversation was less likely to concern trade matters, more likely to reflect common interests,

politics to a certain extent, but more often, sport and entertainment (Stedman Jones 1973–4, p. 487).

In a similar vein, Luria (1979) explains the decline of the Socialist Party of America after 1916 in terms of the migration of its early base to the suburbs, where the class mix dissipated electoral support. He substantiates these claims by correlating degree of suburbanization with election results in 145 US cities from 1900 to 1922, finding that Socialist Party support was lowest in cities experiencing mass suburbanization. Luria makes the interesting, if unsupported, argument that urban densities, and the integration of work and home, politics and culture, were all critical to the success of the Socialist Party. It is not that suburban working-class households deserted the Socialist Party at the polls; instead, suburbanization hurt Party organization and spatially split the Party's electoral support.

The absence of a strong working-class tradition in American urban politics has also been explained by Katznelson (1981) in terms of the spatial and conceptual separation of the residential community from the workplace. His analysis of the processes leading to the separation of the politics of work from the politics of home is noteworthy because it integrates an understanding of spatial effects with a historically grounded analysis of the American political-legal system. Katznelson argues that the effects of the separation of home and work are *mediated* by the political and institutional context.

The separation of the politics of work and home is more extreme in the US than elsewhere because of the institutional context. A comparative tolerance of trade unions by the courts in the 18th century allowed workplace grievances to be organized around the workplace itself. Also in contrast to the European situation, all white male workers in American cities obtained political franchise early and thus a class-based oppositional force did not develop against the state. As a result, class-based politics became localized within the workplace. Residential community came to define the social base and substantive focus of urban politics. Drawing on Gramsci's metaphor, Katznelson argues that the ultimate effect has been the development of 'city trenches' protecting capital from organized working-class resistance. 'In wars of position – like World War One – the system of trenches defines the terrain of battle and thus imparts a logic to the war itself. Because each system of trenches is distinctive, it defines both the place and the context of conflict' (Katznelson 1981, p. 19). Separation of the politics of home and work represents a set of trenches in the United States. Trenches which are the outcome of both the political-legal and geographical history of American cities have the ultimate *effect* of dividing and weakening working-class consciousness in the United States. Katznelson's analysis offers both a demonstration of the importance of the spatial organization of the city for an understanding of class conflict and a warning against any attempt to reduce this understanding to spatial factors alone. The effects of space are mediated by political and legal factors and vary from context to context. This message about reductionism is equally applicable to the study of the effects of residential segregation on class consciousness.

Residential segregation and class conflict

A second trend within the capitalist city was the progressive residential segregation of social classes. Studies of the influence of residential segregation on class consciousness and conflict are extensively reviewed by Harris (1984a). Two seemingly contradictory effects have been attributed to residential segregation. The concentration and isolation of worker households in one quarter of the city is thought to promote consciousness of class and working-class organization, while segregation within and between classes may inhibit class consciousness.

In support of the first claim, there are numerous empirical studies on the effect of spatial concentration on the development of working-class culture (i.e. Fried 1973, Gans 1962, Harris 1984b, Suttles 1968, Stedman Jones 1973–4, Young & Wilmott 1962). There is disagreement, however, about the links between neighbourhood-based working-class culture and class-based political action. Some authors (Luria 1979, Stedman Jones 1973–4) trace the emergence of trade unionism and class-based political parties from a neighbourhood base. Other Marxist scholars view neighbourhood organizations as potentially disruptive to broad-based class struggle, in so far as they direct political organization to consumption and reproduction issues, such as housing, and away from class exploitation in the workplace. The latter view has prompted an extended discussion of the theoretical status of reproduction and consumption struggles and a rich literature has emerged uncovering the theoretical links between struggles in the production and reproduction spheres (Bunge 1977, Castells 1976, 1977, 1978, 1983, Fincher 1984, Harvey 1974, 1976). Nevertheless, Hasson (1986) contrasts theoretical insights and the empirical reality that working-class neighbourhood organizations are often parochial, with a tendency to focus on relatively trivial issues.

Residential segregation may also serve to diffuse class conflict through intra-class fragmentation and the separation of classes. Residential differentiation occurs between, and also within, classes, reflecting the influence of factors like ethnicity, stage in the life-cycle, and position in the division of labour. To the extent that residential differentiation reinforces households' identification with these other factors, it has the potential to fragment class consciousness (Harvey 1975, 1976, Short 1976). For example, Rosenzweig's (1983) study of Worcester, Massachusetts between 1870 and 1920 portrays a city of tight ethnic communities, each with an elaborate organizational infrastructure (including churches, clubs, kinship networks, saloons, and often schools), which led to the almost total isolation of different ethnic groups. The absence of contact fostered animosity that may, at least partially, explain the failure of Worcester unions and the virtual absence of working-class political parties in the city. Though this historical case study is suggestive, it is none the less true, as Harris (1984a) indicates, that the impact of residential differentiation on class fragmentation is largely unsubstantiated and requires further work.

Residential segregation is also thought to diffuse class conflict through the spatial separation of classes, fostering a mutual ignorance. In an oft-quoted passage, Engels outlines the way in which the spatial structure of Manchester, England neatly shields the wealthy from even visual contact with the working class. Writing in 1844, he notes:

> And the finest part of the arrangement is this, that the members of the money aristocracy can take the shortest road through the middle of all the laboring districts without ever seeing that they are in the midst of the grimy misery that lurks to right and left. For the thoroughfares leading from the Exchange in all directions out of the city are lined, on both sides, with an unbroken series of shops . . . they suffice to conceal from the eyes of the wealthy men and women of strong stomachs and weak nerves the misery and grime which form the complement of their wealth (1962, p. 46).

In a reciprocal fashion, the working class is seldom exposed to the suburban landscapes of the wealthy and, therefore, do not as strongly feel their relative deprivation.

Residential segregation thus potentially inhibits and promotes class formation. It may foster intra-class distinctions along the lines of ethnicity, place in the division of labour, stage in the life-cycle, or lifestyle. Spatial separation between classes may promote mutual ignorance and indifference. On the other hand, the concentration of a class within a residential area may encourage contact, promote class consciousness and facilitate political organization.

These arguments are 'blunt tools, incapable of anatomizing the particular significance of segregation in each case' (Harris 1984a, p. 39). There can be no theoretical resolution at an abstract level. The effects of spatial separation and concentration are mediated by a whole range of social processes only understandable in particular historical and situational contexts.

Harris (1984a) suggests several factors mediating the effects of residential segregation in specific contexts. The effects of segregation may be more striking in a city or town with a highly polarized class structure. Continued improvements in communication technology may lessen the effects of spatial segregation. Finally, residential segregation may interact with housing tenure: homeowners may be more sensitive to segregation in so far as they generally have a stronger commitment to their neighbourhoods.

Homeownership and incorporation

It is the last factor, housing tenure, to which we finally turn: homeownership has been considered to be especially significant for the reproduction of labour power and the suppression of class conflict in the postwar period in a number of countries, including the United States, Canada, Britain, Australia, and New Zealand.

A number of scholars working in the Marxist tradition of urban political economy claim that the expansion of homeownership significantly influences the political values of the working class (Bell 1977, Boddy 1980, Castells 1977, Clarke & Ginsburg 1975, Harvey 1976, Luria 1976, Thorns 1981). Homeownership legitimates private property, and promotes 'possessive individualism'. It fragments working-class unity by dividing the working class into housing status groups; common class-based interests are masked and class struggle diffused by intra-class community conflict (Harvey 1976). Further, mortgage debt instills financial conservatism and stability in the workplace: 'A worker mortgaged up to the hilt is, for the most part, a pillar of social stability' (Harvey 1976, p. 272). Homeownership, in other words, has been an extremely effective tool for dampening class conflict and reproducing capitalist class relations.

Incorporation theory, positing that homeownership causes ideological incorporation, has been the focus of a sustained critique (Agnew 1981a, 1981b, Gray 1982, Rose 1980, Saunders 1979). The first wave of criticism addressed the functional nature of incorporation theory, namely, the extent to which homeownership is viewed as a tool of the capitalist class to manipulate working-class consciousness. This is a misleading simplification of the genesis of housing policy, in part because it attributes a scarcely credible degree of intentionality to the state. It minimizes the number of cross-pressures leading to postwar housing policy. In particular, it ignores the extent to which homeownership was actually sought by the working class (Rose 1980, 1981, Mackenzie & Rose 1983). Even if one wishes to maintain that expanded homeownership is not in the long-term interests of the working class, either historically (Edel *et al.* 1984) or in the current context of rising house prices (Ball 1983, Hartman 1983), it mystifies the historical evidence to suggest that homeownership was pressed on the working class by the state in the interests of capital. In general, there is a move towards a more nuanced understanding of the processes that led to expanded homeownership among working-class households, including an appreciation of how working-class households may have unintentionally, but nevertheless actively, reproduced capitalist class relations through attempts to improve living conditions via homeownership.

The effects that incorporation theorists attribute to homeownership are also being reassessed. A difficulty with their analysis is that one has been obliged to accept it on trust; very little empirical evidence is provided.

A number of recent empirical studies indicate the ideological significance of housing tenure (Agnew 1981, Dunleavy 1979, Pratt 1986a, 1986b). These studies also demonstrate, however, the necessity of formulating theories about the ideological impact of homeownership in conjunction with careful empirical work, for they suggest important modifications to the incorporation theory.

Agnew's (1981a, 1981b) comparison of the meaning of homeownership in Britain and the United States shows it to be more clearly viewed as a status symbol and investment and more closely associated with a privatized life-style in the US. This results from a number of factors, including high levels of residential mobility and a weaker tradition of working-class resistance in the United States. The results of this cross-national comparison lead Agnew to conclude that the influence of housing tenure on political consciousness is not strictly determinate but, instead, is mediated by a number of contextual factors.

> Resistance to capitalist social relations, 'struggle' if you will, gives rise to ideas and institutions that reorient the process of self-definition. So in the case of homeownership, the meanings attached to the tenure differ from context to context depending upon the extent to which the dominance of capitalist social relations has been successfully challenged (Agnew, 1981, p. 88).

Research in the Canadian national context (Pratt 1986a, 1986b) suggests another way in which incorporation theory misleadingly oversimplifies the relation between homeownership and ideology. In so far as incorporation theory is used to explain the demise of working-class consciousness, the implication is that homeownership has an extremely potent ideological effect and is capable of overwhelming all other experience, including work. A detailed examination of the relation between social class, housing tenure, and political ideology shows a more complex relationship: the political effects of homeownership are uneven across social classes. The claims of incorporation theory regarding the effects of homeownership on political ideology seem to be true for those in white-collar, but not blue-collar, occupations. The ideological impact of homeownership is, therefore, by no means automatic; experiences of class mediate the influence of housing tenure in important ways. This conclusion leads us back to a central theme of this chapter: there is a complex interaction between different spheres of life, between social reproduction and class relations.

Future research directions

In the past 15 years urban political economists have developed a rich and multifaceted understanding of the ways in which urban form shapes the reproduction of capitalist social relations. Urban space is tied in a number of ways to the reproduction of the capitalist class. Residential segregation creates the preconditions for the extraction of class-monopoly rents. Investment in the built environment is a means of delaying overaccumulation crises. Suburban expansion fosters capitalist accumulation through the expansion of consumer demand. The spatial structure of the city has been influenced by the changed composition of capital in the housebuilding industry.

The spatial organization of the city also affects the reproduction of workers. The separation between home and work fundamentally changes working-class culture and hampers workers' political organization. Distancing of inner-city residents from suburban manufacturing jobs effectively traps them within the secondary sector of labourmarkets. Residential segregation further isolates classes and class fractions. It creates varying social milieux which foster distinctive working-class subcultures, structured along the lines of ethnicity, stage in the life cycle and level of skill. Educational aspirations and resources vary across these milieux, and this has the effect of reproducing skill levels across generations. Finally, expanded suburban homeownership tends to influence the political values of a fraction of North American workers.

What remains to be explored? What new directions of research have been opened up? Three avenues are particularly interesting. One involves epistemological and methodological reorientation. A second, the rethinking of class relations at a less abstract level of analysis. A third addresses increasing crises of social reproduction in many capitalist cities.

Early theories of the relationship between urban form and social class tended to share two characteristics. They were typically functionalist theories drawing on social control metaphors and they were often posited in the absence of careful empirical study. As social control theories, they tended to see the spatial patterns reinforcing class differences as agencies of social control, as tools of manipulation generally forged by finance capital, the state, or capital in general. This relatively static conceptualization misses the point that spatial structures emerge out of a continuous process of class struggle: the working class also shapes urban space and reproduces class relations, and the social relation between capital and labour is under a constant process of renewal and renegotiation. Further, all spatial outcomes should not be seen as directly functional to capital.

The concept of reproduction offers the potential to capture this dynamic process but this opportunity has often been missed. A number of authors (Pickvance 1977, Topalov 1985) note that the reproduction concept is susceptible to the same weaknesses as social control theory if theorists retain a functonalist bias, seeing the process of reproduction as the result of actions taken from above, the results of all urban policies as being functional to the capitalist class, and capitalist class relations as unchanging, static categories. As well, Topalov has criticized the economistic understanding of social reproduction in Marxist analyses. He claims that Marxists have tended to consider workers from the standpoint of capitalists as only producers of surplus value. As such, analyses of social reproduction become one dimensional and the full meaning of conflicts in the reproduction sphere is lost.

Even an understanding that social reproduction results from class conflict can be problematic if reached at too high a level of abstraction. The difficulty with the abstract acceptance that social reproduction both results from class conflict and reproduces class relations is that it is 'too easy to use. According to what is needed in interpreting any specific event, sometimes conflict will be involved, sometimes control' (Topalov 1985, p. 264). There is now renewed appreciation of careful empirical research restoring the situational context, allowing for a more careful assessment to which particular processes have led to historical change and which specific outcomes have secured or impaired social order. This does not entail a return to the empiricism of the past. The current challenge is to devise research questions and methods that integrate the theories developed over the past 15 years with the specifics of particular contexts grounded in space and time. This has led to an interest in locale (e.g. Gregory & Urry 1985, Cooke 1987) and ethnographic methods (Sayer & Morgan 1985).

A recognition of the limitations of abstraction has led to a considerable amount of rethinking of Marxist class categories, as they apply to concrete capitalist societies rather than to abstract modes of production. (For a review of a number of problems that have persistently plagued class analysis, see Walker (1985).) In particular, Marxist class theory failed to account for the increased complexity within the labourmarket, especially the growth of the managerial stratum. In recent years a number of theorists have responded to this gap in Marxist class theory and directed their attention to the theoretical status of the middle classes (Carchedi 1977, Wright 1976, Wright & Perrone 1977). While there is disagreement regarding the appropriate resolution (Wright 1980), Carchedi and Wright have been in some agreement that managers and professionals are in a contradictory class location in so far as the labour time is split between supervising the labour power of others (performing a function of capital) and the production of surplus value (the function of labour).

The discovery of the middle class in theoretical terms opens an avenue of study for urban geographers. Fragmentary evidence shows that the residential environment is of critical importance for the reproduction of middle-class consciousness (Jager 1986, Pratt 1986a, Williams 1986). This, in turn, influences the spatial structure of the city. For instance, the gentrification of inner-city areas has been interpreted in terms of the importance of particular consumption styles for the formation of new middle-class identity (Ley 1980,

Jager 1986). In dynamic interplay, the experience of central city living is thought to act back upon middle-class consciousness and lead to a distancing between the new middle and working classes (Williams 1986). Connections like these between the city and the reproduction of the new middle class are relatively unexplored and warrant further attention.

A third direction for research is prompted by the evident crisis of social reproduction in many North American and European cities. Many urban processes, posited by the theorists reviewed here as critical to social stability in the postwar period, are breaking down. In a number of national contexts, notably the US and Britain, massive cut-backs have been made in national and local provision of urban services. Both national governments are ideologically committed to selling off the existing public housing stock. Yet housing access and affordability have been increasing problems since the mid-1970s. In the US, renters are paying more of their incomes for housing, homeownership is becoming less accessible to first-time home purchasers, rates of homeowner mortgage default are rising, and homelessness of young people, the elderly, and families appears to be an increasing problem (e.g. Hartman 1983, Rudel 1985). The situation is exacerbated by the spatial restructuring of employment, both regionally and internationally, a theme covered in considerable detail by a number of authors in this volume. Even in cities experiencing economic growth, there is clear evidence that such growth will not produce a massive expansion of middle-income groups, as has been typical of the postwar period. Instead, there are signs of polarization in the occupational structure, with an expansion in the supply of low-wage jobs and a contraction in the supply of middle-income jobs (Sassen-Koob 1984, Stanback & Noyelle 1982). Further, work by Pahl (1984) dispels illusions that households which have become unemployed or underemployed as a result of this restructuring will manage to get by through an informal economy of barter and self-help.

Several questions present themselves. How will the growing number of low-income households maintain themselves? What new strategies for reproduction will households develop? What types of collective action will result? How will the spatial structure of the city come to reflect new strategies for reproduction and mediate the growing polarization of social classes?

Notes

I would like to thank the editors, and Trevor Barnes for their comments on a draft of this chapter.

1 In this chapter I focus on the positive theoretical contributions developed within the radical political-economy tradition over the last 15 years. Also of importance are the critiques of mainstream perspectives developed by Marxist theorists. See Castells (1977) and Harvey (1973) for what have become classic critiques of the Chicago School and ecological traditions in urban geography. For cogent criticism of neoclassical location rent models, see Harvey (1973) and Walker (1978).

2 A study by Dickens *et al.* (1985) indicates the specificity of Ball's findings. Dickens *et al.* draw an extremely effective comparison between housebuilding industries in Britain and Sweden. They find that Swedish housebuilders have been more successful in the introduction of mass-production techniques to residential construc-

tion and attribute this success to the fact that Swedish housebuilders have been unable to capture profits through efficient land banking. This reflects the active rôle that the Swedish government has taken in controlling the land market, which, in turn, results from the relative weakness of Swedish landed capital.

3 The latter factor, the effects of negotiations between capitalists and workers, is often neglected but nevertheless of considerable importance to the determination of skill levels. The feminist literature on the occupational segregation of women highlights the negotiated quality of skill definitions and the extent to which the low skill levels of 'women's jobs' in some cases reflect the handiwork of both employers and male union members (Philips & Taylor 1980, Walby 1985).

4 The current applicability of the theory of monopoly capitalism has been questioned (Gordon *et al.* 1982, Trachte & Ross 1985). The segmentation of the labourforce into competitive and monopoly sectors may decrease in the current phase of capitalist development, as monopoly sector workers lose many of the benefits gained through the postwar period.

5 Ward (1975) argues against a historically linear understanding of residential segregation, in so far as he detects less segregation in the mercantile city than in earlier city forms.

References

Aglietta, M. 1979. *A theory of capitalist regulation: the U.S. experience.* London: New Left Books.

Agnew, J. 1981a. Home-ownership and identity in capitalist societies. In *Housing and identity: cross-cultural perspectives*, J. S. Duncan (ed.), 60–97. London: Croom Helm.

Agnew, J. A. 1981b. Home-ownership and the capitalist social order. In *Urbanization and urban planning in capitalist society*, M. Dear & A. J. Scott (eds), 457–80. London: Methuen.

Althusser, L. 1971. *Lenin and philosophy and other essays.* London: New Left Books.

Badcock, B. 1984. *Unfairly structured cities.* Oxford: Basil Blackwell.

Ball, M. 1983. *Housing policy and economic power: the political economy of owner occupation.* London: Methuen.

Baran, P. A. & P. M. Sweezy 1966. *Monopoly capital.* New York: Monthly Review Press.

Belec, J., J. Holmes & T. Rutherford 1986. The rise of fordism and the transformation of consumption norms: mass consumption and housing in Canada, 1930–1945. Paper presented at International Research Conference on Housing Policy, Gävle, Sweden.

Bell, C. 1977. On housing classes. *Australian and New Zealand Journal of Sociology* **13**, 36–40.

Benjamin, W. 1973. *Charles Baudelaire: a lyric poet in the era of high capitalism*, trans. H. Zohn. London: New Left Books.

Berry, M. 1981. Posing the housing question in Australia: elements of a theoretical framework for a Marxist analysis of housing. *Antipode* **13**, 3–14.

Blaut, J. M. 1974. The ghetto as an internal neo-colony. *Antipode* **6**, 37–41.

Blaut, J. M. 1975. Imperialism: the Marxist theory and its evolution. *Antipode* **7**, 1–19.

Blaut, J. M. 1983. Assimilation versus ghettoization. *Antipode* **15**, 35–41.

Blaut, J. M. 1985. Endnote, 1985. *Antipode* **17**, 162.

Boddy, M. 1980. *Building societies.* London: Macmillan.

Bourdieu, P. 1974. Cultural reproduction and social reproduction. In *Power and ideology in education*, J. Karabel & A. H. Halsey (eds), 487–511. New York: Oxford University Press.

Bowles, S. & H. Gintis 1977. *Schooling in capitalist America.* New York: Basic Books.

Braverman, H. 1974. *Labour and monopoly capitalism.* New York: Monthly Review Press.

Buder, S. 1969. *Pullman: an experiment in industrial order and community planning, 1880–1930*. New York: Oxford University Press.

Bunge, W. 1977. The point of reproduction: a second front. *Antipode* **9**, 60–76.

Carchedi, G. 1977. *On the economic identification of social classes*. London: Routledge & Kegan Paul.

Castells, M. 1976. Theoretical propositions for the experimental study of urban social movements. In *Urban sociology: critical essays*, C. Pickvance (ed.). London: Tavistock.

Castells, M. 1977. *The urban question: a Marxist approach*, trans. A. Sheridan. Cambridge, Mass.: MIT Press.

Castells, M. 1978. *City, class, power*, trans. E. Lebas. New York: St Martin's Press.

Castells, M. 1983. *The city and the grassroots*. Berkeley: University of California Press.

Checkoway, B. 1980. Large builders, federal housing programmes, and postwar suburbanization. *International Journal of Urban and Regional Research* **4**, 21–45.

Child Hill, R. 1984. Separate and unequal: governmental inequality in the metropolis. *American Political Science Review* **68**, 1563–8.

Clarke, S. & N. Ginsburg 1975. The political economy of housing. In *Political Economy of Housing Workshop, Political Economy and the Housing Question*, 3–33. London: PEHW.

Cooke, P. 1987. Individuals, localities and postmodernism. *Environment and Planning D, Society and Space* **5**, 408–12.

Deskins, D. R. 1970. Residence workplace interaction vectors for the Detroit metropolitan area: 1958 to 1965. Northwestern University, Department of Geography, Special Publication 3, 1–24.

Dickens, P. S. Duncan, M. Goodwin & F. Gray 1985. *Housing, states and localities*. London: Methuen.

Doherty, J. 1973. Race, class and residential segregation in Britain. *Antipode* **5**, 45–51.

Dublin, T. 1979. *Women at work*. New York: Columbia University Press.

Dunleavy, P. 1979. The urban basis of political alignment: social class, domestic property ownership, and state intervention in consumption processes. *British Journal of Political Science* **9**, 409–43.

Edel, M. 1981. Capitalism, accumulation and the explanation of urban phenomena. In *Urbanization and urban planning in capitalist society*, M. Dear & A. J. Scott (eds), 19–44. London: Methuen.

Edel, M., E. D. Sclar & D. Luria 1984. *Shaky palaces: homeownership and social mobility in Boston's suburbanization*. New York: Columbia University Press.

Edwards, R. 1979. *Contested terrain: the transformation of the work place in the twentieth century*. New York: Basic Books.

Elgie, R. 1970. Rural immigration, urban ghettoization and their consequences. *Antipode* **2**, 35–54.

Engels, F. 1962. *The condition of the working class in England in 1844*. London: Allen & Unwin.

Ewen, S. 1976. *Captains of consciousness: advertising and the social roots of consumer culture*. New York: McGraw-Hill.

Feldman, M. 1977. A contribution to the critique of urban political economy: the journey to work. *Antipode* **9**, 30–50.

Fincher, R. 1984. Identifying class struggle outside commodity production. *Environment and Planning D, Society and Space* **2**, 309–27.

Fried, M. 1973. *The world of the urban working class*. Cambridge, Mass.: Harvard University Press.

Gale, S. & E. G. Moore 1975. *The manipulated city*. Chicago: Maaroufa Press.

Gans, H. 1962. *The urban villagers*. New York: The Free Press.

Giddens, A. 1979. *Central problems in social theory: action, structure and contradiction in social analysis*. Berkeley: University of California Press.

Giddens, A. 1984. *The constitution of society: outline of the theory of structuration*. Oxford: Polity Press.

Gordon, D., R. Edwards & M. Reich 1982. *Segmented work, divided workers*. New York: Cambridge University Press.

Gottdiener, M. 1985. *The social production of urban space*. Austin: University of Texas Press.

Gray, F. 1976. Racial geography and the study of education. *Antipode* **8**, 33–44.

Gray, F. 1982. Owner occupation and social relations. In *Owner occupation in Britain*, S. Merrett (ed.), 267–91. London: Routledge & Kegan Paul.

Gregory, D. & J. Urry (eds) 1985. *Social relations and spatial structures*. New York: St Martin's Press.

Harris, R. 1984a. Residential segregation and class formation in the capitalist city: a review and directions for research. *Progress in Human Geography* **8**, 26–49.

Harris, R. 1984b. A political chameleon: class segregation in Kingston, Ontario, 1961–1976, *Annals of the Association of American Geographers* **74**, 454–76.

Hartman, C. (ed.) 1983. *America's housing crisis: what is to be done?* Boston: Routledge & Kegan Paul.

Harvey, D. 1972. A commentary on the comments. *Antipode* **4**, 38.

Harvey, D. 1973. *Social justice and the city*. Baltimore: Johns Hopkins University Press.

Harvey, D. 1974. Class-monopoly rent, finance capital and the urban revolution. *Regional Studies* **8**, 239–55.

Harvey, D. 1975. Class structure in a capitalist society and the theory of residential differentiation. In *Processes in physical and human geography*, R. Peel, M. Chisholm & P. Haggett (eds), 354–69. London: Heinemann.

Harvey, D. 1976. Labor, capital and class struggle around the built environment in advanced capitalist societies. *Politics and Society* **6**, 265–95.

Harvey, D. 1978. Urbanisation under capitalism: a framework for analysis. *International Journal of Urban and Regional Research* **2**, 101–31.

Harvey, D. 1982. *The limits to capital*. Oxford: Basil Blackwell.

Harvey, D. & L. Chatterjee 1974. Absolute rent and the structuring of space by governmental and financial institutions. *Antipode* **16**, 22–36.

Hasson, S. 1985. The neighbourhood organisation as a pedagogic project. *Environment and Planning D, Society and Space* **3**, 337–55.

Hayden, D. 1981. *The grand domestic revolution: a history of feminist designs for American homes, neighbourhoods, and cities*. Cambridge, Mass.: MIT Press.

Herbert, D. T. 1979. Urban crime: a geographical perspective. In *Social problems and the city: geographical perspectives*, D. T. Herbert & D. M. Smith (eds). Oxford: Oxford University Press.

Hoch, C. 1979. Social structure and suburban spatio-political conflicts in the U.S. *Antipode* **11**, 44–55.

Hoch, C. 1984. City limits: municipal boundary formation and class segregation. In *Marxism and the metropolis*, 2nd edn, W. K. Tabb & L. Sawers (eds), 101–19. New York: Oxford University Press.

Jager, M. 1986. Class definition and the aesthetics of gentrification: Victoriana in Melbourne. In *Gentrification of the city*, N. Smith & P. Williams (eds), 78–91. Boston: Allen & Unwin.

Johnston, R. J. 1984. *Residential segregation, the state and constititional conflict in American urban areas*. London: Academic Press.

Katznelson, I. 1979. Community, capitalist development, and the emergence of class. *Politics and Society* **9**, 203–37.

Katznelson, I. 1981. *City trenches: urban politics and the patterning of class in the United States*. New York: Pantheon.

Ley, D. 1980. Liberal ideology and the post industrial city. *Annals of the Association of American Geographers* **70**, 238–58.

Luria, D. 1976. Suburbanization, homeownership and working class consciousness. PhD dissertation. University of Massachusetts, Amherst, Department of Economics.
Luria, D. 1979. Suburbanization, ethnicity and party base: spatial aspects of the decline of American socialism. *Antipode* **11**, 76–80.
Mackenzie, S. & D. Rose 1983. Industrial change, the domestic economy and home life. In *Redundant spaces in cities and regions*, J. Anderson, S. Duncan & R. Hudson (eds), 155–99. London: Academic Press.
Markusen, A. R. 1984. Class and urban social expenditure: a Marxist theory of metropolitan government. In *Marxism and the metropolis*, 2nd edn, W. K. Tabb & L. Sawers (eds), 82–100. New York: Oxford University Press.
Marx, K. 1967. *Capital* (3 vols). New York: International Publishers.
Myers, D. 1985. Wive's earnings and rising costs of homeownership. *Social Science Quarterly* **66**, 319–29.
Neenan, W. 1970. Suburban-central city exploitation thesis: one city's tale. *National Tax Journal.*
Neenan, W. 1974. *Political economy of urban areas.* Chicago: Markham.
O'Connor, J. 1973. *The fiscal crisis of the state.* New York: St Martin's Press.
O'Connor, J. 1984. *Accumulation crisis.* New York: Basil Blackwell.
Pahl, R. E. 1984. *Divisions of labour.* Oxford: Basil Blackwell.
Peet, R. 1975. Inequality and poverty: a Marxist-geographic inquiry. *Annals of the Association of American Geographers* **65**, 564–71.
Philips, A. & B. Taylor 1980. Sex and skill: notes towards a feminist economics. *Feminist Review* **6**, 79–88.
Pickvance, C. 1977. Some aspects of the political economy of French housing or why the reproduction of labour power theory of housing is invalid. Paper presented at Political Economy of Housing Workshop, Birmingham.
Popenoe, D. 1977. *The suburban environment: Sweden and the United States.* Chicago: University of Chicago Press.
Pratt, G. 1986a. Housing tenure and social cleavages in urban Canada. *Annals of the Association of American Geographers* **76**, 366–80.
Pratt, G. 1986b. Housing consumption sectors and political response in urban Canada. *Environment and Planning D, Society and Space* **4**, 165–82.
Rose, D. 1980. Toward a re-evaluation of the political significance of home-ownership in Britain. In *Political Economy of Housing Workshop, Housing, Construction and the State*, Conference of Socialist Economists. London: CSE.
Rose, D. 1981. Home-ownership and industrial change: the struggle for a 'separate sphere'. University of Sussex Working Papers in Urban and Regional Studies, 25.
Rosenzweig, R. 1983. *Eight hours for what we will: workers and leisure in an industrial city, 1870–1920.* Cambridge: Cambridge University Press.
Rudel, T. K. 1985. Changes in access to homeownership during the 1970's. *Annals of Regional Science* **19**, 37–49.
Sassen-Koob, S. 1984. The new labor demand in global cities. In *Cities in transformation: class, capital and the state, vol. 26: Urban affairs annual review*, M. P. Smith (ed.), 139–71. Beverly Hills: Sage Publications.
Saunders, P. 1979. *Urban politics: a sociological approach.* London: Hutchinson.
Saunders, P. 1981. *Social theory and the urban question.* London: Hutchinson.
Sayer, A. & K. Morgan 1985. A modern industry in a declining region: links between method, theory and policy. In *Politics and method*, D. Massey & R. Meegan (eds), 144–74. London: Methuen.
Scott, A. J. 1980. *The urban land nexus and the state.* London: Pion.
Short, J. 1976. Social system and spatial patterns. *Antipode* **8**, 77–87.
Smith, N. 1979. Gentrification and capital: practice and ideology on society hill. *Antipode* **11**, 24–35.
Smith, N. 1982. Gentrification and uneven development. *Economic Geography* **58**, 139–55.

Smith, N. 1986. Gentrification, the frontier and the restructuring of urban space. In *Gentrification of the city*, N. Smith & P. Williams (eds), 15–34. Boston: Allen & Unwin.
Stanback, T. M. & T. Noyelle 1982. *Cities in transition*. Totowa, NJ: Allanheld, Osmun.
Stedman Jones, G. 1971. *Outcast London*. London: Oxford University Press.
Stedman Jones, G. 1973–4. Working-class culture and working-class politics in London, 1870–1900; notes on the remaking of a working class. *Journal of Social History* **7**, 460–508.
Stone, M. 1974. The housing crisis, mortgage lending, and class struggle. *Antipode* **6**, 22–37.
Stone, M. 1983. Housing and the economic crisis: an analysis and emergency program. In *America's housing crisis: what is to be done?* C. Hartman (ed.), 99–150. Boston: Routledge & Kegan Paul.
Suttles, G. 1968. *The social order of the slum*. Chicago: University of Chicago Press.
Thorns, D. C. 1981. Owner-occupation: its significance for wealth transfer and class formation. *Sociological Review* **29**, 705–28.
Topalov, C. 1985. Social policies from below: a call for comparative historical studies. *International Journal of Urban and Regional Research* **9**, 254–71.
Trachte, K. & R. Ross 1985. The crisis of Detroit and the emergence of global capitalism. *International Journal of Urban and Regional Research* **9**, 186–217.
Walby, S. 1985. Theories of women, work, and unemployment. In *Localities, class and gender*, Lancaster Regionalism Group (eds), 145–60. London: Pion.
Walker, R. A. 1978. The transformation of urban structure in the nineteenth century and the beginning of suburbanization. In *Urbanization and conflict in market societies*, K. R. Cox (ed.), 165–211. London: Methuen.
Walker, R. A. 1981. A theory of suburbanization: capitalism and the construction of urban space in the United States. In *Urbanization and urban planning in capitalist society*, M. Dear & A. J. Scott (eds), 383–429. London: Methuen.
Walker, R. A. 1985. Class, division of labour and employment in space. In *Social relations and spatial structures*, D. Gregory & J. Urry (eds), 164–89. New York: St Martin's Press.
Ward, D. 1975. Victorian cities: how modern? *Journal of Historical Geography* **1**, 135–52.
Williams, P. 1986. Class constitution through spatial reconstruction? A re-evaluation of gentrification in Australia, Britain, and the United States. In *Gentrification of the city*, N. Smith & P. Williams (eds), 56–77. Boston: Allen & Unwin.
Willis, P. 1977. *Learning to labour*. Farnborough: Saxon House.
Wright, E. O. 1976. Class boundaries in advanced capitalist societies. *New Left Review* **98**, 3–41.
Wright, E. O. 1980. Varieties of Marxist conceptions of class structure. *Politics and Society* **9**, 299–322.
Wright, E. O. & L. Perrone 1977. Marxist class categories and income inequality. *American Sociological Review* **42**, 32–56.
Wright, G. 1981. *Building the dream: a social history of housing in America*. New York: Pantheon.
Yago, G. 1978. Current issues in U.S. transportation politics, *International Journal of Urban and Regional Research* **2**, 351–9.
Young, M. & P. Wilmott 1962. *Family and kinship in east London*. Harmondsworth: Penguin.
Zunz, O. 1982. *The changing face of inequality: urbanization, industrial development, and immigrants in Detroit, 1880–1920*. Chicago: University of Chicago Press.

5 *Women in the city*

Suzanne Mackenzie

Change in gender rôles and gender definitions has been an insistent force shaping social and economic life over the past two decades. As women and men transformed their activities, so they reshaped social institutions, personal relationships and human self-definitions. These are changes that have deep personal and institutional significance, and they have affected all social actors.

Throughout this period, the woman's movement has articulated and furthered these changes. Feminism moved from expressions of unease – Friedan's 'vague discontent' (Friedan 1963) – to an energetic and comprehensive politics, audaciously claming to move 'beyond the fragments' which constitute a society and social understanding in transition (Rowbotham *et al.* 1979). As gender rôle change becomes a focus of progressive movements, it also forms a focus of mobilization for the anti-feminist New Right.[1] Gender politics are taking on an increasingly pivotal rôle in the political agenda.

Over the last two decades, there has been growing recognition that these are issues of fundamental importance to geographers. The meeting of feminism and geography has not been a simple matter of introducing new content into the study of human-environmental relations; it has involved recognizing that gender is an integral parameter of environmental change, and that feminism – as the politics and analysis of gender change – is integral to geographical studies.

Geographers are coming to understand that gender and environment have altered in the process of interaction. Women as well as men have always organized to extend the resources available for their work. But because gender has always been a criterion for differential allocation of resources, women's and men's organization has generally taken different forms within this human endeavour.[2] The difference has been in large part a function of the kinds of environments in which women and men lived and worked. The nature of women's organization has been propelled and constrained by particular social environments, defined as 'women's space'. This space has offered an historically varying set of social resources for performing those tasks defined as women's work. The feminist tradition has long organized to improve, and often to extend, women's space, to facilitate and often alter women's work. Contemporary feminism reflects both the social environment from which it springs – the context and content of women's space – and the attempts to create an altered, more adequate context and content.

This chapter traces how this complex and dynamic relation between gender and environment has been integrated into Anglo–North American geography within the last two decades. It examines first the social and environmental context within which urban women were organizing in this period, outlining the problems they encountered and the responses they made. The following sections discuss the initial encounters between these problems and responses and urban geographic analysis, and outline the gradual development of feminist-informed geographical concepts and the relations between feminist analysis and other non-positivist geographical perspectives. In conclusion, the chapter indicates how feminist analysis is becoming increasingly central to geographical analysis in a period of social and economic restructuring, and suggests some of the implications of this for future research and for urban politics.

The political motivation – retrospective: women as urban activists and feminism as urban analysis

The social environment within which contemporary feminism emerged – the mid-20th century city – was one which expressed and reinforced sharply differentiated gender rôles. Since the early part of the century, women's work has been increasingly separated from productive activity and confined to activities of domestic work and family nurture. Women's space was delimited as home and community. The resources available in this space were planned and arranged to facilitate the reproduction and leisure of current and future wage workers. Women worked from a material base which was defined as private and was geographically separated from the public workplaces of men. Feminist work in geography reflects and is articulating the disintegration of this gender-specific spatial separation.

Alterations in the social relations reflected in this type of city took a number of forms. By the 1950s, maintaining the private family had become more and more of a social issue. Demographic issues became public policy. The baby boom became the population bomb, with its attendant neo-Malthusian concern to find sufficient resources to feed, house, and employ all the products of millions of private reproductive decisions (Brooks 1973, Ehrlich 1971, Leathard 1980). Similarly, reproductive control and abortion became social issues for mankind, as did care of the growing proportion of the elderly. Rising numbers of divorces, separations, and households headed by single working mothers led to growing legal and welfare intervention into those areas of family life which had hitherto remained outside the domain of public agencies.[3] Perhaps most striking was the fact that, by the 1960s, purchasing and maintaining a home for family life often required two incomes, and a growing number of women with families moved into public wage work. The single-earner nuclear family which had dominated the ideology of most people and the lives of some in the 1950s, became a rarity.

These changes were not unproblematic. They were the outcomes of millions of people responding to distinct pressures and making constrained but often unexpected and creative choices. For many women, these changes were experienced as living a double life – attempting to fulfil their responsibilities for maintaining a home and community while at the same time performing public, economic rôles.

Neither women's relation to the domestic environment nor to the public wage-work environment was the same as that of their male counterparts. In the domestic environment, women continued to be the primary workers. In the wage-work environment, they came numerically to dominate the expanding service sector, often performing public equivalents of what they did at home. Most women worked in female occupational ghettos characterized by low pay, low status, insecure tenure, and part-time hours. The postwar expansion of the service sector was built around this kind of wage labour, and the predominantly domestic type of tasks involved fuelled the assumption that because their jobs mirrored their continuing responsibilities in the home, women were not 'real workers'. The home and community were still 'women's place', ensuring that their ventures into the public sphere were both constrained and badly rewarded.

The difficulties of dual rôles were exacerbated by the form of the urban environment. The design of homes and communities assumed someone was working full time to maintain and organize domestic life. This created pressures on the growing number of women who were living independently, or who were members of dual-income or single-parent families. The organization and location of wage jobs assumed their workers had minimal domestic responsibilities. There were few bridging services between women's work at home and their work at work. Affordable and accessible childcare was virtually non-existent. Transit routes were radial and geared to business hours, while women's wage work often required lateral movement at odd and irregular hours (Cichocki 1980, Lopata 1981, Wekerle 1981).

Changes in women's rôles created needs for new forms of urban organization and new urban services. In performing their new rôles, women created new urban movement patterns, new demands for public services, new patterns of wage work, new shift schedules, and altered hours in the shops and services where they carried out part of their domestic work.

Contemporary feminist activity was an articulation of these problems, and of women's attempts to gain new resources, including adequate social constructs, for their altered lives. The required resources were often those defined as inappropriate for women: access to secure, remunerative jobs, and ownership of property, or resources which did not yet exist: accessible childcare, facilities for reproductive control and job retraining.[4]

Feminism became socially evident because it articulated the process by which women asked for new resources and acted in new ways. It became increasingly recognized as political because demands were voiced and actions were taken more and more in the public sphere.

By the late 1960s, it became difficult for academia to ignore women's unforeseen, autarchic actions. Academics, largely male, manfully attempted to encompass them within frameworks designed to explore man's social relations. The results, predictably, were learned journals carrying studies of neglected latchkey children juxtaposed with studies of the growing need for cheap female labour. By the early 1970s, feminism could no longer be denied candidacy as a social, if not yet as a scholarly framework, although it was rarely recognized that it could connect empirically the latchkey children with the female workers.[5]

By the mid-1970s, the demands which women, in carrying out their dual rôles, made on cities – for new transit routes and resources, childcare, new home and neighbourhood design – could no longer be wholly ignored by geographers. With the seminal discussions of Bruegal (1973), Burnett (1973), and Hayford (1974), the discipline of geography was breached.

Initial exploration: discovering and measuring women in the city

The initial incursion of women's issues and feminist analysis into urban geography was an angry one. Early writers argued that geographers had ignored women's activities and that this distorted both the reality of women's lives and the understanding of human–environmental relations. They said that geographers were building models on the basis of family forms and gender-based movement patterns which no longer existed (Burnett 1973, Enjeu & Save 1974, Hayford 1974). Yet this incursion was also a cautious one. As with any attempt to introduce new content into a discipline developed in its absence, people moved warily. They looked for footholds, places in existing frameworks for women's new and unratified patterns.

Existing frameworks did not appear to offer many such footholds. Despite the pioneering efforts of a few humanists and historical materialists, at this point still zealously and often dogmatically explicating their respective philosophies, urban geography was largely a spatial science. It focused on attempts to explain and predict patterns of movement using models which extrapolated from empirical evidence of human and commodity movement in the past. The patterns women created were unpredictable and even inexplicable in terms of this kind of science. Demographic predictions foundered, victims of the failure to acknowledge the growing campaigns around fertility control and new family forms. The 'units' of demographic change acquired unexpected political convictions and made themselves known as complex, conscious beings. Similarly, discussion of residential location and the categorization of socio-economic status were disordered by unforeseen, two-income families. Journey-to-work researchers were perplexed by women's apparently erratic non-maximizing movements which were punctuated by trips to childcare, shops, and childrens' teachers (Bowlby *et al.* 1981, Burnett 1973, Monk & Hanson 1982, Hayford 1974, Tivers 1978).

Initially, such problems were seen as empirical perplexities. The first reaction involved 'adding women on', trying to find models powerful enough to disaggregate families and movement patterns by gender. The application of unchanged methods to women's specific activities led to the growth of a geography of women as a separate subdiscipline. Sufficient information was amassed about women's perceptions and movement patterns in cities to document their definition as an urban subgroup, deviating from the assumed male norm by virtue of occupying less space, having access to fewer resources, travelling less, and generally suffering specific spatial constraints.[6]

While providing a new dimension on urban form and development, initial explorations were constrained by the need to engage an entrenched positivism. They were also constrained by the limitations of feminist analysis which was only beginning to develop comprehensive concepts and categories.[7] In geography, as elsewhere in academia, the initial meeting was a limited and necessarily empirical one.

However, in geography, as elsewhere, there was increasing frustration with frameworks which confused rather than enlightened. A generation of geographers who had come to maturity with feminism as an evolving parameter of their lives, and who now encountered humanism and historical materialism simultaneously with spatial science, began to search for new frameworks.

Theoretical explorations: women against the divided city

The apparently simple question as to why women suffer these spatial constraints was the basis of a challenge to positivist urban analysis. The explanations offered by correlative spatial and behavioural science provided some direction: women did less travelling because they made more stops; they were less mobile because they had lower than average incomes, or had less access to cars. Such correlations, while empirically true, were only a first step toward explanation. Like many other urban analysts in this period, feminists recognized that the end point of positivist analysis was 'for us an indicator of what the problem is' (Harvey 1973, p. 137). The obvious questions were, Why are women's lives constrained in these ways? What social processes underlie these spatial patterns and this differential access to resources?

Suggestions were made that such questions could not be answered using standard frameworks or that they were outside the proper realm of spatial science. While this was arguably true, they were nevertheless well within the realm of human–environmental relations, increasingly advocated as a basis for geography by humanists and historical materialists. Humanism and historical materialism appeared to provide potential support for the feminist proposal that it was necessary to go beyond positivism in order to examine gender-based questions. A feminist voice was added to the search for and development of new theory and method in geography.

Within this wide rubric of human–environmental relations, feminists began to look beneath spatial problems, searching for their social roots. They found the divided city.

The divided city was first seen as a city separated into men's spaces and women's spaces (Enjeu & Save 1974, Loyd 1975). Examining these divisions allowed other labels to be added. Men's spaces were public and economic; women's spaces private and social (Saegert 1981, Wekerle 1981). A seminal article by Wekerle, Peterson and Morley argued that the divided city, and 'women's traditional association with the home has been a major stumbling block to their access to the wider opportunities of the city. . . . This sexually segregated, public–private dichotomy is fundamental to modern capitalist societies and is reinforced by urban planning and design decisions' (Wekerle *et al.* 1980, pp. 8–9).

Socialist feminist literature allowed these separate spaces to be labelled in terms of the activities which went on in them. Men's public spaces were sites for production, women's private spaces were sites for reproduction. Socialist feminists argued that production and reproduction were integrated, forming a social and historical dialectic (Bridenthal 1976, Kuhn & Wolpe 1978). Feminists within geography and other environmental disciplines found that women worked in both, and created an intersection between them. The problem was not simply that women were different. Moving between work at home and work in public places, they bridged private and public spaces and activities. Women's daily activities were carried out in opposition to a city made up of distinct work spaces and home spaces. Women were altering the nature of these spaces and the relations between them. In effect, women's daily activities were rendering current urban form obsolete.

Women's activities were also rendering obsolete a geographical analysis based on this dichotomy. If women's lives merged the dichotomies of the divided city, a focus on the relations between production and reproduction was necessary to understand these lives. In terms adopted from Sayer's (1982) seminal article on the nature of abstraction, the division of production and reproduction led to chaotic abstractions in urban analysis; feminists added that it also led to bad planning (Haar 1981, Matrix 1984, Wekerle *et al.* 1980).

An analytical focus on the intersection of production and reproduction allowed geographers to incorporate gender as a parameter in studying environmental processes. Understanding conflicts in the movement patterns of dual-rôle women provided a new perspective on urban revitalization (Holcombe 1981, Rose 1984), and on processes of relocating jobs to the suburbs (Nelson 1986). The constraints placed on women's earning power by their limited mobility were seen to reinforce their poverty, especially that of single parents. Restricted access to urban resources, especially housing, often led to a vicious cycle of deprivation (Christopherson 1986, Klowdawsky *et al.* 1985).

The focus on the intersection was simultaneously liberating and confusing. It broke down many of the categories geographers had used, leaving only a mass of conceptually unclad and immediate content. One could not assume a neat division between economic and social activities, nor between economic and social geographies; activities and forms had to be seen anew. It was necessary to watch them being actively produced, reproduced, and altered.

Activity became the concept linking gender rôles and environmental change. It became evident that change of the gender category 'woman' was effected through alterations in the way women used, appraised, and created resources in space and time. Women were changing their space and time patterns in taking on dual rôles. They were appropriating new spaces: moving into waged workplaces and public political venues. They were appropriating and creating new resources: purchasing goods rather than manufacturing them at home; redefining unused church basements as childcare drop-in centres. All of these innovations were contributing to a redefinition of gender relations. The availability, accessibility, and form of resources enabled or constrained women's capacity to change their lives.[8] Women changing their lives in turn altered needs for and the form of resources. The question was no longer simply, How do women use space? but, What is the relation between the form and change of gender and the form and change of the environment?

This broadening of the question's scope involved replacing an unexamined concept of woman with a concept of historically specific and mutable gender. This led feminist geographers to suggest that gender was a human characteristic, not just a female one. The proposal that men's activities should also be analyzed in terms of their gender – a reference point hitherto reserved solely for women – while confusing many pre-feminist geographers, directly challenged the concept of 'human' used in geographical analysis. Feminists argued that the term human was empirically androgynous and should be conceptually recognized as such. In stating that gender change was a fundamental parameter of environmental change, they also redefined the concept of environment. Environments were human creations, produced and altered in a dialectical relation with the production and alteration of gender categories and rôles. In making these suggestions, geographers were drawing upon an increasingly commodious and discerning feminism, which contained both the radical feminist concern for the development of an androgynous humanity and the socialist feminist concern for the historical and political process of change in this humanity.

By the early 1980s, feminist work in geography had a theoretical structure and contained within it methodological guidelines.[9] As its concerns broadened, it began to draw increasingly upon the concepts of human–environmental relations being proposed and elaborated by humanists and historical materialists.

Many feminist geographers had realized the significance of gender through historical materialist study of human–environmental relations. Others had come to historical materialism from their feminist concerns. The merger – socialist feminism – was a specific kind of historical materialism from the beginning. It was one imbued with the content of human life and the nature of interpersonal relations. Both this content and the organizational structures of feminism encouraged scepticism toward received categories and laws in any form, including those of historical materialism (Rowbotham *et al.* 1979, Wandor 1972).

Feminists were especially sceptical of Althusserian structuralism. Its concepts and laws – while offering a certainty which was potentially tempting to people struggling with an overflow of untidy and restless human content – appeared depopulated and abstract as gravity models and nearly as inhospitable to feminist questions. More hospitable was the work of critical theorists, with their emphasis on the internal unity and dialectical change of nature and human nature.[10] Critical theory provided a basis for understanding the articulation of gender and environment. It also provided a basis for integrating emphasis on human agency and the constitution of meaning which feminists shared with humanist geographers.

Feminism incorporated and extended elements of both humanism and historical materialism and contributed to the evolution of both within the discipline. But it did more than that. In discussing the divided city and focusing on the intersection of production and reproduction, it offered elements of a new model of urban structure. This is a model which is becoming increasingly central to urban analysis as we experience, and attempt to understand, a restructured capitalism.

Crossing the boundaries: women restructuring the city and some suggestions for future research

As the recession of the late 1970s metamorphosed into the economic and social restructuring of the 1980s, geographers began to recognize some fundamental changes in the relations between family life and the economy. The form of the wage labour economy was altering, leading to rising rates of structural unemployment in capitalist industrial cities and to new ways of organizing social production and reproduction.[11] These changes had different implications for women and men.

Feminization of the wage labourforce and of work processes had been an element of the postwar boom. It also appeared to be a correlate of restructuring, in both quantitative and qualitative terms. In the cities of developed countries, a primary aspect of restructuring was the declining number of jobs in the male-dominated primary and secondary sectors and the continued growth, or at least stability, in numbers of jobs in the female-dominated tertiary sector. In most industrialized countries, the male participation rate in the workforce declined slowly, while the female rate continued to climb. At the same time, the low pay, short-term contracts, casual and part-time hours which had characterized many service sector jobs and thus most women's wage work appeared to be extending beyond the female ghettos.

Gross figures on labourforce participation rates belied both the deteriorating conditions and security of women's wage jobs and their rising rates of unemployment (Armstrong 1984, Murgatroyd & Urry 1985, Walby 1985). Not only were jobs disappearing or being reorganized, but women's tentative hold on public-sphere jobs was weakened by cuts and threats to social services which had facilitated the carrying out of dual rôles in a divided city. Socially provided childcare, retraining, and pay equity – services which feminists had defined as conditions for integrating human work and life – were increasingly defined as special, and expensive, as concessions to women workers. Women's wage work was still valued for its cheapness and flexibility. But their responsibilities for carrying out free and invisible reproductive work in the home and community were also given new ideological and practical prominence as social responsibility for people declined.

Women responded through heightening public pressure for services and employment equity. At the same time, partly in response to growing unemployment and declining services, many women altered the sites of their waged activity. Women became the fastest growing section of the expanding self-employed sector. Many of these new businesses were based in the home, as was a growing amount of industrial production, especially in the garment industry (Armstrong 1984, Johnson & Johnson 1982, Johnson 1984, Levesque 1985, Urry 1985). It appeared that restructuring was leading to the creation and expansion of an informal economy, that social space where people combined the resources of their homes and communities with networks of casual work to develop survival strategies.

The informal economy had long been central to analysis of urban economies in the Third World where women in neocolonial societies had developed strategies for household and community survival and filled the labour demands of multinational corporations (Bromley & Gerry 1979, Connolley 1985, Oppong 1983, Roldan 1985, Steyn & Uys 1983). The informal economy was now becoming empirically central in First World cities, and social scientists, including geographers, were attempting to find concepts for analyzing it.

Analysis based on the formal economy was inadequate, even misleading for an understanding of the processes of restructuring. Concepts for analyzing political and economic processes and delimiting the appropriate spatial scale for study had to be developed in terms of the interaction between formal and informal economies and the 'characteristic social relations and social practices within and between households' (Urry 1985, p. 22). In short, understanding restructuring required a focus on the relations between production and reproduction, and study at a scale defined by these relations.[12]

Within analysis of capitalist industrial cities, the most readily available set of concepts was that developed by feminists. Women's relation to the wage labourforce not only differed from men's, but had always involved the development of bridging services to help carry out their domestic work. As fiscal restraint threatened the few public bridging services available, women both resisted and redoubled their efforts to compensate. In sustaining their dual rôles, women had been creating an informal economy for decades, and feminists had been studying this process. Feminist geographers had located analytically the informal economy at the intersection of production and reproduction and indicated that it had a concrete form in community childcare and health networks, housing developments which integrated living space with a range of childcare, counselling and educational services, and in networks of women earning money at home.[13] Feminist geographers had documented that this was not a retreat from the economy, but an alteration and extension of the domestic workplace to incorporate economic activities. Women's space now encompassed not only resources for private family life, but also resources for providing public services and for waged work. These studies had indicated that attempts to develop strategies had been constrained by obsolete environments in the divided city and that women had tried to overcome their problems and meet their new needs by developing new forms and uses of space. As ever, feminist activity was both confounded by 'women's space' and impelled to create more adequate spaces.

In a period of restructuring this creativity takes on new prominence. With the extension of remote communications technology and piecework in many industries and rising levels of unemployment, the home and community become sites both of private life and economic subsistence for more and more people. While environmental scientists document that women still have more intense working relations to the home and community than do men (Saegert 1981, Saegart & Winkler 1980, Williams 1986), the resources of the home and community are forming an increasingly important part of the economic subsistence of a growing number of people. Similarly, the ways in which women use these resources, in combination with resources from the public sphere – stretching the wage through home manufacture, earning money at home, trading and bartering goods and services with neighbours, creating community networks for childcare, self-education and wholesale purchases – are becoming more central to the survival strategies of a growing number of families (Nicholls & Dyson 1983, Mackenzie & Rose 1983). The intersection, and the feminist analysis which had discovered and discussed it, is becoming increasingly central to understanding urban form.

Geographers' ability to comprehend these processes will depend upon extending the research agenda which feminists have initiated. A necessary starting point is the universal incorporation of feminist concepts within geographical analysis as a whole, simultaneously with further investigation of the empirical conditions which define the parameters and make up the content of these concepts. At the most general level, the feminists' reconceptualization of the terms 'human' and 'environment' can be extended to the study of restructuring. Feminist geographers have shown that economic restructuring is a dialectical process, involving the actions of corporate forces at a global and national scale interacting with the responses of individuals at a local scale.[14] We require more feminist-informed studies of global, systemic changes which recognize the necessary androgyny of humans. We also require studies of how these people, women and men, are responding to restructuring. People are both resisting the disappearance of jobs and services and developing survival strategies in an informal economy. These two forms of response are interrelated, and our empirical examination must take account of how public resistance interacts with new forms of work in the home, new household and community networks, the renovation and redefinition of women's space, and the creation of new places. All of these responses are important and it appears that all of them may be tending to merge productive and reproductive activities and spaces. Our empirical studies could monitor the ways in which restructuring is producing the rudiments of a new urban form centred on the sites where production and reproduction interact.

One of the implications of this may be a new form of divided city. As the wage economy alters, it may be that proportionately fewer people move between public and private spaces and fewer goods find their way from the productive to the reproductive sphere (Gershuny 1978, 1985, Pahl 1984, Pahl & Wallace 1985). The city may be redividing along lines defined by the employed – those who still have rôles within the public sphere, and the unemployed – those whose time and space is structured primarily around their homes and communities. This is a division which would, in many respects, cut across the gender lines which delineated the city divided into productive and reproductive spaces. Existing research indicates that increasing numbers of employed, often dual-income families live and work in revitalized inner-city areas or in suburbs where their homes are juxtaposed with industrial–commercial estates. There is some evidence that more and more unemployed households live and develop survival strategies in other areas, urban and, increasingly, suburban (Boles 1986, Ley 1986, Nelson 1986, Rose 1984, Social Planning Council of Metropolitan Toronto 1979). Understanding and responding to restructuring requires that we assess the extent to which gentrification, the formation of new ghettos, and industrial relocation may be aspects of a new form of city divided into a patchwork of communities containing both public and private resources.

As yet we have little concrete information about how people are adapting their use of time and space to respond to changes in the wage economy. The work of feminist geographers has indicated, however, that restructuring gender activities is a central component of this. Women's patterns of activity have been geographically extended and the concepts feminists have developed to study these activities are being expanded to cover urban analysis a a whole. Feminist work also indicates that change in gender relations has been fundamental in bringing about these alterations in urban form. Over the past three decades, significant elements of urban change have been the outcomes of the strategies of dual-income families and female-headed single-parent families. This new demographic and geographic structure and the attendant changes in resource use have contributed to new modes of subsistence involving the use of a variety of resources which are both public and private. It has also altered women's space, the home and community. Just as it is not possible fully to understand the development, temporary success, and subsequent alteration of a city divided into separate industrial–commercial and residential areas without understanding gender as a parameter, so it is impossible to comprehend, or even to see clearly, current changes without recognizing that they result from and extend gender rôles.

But feminism is more than an analysis, and it does more than provide us with new empirical insights and new theoretical vision. Feminism is also inextricably and irreducibly a politics of change, including environmental change. It works in prefigurative ways, creating the basis of new social relations on a day-to-day basis. The content is fundamental and deeply embedded in everyday life – in sexuality, biological reproduction, relations to children and other adults, use of resources in the immediate environment, self-definitions. Change comes incrementally, as we live out and reproduce new patterns. As a consequence, the effectiveness of feminism is often invisible, at the same time it is often irreversible. It is, above all, played out in and through the process of incrementally changing our social environment. The research agenda suggested above, with its emphasis on the newly visible resistance and survival strategies women and men are creating to combat the divided city, will be carried out in a highly charged political context, focused around the definition of gender.

The political motivation – prospects: women as urban actors and feminism as urban politics

One of the implications of economic restructuring and the effectiveness of feminism in articulating and extending changes in gender rôles is the rise of the new right, 'the convergence of economic and social conservatism' (Dubinsky 1985, p. 30). This has helped propel gender politics to the centre of the political stage, and created a new political climate which will be expressed and altered through environmental change.

As Dubinsky points out, the new right is new in that

> [it has] succeeded in combining . . . two conservative impulses: an anti-feminist backlash, which is manifested in opposition to abortion, birth control, homosexuality, and alternatives to the family, and an anti-social-welfare backlash, which is manifested as opposition to the post-war notion of the state as the provider of social welfare and equal opportunity. What brings these two impulses together and gives new right ideology its 'coherence' and wide-spread appeal is its defence of the patriarchal family form (Dubinsky 1985, p. 33).

To some extent, the political agenda of the First World is centred on the politics of conscious gender change, whether feminist – aiming to extend choices and remove gender-based constraints, or anti-feminist – aiming to extend gender differentiation as a basis for resource allocation.[15] The latter politics, that of the new right, is as radical and active as the former, despite the rhetoric about returning to tradition. Processes of production and reproduction have altered too much to allow a retreat into a tradition of patriarchal nuclear families; both feminists and anti-feminists are constrained to build something new.

People will meet this challenge in concrete ways which should be visible to geographers and which will provide a context for, and some of the content of, future research. This building will depend upon assembling resources in new ways, and creating new resources. The anti-feminist vision relies upon the home, as a separate and private place, occupied by a single male economic provider, a single female domestic worker, and their children. Realizing this ideal will require a wide-ranging social and economic reordering. Most households will have to alter their demographic structure. New forms of home and community will have to be provided and old ones rebuilt. Sustaining family life on one income, with diminished social welfare support will require developing radically new forms of support to meet the social needs generated by private family life (Eichler 1985). Just as the development of suburban domestic workplaces in late 19th-century and early 20th-century cities depended upon and reinforced particular gender rôles (Mackenzie & Rose 1983), so the attempts to emulate this environment in a new historical context will depend upon and reinforce new gender rôles. Feminist proposals, in contrast, advocate breaking down gender and environmental separations, unifying public and private domains, work and home life. This is not just a vision of future possibilities, it is a recognition of continuous processes.

The confrontation between these two groups will accelerate, and involve a growing number of people. Men will find it increasingly difficult to distance themselves from gender issues. As the conditions of biological and social reproduction alter in dialectical relation to changes in production, so women's and men's rôles alter. Over history, this dialectic has been more or less conscious (Beard 1962, Ehrenreich & English 1979, Rowbotham 1974); it appears that it will become increasingly conscious in our own lifetimes.

The possibility of anyone remaining neutral, passive, or ignorant with respect to gender issues is effectively removed. Pre-feminist consciousness is doomed. The processes of creating our futures will not be settled, however, at the level of rhetoric, nor at the level of consciousness, feminist or anti-feminist. It will develop at the material level, in the activities by which women and men create and reproduce their daily lives. Analysis and ideology can both guide and articulate this process, but the major task is to see it, to see the continuous, ephemeral, and unpredictable relation between change towards androgynous human rôles and change in the environment.

Notes

1 The new right is an American political force, although elements of anti-feminism are evident in Canada and Britain. For discussion of the new right in America, see Dworkin (1983), Eisenstein (1981), and Petchesky (1981). On REAL women in Canada, see Dubinsky (1985) and Eichler (1985).

2 Women's organization has not always been a devalued part of this endeavour. The extent to which women's organization and general contribution to society has been devalued or rendered invisible is correlated to women's degree of social isolation – the extent to which they have been separated from those aspects of society defined

as central. In capitalist society, public, economic rôles are defined as central, and women have been defined in terms of private, domestic work.

3 On issues of reproductive control in Canada, see Dubinsky (1985), in the United States see Gordon (1977), and in Britain see Leathard (1980), and Walsh (1980). On the changing nature of the family, see for example, Brophy & Smart (1981), Eichler (1983), and Wilson (1977, 1980).

4 These conflicts and needs, and this potential creativity were articulated within the wider set of material changes and the climate of resistance in the late 1960s and early 1970s. The postwar generation of the First World matured in conditions of relative material security and rising expectations. For the first time, a large minority of people – including a substantial number of women – was assured subsistence and some education, and was relatively free to question the conditions under which subsistence was produced and reproduced. The cultural revolution of the 1960s and 1970s was, in many respects, an assertion of control. For many, this took the form of attempts to find more ecologically sensitive, self-sufficient modes of survival, new ways of producing and utilizing resources, and, concomitantly, of attempting to define new racial, gender, and sexual rôles which more adequately reflected the new material conditions.

This social climate had profound, if somewhat delayed repercussions for academic social science. It rendered the projections of the current structuralist–functionalist social science somewhat inaccurate and a little absurd. People's actions were implicitly or explicitly rejecting the human engineering ethos of positivism as a whole (on the human engineering implications and limitation of positivist social science, see Fay (1975)). Some actively challenged the idea that their actions could be predicted on the basis of past actions and rebelled against the limits which positivist views implied for imagination and action. Unratified and unpredictable change – successfully resisting a war, resisting given career trajectories, resisting the 'natural' inferiority of women and ethnic minorities – was seen to be possible.

5 For some discussions of these early challenges to social science as a whole in Canada, see Canadian Women's Educational Press Collective (1972) and Hamilton (1985), in America, Morgan (1970), and in Britain, Wandor (1972).

6 The geography of women and its documentation of constraints makes up much of the empirical literature reviewed in Mazey & Lee (1983) and in Zelinsky *et al*. (1982). One of the most comprehensive discussions of geographic constraints is Tivers's (1986) extensive study of women in Britain.

7 Feminism as a whole was still bound by two sets of constraints: the liberal feminist attempt to make women 'equal' to men within an unquestioned social system, and the concern of radical and socialist feminists to discover the potential creativity of a basically unquestioned 'woman'. For geographic discussion of feminist theoretical development, see Mackenzie (1984) and Women and Geography Study Group of the Institute of British Geographers (1984). For general discussion, see Beechey (1979) and Rowbotham (1972) on Britain, Eisenstein (1979) on the United States, and Teather (1976) on Canada.

8 Childcare was the most obvious example. Its affordability, hours, and location influence the hours and locations of parents' (especially mothers') wage jobs. Childcare also provides a focus of community-based feminist organization and a source of employment.

9 On the relation of feminist philosophy and methodology, see McRobbie (1982), Roberts (1981) (especially Oakley (1981)), and Stanley & Wise (1983).

10 For general discussions of critical and realist theory, see Bhaskar (1978), Giddens (1977), Keat & Urry (1975). For specific discussion of human–environmental relations, see Ollman (1976), Sayer (1979).

11 There is a growing international, interdisciplinary literature on restructuring, ranging from Gershuny's early arguments on post-industrial society (Gershuny,

1978) through more recent detailed studies on the implications of restructuring for household strategies (Nicholls & Dyson 1983, Pahl 1984, Roldan 1985) to theoretically informed overviews (Gill 1985, Handy 1984, Redclift & Mingione 1985). Some recent geographical analysis includes Rigby (1986) and articles in Scott & Storper (1986).

12 See discussions in Connolly (1985), Murgatroyd *et al.* (1985).

13 There is a growing literature on the development and importance of integrated communities and housing developments. For historical discussions see Hayden (1981b, 1982); for contemporary examples and proposals, see France (1985), Hayden (1981a), Hitchcock (1985), Klodawsky & Spector (1985), Leavitt (1985); on homeworkers, see Mackenzie (1986a, 1986b) and Netting (1985).

14 This is also a strong theme in locality studies (see Murgatroyd *et al.* (1985)).

15 The feminist concept of choice extends far beyond reproductive choice to include all areas of personal and social life, although reproductive choice remains the dominant issue.

References

Armstrong, P. 1984. *Labour pains: women's work in crisis.* Toronto: The Women's Press.

Beard, M. 1962. *Women as a force in history: a study in traditions and realities.* New York: Collier.

Beechey, V. 1979. On patriarchy. *Feminist Review* **3**, 66–82.

Bhaskar, R. 1978. On the possibility of social scientific knowledge and the limits of naturalism. *Journal for the Theory of Social Behavior* **8**, 1–28.

Boles, J. (ed.) 1986. *The egalitarian city: issues of rights, distribution, access and power.* New York: Praeger.

Bowlby, S., J. Foord & S. Mackenzie 1981. Feminism and geography. *Area* **13**, 711–16.

Bridenthal, R. 1976. The dialectic of production and reproduction in history. *Radical America* **10**, 3–11.

Bromley, R. & C. Gerry (eds) 1979. *Casual work and poverty in Third World cities.* Chichester: Wiley.

Brooks, E. 1973. *This crowded kingdom: an essay on population pressure in Great Britain.* London: Charles Knight.

Brophy, J. & C. Smart 1981. From disregard to disrepute: the position of women in family law. *Feminist Review* **9**, 3–16.

Bruegal, I. 1973. Cities, women and social class: a comment. *Antipode* **5**, 62–3.

Burnett, P. 1973. Social change, the status of women and models of city form and development. *Antipode* **5**, 57–62.

Canadian Women's Educational Press Collective 1972. *Women unite: an anthology of the Canadian women's movement.* Toronto: Canadian Women's Educational Press.

Christopherson, S. 1986. Parity or poverty? The spatial dimension of income inequality. Working Paper 21, Southwest Institute for Research on Women, Tucson, Arizona.

Cichocki, M. 1980. Women's travel patterns in a suburban development. In *New space for women*, G. Wekerle, R. Peterson & D. Morley (eds), 151–64. Boulder, Col.: Westview Press.

Connolly, P. 1985. The politics of the informal sector: a critique. In *Beyond employment: household, gender and subsistence*, N. Redclift & E. Mingione (eds), 55–91. London: Basil Blackwell.

Dubinsky, K. 1985. *Lament for a 'Patriarchy Lost'? Anti-feminism, anti-abortion and R.E.A.L. Women in Canada.* Ottawa: Canadian Research Institute for the Advancement of Women/Institut Canadien de Recherches sur les Femmes.

Dworkin, A. 1983. *Right wing women.* New York: Putnam.

Ehrenreich, B. & D. English 1979. *For her own good: 150 years of the experts' advice to women*. London: Pluto Press.
Ehrlich, P. 1971. *The population bomb*. London: Ballantine.
Eichler, M. 1983. *Families in Canada today*. Toronto: Gage.
Eichler, M. 1985. *The pro-family movement: are they for or against families?* Ottawa: Canadian Research Institute for the Advancement of Women/Institut Canadien de Recherches sur les Femmes.
Eisenstein, Z. 1979. Developing a theory of capitalist patriarchy and socialist feminism. In *Capitalist patriarchy and the case for socialist feminism*, Z. Einsenstein (ed.), 5–40. New York: Monthly Review Press.
Eisenstein, Z. 1981. The sexual politics of the new right – understanding the crisis of liberalism, In *Feminist theory – a critique of ideology*, N. Keohane (eds). Chicago: University of Chicago Press.
Enjeu, C. & J. Save 1974. The city: off limits to women. *Liberation* **18**, 9–13.
Fay, B. 1975. *Social theory and political practice*. London: Allen & Unwin.
France, I. 1985. Hubertusvereniging: a transition point for single parents. *Women and Environments* **7**, 20–2.
Friedan, B. 1963. *The feminine mystique*. New York: Dell.
Gershuny, J. 1978. *After industrial society: the emerging self-service economy*. London: Macmillan.
Gershuny, J. 1985. Economic development and change in the mode of provision of services. In *Beyond employment: household, gender and subsistence*, N. Redclift & E. Mingione (eds). London: Basil Blackwell.
Giddens, A. 1977. *Studies in social and political theory*. London: Hutchinson.
Gill, C. 1985. *Work, unemployment and the new technology*. Cambridge: Polity Press.
Gordon, L. 1977. *Woman's body, woman's right: a social history of birth control in America*. New York: Penguin.
Haar, C. 1981. Foreword. In *Building for women*, S. Keller (ed.), vii–viii. Lexington, Mass.: Lexington Books.
Hamilton, R. 1985. Feminists in the academy: intellectuals or political subversives? *Queen's Quarterly* **92**, 3–20.
Handy, C. 1984. *The future of work: a guide to a changing society*. London: Basil Blackwell.
Harvey, D. 1973. *Social justice and the city*. Baltimore: Johns Hopkins University Press.
Hayden, D. 1981a. What would a non-sexist city be like? Speculations on housing, urban design and human work. In *Women and the American city*, C. Stimpson, E. Dixler, M. Nelson & K. Yatrakis (eds), 170–87. Chicago: University of Chicago.
Hayden, D. 1981b. Two utopian feminists and their campaigns for kitchenless houses. In *Building for women*, S. Keller (ed.), 3–19. Lexington, Mass.: Lexington Books.
Hayden, D. 1982. *The grand domestic revolution: a history of feminist designs for American homes, neighbourhoods and cities*. Cambridge, Mass.: MIT Press.
Hayford, A. 1974. The geography of women: an historical introduction. *Antipode* **6**, 1–19.
Hitchcock, P. 1985. St. Clair O'Connor Community: an extended family. *Women and Environments* **7**, 18–19.
Holcombe, B. 1981. Women's roles in destressing and revitalizing cities. *Transition* **11**, 1–6.
Johnson, L. & R. Johnson 1982. *The seam allowance: industrial home sewing in Canada*. Toronto: The Women's Press.
Johnson, S. 1984. When home becomes your corporate headquarters. *Working Woman* (October), 75–7.
Keat, R. & J. Urry 1975. *Social theory as science*. London: Routledge & Kegan Paul.
Klodawsky, F. & A. Spector 1985. Mother-led families and the built environment in Canada. *Women and Environments* **7**, 12–17.

Klodawsky, F., A. Spector & D. Rose 1985. *Single parent families and Canadian housing policies: how mothers lose.* Ottawa: Canada Mortgage and Housing Corporation.

Kuhn, A. & A.-M. Wolpe 1978. Feminism and materialism. In *Feminism and materialism: women and modes of production*, A. Kuhn & A.-M. Wolpe (eds), 1–10. London: Routledge & Kegan Paul.

Leathard, A. 1980. *The fight for family planning: the development of family planning services in Britain, 1921–1974.* London: Macmillan.

Leavitt, J. 1985. A new American house. *Women and Environments* **7**, 14–16.

Levesque, J.-M. 1985. Self-employment in Canada: a closer examination. *Labour Force* (February), 91–105. Ottawa: Statistics Canada.

Ley, D. 1986. *Gentrification in Canadian inner cities: patterns, analysis, impacts and policy.* Ottawa: Canada Mortgage and Housing Corporation (External Research Grant 6583/L31).

Lopata, H. 1981. The Chicago woman: a study of patterns of mobility and transportation. In *Women and the American city*, C. Stimpson, E. Dixler, M. Nelson & K. Yatrakis (eds), 158–66. Chicago: University of Chicago.

Loyd, B. 1975. Women's place, man's place. *Landscape* **20**, 10–13.

Mackenzie, S. 1984. Editorial introduction. *Antipode* **16**, 3–10.

Mackenzie, S. 1986a. Women's responses to economic restructuring: changing gender, changing space. *The politics of diversity: feminism, Marxism and Canadian society*, M. Barrett & R. Hamilton (eds). London: Verso.

Mackenzie, S. 1986b. Restructuring the local community: the case of self-employed homeworkers. Paper presented to the Annual Meeting of the Canadian Association of Geographers, Calgary.

Mackenzie, S. & D. Rose 1983. Industrial change, the domestic economy and home life. In *Redundant spaces in cities and regions? Studies in industrial decline and social change*, J. Anderson, S. Duncan & R. Hudson (eds), 155–99. London: Academic Press.

McRobbie, A. 1982. The politics of feminist research: between talk, text and action. *Feminist Review* **12**, 46–57.

Matrix 1984. *Making space: women and the man-made environment.* London: Pluto Press.

Mazey, M. & D. Lee 1983. *Her space, her place: a geography of women.* Washington, DC: Association of American Geographers.

Monk, J. & S. Hanson 1982. On not excluding half the human in geography. *Professional Geographer* **34**, 11–23.

Morgan, R. 1970. *Sisterhood is powerful: an anthology of writings from the women's liberation movement.* New York: Vintage.

Murgatroyd, L. & J. Urry 1985. The class and gender restructuring of the Lancaster economy, 1950–1980. In *Localities, class and gender*, L. Murgatroyd, M. Savage, D. Shapiro *et al.*, 30–53. London: Pion.

Murgatroyd, L., M. Savage, D. Shapiro, J. Urry, S. Walby, A. Warde with J. Mark-Lawson 1985. *Localities, class and gender.* London: Pion.

Nelson, K. 1986. Labor demand, labor supply and the suburbanization of low-wage office work. In *Production, work, territory: the geographical anatomy of industrial capitalism*, A. Scott & M. Storper (eds), 149–71. Boston: Allen & Unwin.

Netting, N. 1985. Women and work: the job-free alternative. Paper presented to the Women and the Invisible Economy Conference, Simone de Beauvoir Institute, Montréal.

Nicholls, W. & W. Dyson 1983. *The informal economy: where people are the bottom line.* Ottawa: Vanier Institute for the Family.

Oakley, A. 1981. Interviewing women: a contradiction in terms. In *Doing feminist research*, H. Roberts (ed.). London: Routledge & Kegan Paul.

Ollman, B. 1976. *Alienation: Marx's conception of man in capitalist society.* Cambridge: Cambridge University Press.

Oppong, C. 1983. Women's roles and conjugal family systems in Ghana. In *The changing position of women in family and society: a cross-national comparison*, E. Lubri (ed.), 331–43. Leiden: E. J. Brill.

Pahl, R. 1984. *Divisions of labour*. London: Basil Blackwell.

Pahl, R. & C. Wallace 1985. Household work strategies in economic recession. In *Beyond employment: household, gender and subsistence*, N. Redclift & E. Mingione (eds), 189–227. London: Basil Blackwell.

Petchesky, R. 1981. Anti-abortion, anti-feminism and the rise of the new right. *Feminist Studies* **10**, 21–32. Summer.

Redclift, N. & E. Mingione 1985. Introduction: economic restructuring and family practices. In *Beyond employment: household, gender and subsistence*, N. Redclift & E. Mingione (eds), 1–11. London: Basil Blackwell.

Rigby, D. 1986. Investment, employment and the age structure of capital stock in Canada, 1955–1981. Paper presented to the Annual Meeting of the Canadian Association of Geographers, Calgary.

Roberts, H. (ed.) 1981. *Doing feminist research*. London: Routledge & Kegan Paul.

Roldan, M. 1985. Industrial outworking, struggles for the reproduction of working-class families and gender subordination. In *Beyond employment: household, gender and subsistence*, N. Redclift & E. Mingione (eds), 248–85. London: Basil Blackwell.

Rose, D. 1984. Rethinking gentrification: beyond the uneven development of Marxist urban theory. *Environment and Planning D, Society and Space* **2**, 47–74.

Rowbotham, S. 1972. The beginnings of women's liberation in Britain. In *The Body Politic: Women's Liberation in Britain 1959–1972*, M. Wandor (ed.), 91–102. London: Stage 1.

Rowbotham, S. 1974. *Women, resistance and revolution*. Harmondsworth: Penguin.

Rowbotham, S., L. Segal & H. Wainwright 1979. *Beyond the fragments: feminism and the making of socialism*. London: Merlin Press.

Saegart, S. 1981. Masculine cities and feminine suburbs: polarized ideas, contradictory realities. In *Women and the American city*, C. Stimpson, E. Dixler, M. Nelson & K. Yatrakis (eds). Chicago: University of Chicago Press.

Saegart, S. & G. Winkeler 1980. The home: a critical problem for changing sex roles. In *New space for women*, G. Wekerle, R. Peterson & D. Morley (eds), 41–63. Boulder, Colorado: Westview.

Sayer, A. 1979. Epistemology and conceptions of people and nature in geography. *Geoforum* **10**, 19–43.

Sayer, A. 1982. Explanation in economic geography: abstraction versus generalization. *Progress in Human Geography* **6**, 68–88.

Scott, A. & M. Storper (eds) 1986. *Production, work, territory: the geographical anatomy of industrial capitalism*. Boston: Allen & Unwin.

Social Planning Council of Metropolitan Toronto 1979. *Metro's suburbs in transition*. Toronto: Social Planning Council.

Stanley, L. & S. Wise 1983. *Breaking out: feminist consciousness and feminist research*. London: Routledge & Kegan Paul.

Steyn, A. & J. Uys 1983. The changing position of black women in South Africa. In *The changing position of women in family and society: a cross-national comparison*, E. Lupri (ed.), 344–70. Leiden: E. J. Brill.

Teather, L. 1976. The feminist mosaic. In *Women in the Canadian mosaic*, G. Matheson (ed.), 300–46. Toronto: Peter Martin.

Tivers, J. 1978. How the other half lives: the geographical study of women. *Area* **10**, 302–6.

Tivers, J. 1986. *Women attached: the daily lives of women with young children*. London: Croom Helm.

Urry, J. 1985. Deindustrialization, households and politics. In *Localities, class and gender*, L. Murgatroyd, M. Savage, D. Shapiro *et al.*, 13–29. London: Pion.

Walby, S. 1985. Spatial and historical variations in women's unemployment and employment. In *Localities, class and gender*, L. Murgatroyd, M. Savage, D. Shapiro *et al.*, 161–76. London: Pion.

Walsh, V. 1980. Contraception: the growth of a technology. In *Alice through the microscope: the power of science over women's lives*, Brighton Women and Science Group (eds), 182–207. London: Virago.

Wandor, M. (ed.) 1972. *The body politic: women's liberation in Britain, 1968–1972.* London: Stage 1.

Wekerle, G. 1981. Women in the urban environment. In *Women and the American city*, C. Stimpson, E. Dixler, M. Nelson & K. Yatrakis (eds), 185–211. Chicago: University of Chicago Press.

Wekerle, G., R. Peterson & D. Morley 1980. Introduction. In *New space for women*, G. Wekerle, R. Peterson & D. Morley (eds), 1–34. Boulder, Col.: Westview Press.

Williams, D. 1986. Gender differences in perception of neighbourhood. Paper presented to the Annual Meeting of the Canadian Association of Geographers, Calgary.

Wilson, E. 1977. *Women and the welfare state.* London: Tavistock Publications.

Wilson, E. 1980. *Only halfway to paradise: women in post war Britain, 1945–1968.* London: Tavistock Publications.

Women and Geography Study Group of the Institute of British Geographers 1984. *Geography and gender: an introduction to feminist geography.* London: Hutchinson.

Zelinsky, W., J. Monk & S. Hanson 1982. Women and geography: a review and prospectus. *Progress in Human Geography* **6**, 317–66.

6 *Third World cities*

Lata Chatterjee

The analysis of contemporary Third World cities is important for cities reflect a conjuncture of processes occurring at the global and the concrete levels. At the global level, urbanization is important for the expansion of capitalism and thus for the reproduction of the capitalist system. At the concrete level, urbanization relates to reproduction of the human species and focuses on issues such as employment and housing. While there is almost universal consensus about the empirical characteristics of Third World cities, numerous, competing theoretical approaches attempt to explain the causal mechanisms and underlying dynamics. McGee (1979) provides a useful review of the various stages in the interpretation of Third World urbanization ranging from the initial application of the Western model, to the later dualistic model, to the more recent interpretations of urbanization under conditions of peripheral capitalism. Only explanations drawn from the perspective of political economy are discussed here, however.

At one level the perspective of political economy appears unified in its explanation, particularly in its emphasis on the need to situate the interpretation of Third World cities within the logic of the expansionary nature of the world capitalist system. Its basic premise is that the empirical characteristics of Third World cities are dependent on, and reflect, wider changes in the social formations of Third World countries. As such, urban phenomena cannot be deciphered independently of the general dynamics of change in society. In this non-economic factors such as value transformations, political ideologies, class struggles, the rôle of the state, and like issues must be considered in conjunction with the logic of capitalist accumulation. In its basic tenets, then, the interpretation of urbanization from the political-economic perspective can be differentiated from the more conventional, choice theoretic, individualistic and functionalist analyses of urban processes.

However, the last three decades have witnessed complex debates within the political-economy perspective with respect to the dynamics of the incorporation of Third World societies into the capitalist system, particularly: the relevance of the concept of the mode of production; the relative primacy of economic versus political factors; the logic of accumulation *vis-à-vis* class struggle; and so on. We do not as yet have a single comprehensive theory that pays adequate justice to the various processes and subprocesses interacting to produce the complex phenomena of Third World development and urbanization. The theories vary in their emphasis on the different processes, so it is useful to discuss Third World cities in the context of the major competing theories of development. As the literature is complex and sophisticated, this discussion is, at best, introductory, exploratory, and non-exhaustive.

We can broadly divide Third World urbanization in market societies into two phases: (a) urbanization under conditions of colonialism; and (b) urbanization under conditions of dependent capitalism. It is difficult to draw precise boundaries between these two phases with respect to countries; for example in India there has been a gradual transition between the two periods of capitalist expansion. Indigenous industrialists developed characteristics of dependent capitalism even during the colonial period. In fact, the contradiction between the two classes of capitalists – foreign colonial and emerging domestic – reinforced the movement for political independence and the creation of a nation–state favouring the interests of domestic capital. The attempted resolution of this contradiction has important consequences for Indian urbanization and for her cities. For other countries, particularly in Africa, the boundary between the two phases coincides with that of political independence. Currently there are considerable differences between and within Third World cities resulting from the relative impacts of these two stages of capitalist development in their social formation.

Contemporary Third World socialist countries share a common experience of colonialism with contemporary market societies, hence many of their urban characteristics are similar to those of market societies. However, socialist countries have sought to modify their urban systems through spatial planning policies. Since there are approximately 32 socialist countries in the Third World (Forbes & Thrift 1987), their experience is also briefly discussed.

The observable manifestations of urban processes in Third World cities are described in the next section. This is brief since several books describing Third World cities are available (Bromley & Gerry 1979, Gilbert *et al.* 1982, Brunn & Williams 1983, Gilbert & Gugler 1983, Timberlake 1985, Eisenstadt & Shachar 1987). The explanatory sections that follow contain a discussion of the conceptual underpinnings common in the political-economic interpretation of Third World cities. This is followed by a discussion of Third World cities in market societies, drawing on several differing approaches that belong to the political-economy perspective. These approaches vary in the primacy they place on causal mechanisms. In the following section there is a discussion of cities in the socialist Third World. This chapter provides a brief evaluation of explanatory strengths and weaknesses of the various approaches relating to the phenomena that collectively define the Third World city.

Empirical characteristics of Third World cities

The literature has been dominated by discussions of problems caused by excessive rates of urban growth in resource-poor societies and imbalanced city systems accompanying uneven spatial development at the national and regional levels (Abu-Lughod & Hay 1977, Safa 1982, Smith 1985, Timberlake 1985, Walton 1982, Smith & Feagin 1987). The classic features are hyperurbanization and the explosive growth of megacities, industrial concentration in a few metropolises, deteriorating urban income distribution, increase in urban unemployment and poverty, a burgeoning urban informal sector, inadequate provision of basic goods and services, speculation in land and housing markets with rapid proliferation of slums and squatter settlements, and various expressions of social and political conflict resulting from these negative trends. Third World cities are also the administrative centres for the public and private sectors, have the highest concentration of the economic and political élites of a nation, are the location of the superstructural institutions, account for the largest increase in industrial output, and have been the beneficiaries of urban bias in development, at least in market societies. Scarce infrastructural resources have been channelled to the national capitals and major cities. All these characteristics reflect a milieu of economic decline for a large segment of the population with increasing conspicuous consumption for the minority.

Numerous problems result from the merging of several trends some of which are primary and structural; others, being secondary, are thus more easily identifiable. The structural causes are discussed in the next section. Among the secondary factors, the acceleration in the growth of urban populations and their sheer magnitude is problematic. For example, in Asia, in the period between 1950 and 1980, the urban population increased from 216·3 million to 689·3 million (Hauser & Gardner 1982). This 3-fold increase in size would strain the budgets and management capacities of the richest of countries. The maldistribution of the urban population is another problem. For example, in Paraguay 92 per cent of the urban population is concentrated in one or two cities, comparable figures for Argentina and Uruguay are 82 per cent and 75 per cent (Portes 1985b). Part of the urban growth is fuelled by rural–urban migrations due to unbalanced regional and sectoral development. Any solution to the urban growth problem has to confront head on the rural–urban issue, particularly the equitable and efficient management of the geographic distribution of the national population and resources.

In terms of the internal structure of cities, the trends relate to labour and housing markets. Due to policy choices made by the public and private sectors at various spatial scales, the urban labourforce is growing at a faster rate than employment opportunities. At current rates it takes approximately 35 years for the labourforce to double in the Third World. Industrial investment policies have not been responsive to these trends. Capital-intensive industrialization is favoured by the capital-owning sector, for capital intensity with low wages permits high rates of surplus value appropriation. The state supports this with fiscal and monetary policies. The result of sluggish employment growth in the urban industrial sector gives rise to the class of casual workers, working irregularly with short-term contracts, or in the informal sector with a high degree of self-provisioning. Such labourmarket conditions permit industrial wages to be depressed. Whereas the urban poor are compelled to work even for minimal incomes, there are high rates of unemployment among the educated youth in the majority of Third World cities.

The concentration of capital-intensive industrial development in Third World cities allows the urban industrial, finance, and rentier classes to benefit disproportionately from national development policies. This is expressed in expensive, well serviced neighbourhoods containing luxury housing. An increasing national bourgeoisie capable of expressing its preference for high standards of living in the major urban areas aggravates the shortages of real resources for urban development. Scarce resources such as urbanized land and building materials are diverted to luxury housing, due to market forces (Chatterjee 1985). The adoption of the ability to pay criterion for access to basic goods and services, such as housing and transportation, has caused land rents and materials prices to escalate with demand and supply imbalances.

This distortion negatively affects the majority of the urbanites in Third World cities who have to settle for minimal and below-minimal standards of consumption. High rates of unemployment and underemployment combine with urban inflation to cause widespread poverty, proliferation of slums, inadequate infrastructure provision, and severe environmental degradation. These conditions are functional for industrial capitalism in Third World cities. The depression of the living conditions of workers reduces the cost of subsistence of the labourforce and thus the acceptable minimum wage. There are, however, considerable variations among Third World cities. Hong Kong and Singapore have adopted policies which successfully reduce their urban problems (Pryor 1983, Yeh 1975). China also provides a notable counter example: policies adopted by the state are able to reduce rural–urban differences in consumption and eradicate homelessness, even though severe urban housing shortages continue (Demko & Regulska 1987).

The parameters of the lack of basic service provision in most Third World cities are described in numerous academic and planning studies (e.g. Pacione 1981, Angel *et al.* 1983, Skinner & Rodell 1983, Lowder 1986). Explanations for these attributes are summarized in the following sections.

The common conceptual underpinning

Any interpretation of Third World cities cannot be divorced from an analysis of production and reproduction relations in the society. Considerable variations between Asian, African, and Latin American cities reflect the specific dynamics of their social formations. The social formation, an historically determined configuration in a geographic area, reflects an intertwining of social, cultural, economic, and political processes, some of which are common to the Third World, while others are specific to the locality. The common characteristics derive from the peripheral location of Third World countries in an expansionary global capitalist system. The particular ones derive from the characteristics of the pre-capitalist societies partially dissolved by the hegemonic penetration of world capitalism. Since there are many pre-capitalist societies, and their articulation with global capitalism varies, this gives rise to many differing urban expressions of the common processes underlying the incorporation of pre-capitalist societies into the capitalist system. That is, the attributes of Third World cities are not universal, rather they reflect the contradictions internal to a social formation or arising from the articulation of different social formations. Only the general attributes are discussed in this chapter, due to feasibility constraints. However, it is important to emphasize that the attributes of any specific Third World city can be interpreted only when the general is combined with the conjunctural (Cardoso & Falletto 1978, Palma 1978).

The political-economic perspective views the conditions of Third World urbanization as empirical manifestations of processes rooted in relations of domination and subdomination. At a global level these processes lead to uneven development among countries (Smith 1984). The characteristics of Third World cities express, to a large extent, the interlinkages of a global production and consumption system in which the interests of international capital are maximized. Core–periphery, dependency, and world systems theories are relevant in this regard. Each looks at global linkages through different lenses and thus emphasizes different subprocesses. While each perspective seeks to establish its theoretical supremacy, this chapter will discuss the strengths of each for they represent different phases of an evolutionary schema, in which the last word has yet to be said. At the national level the twin processes of domination and subdomination have urban expressions in regional inequalities, urban bias in infrastructure investment and processes such as rural–urban migration, and like responses. At the local level, manifestations of poverty such as slums and squatter settlements are all expressions of domination by economically and politically powerful groups seeking to maximize their class interests either through ownership of urban space or privileged access to urban infrastructure investments. At each of these geographical levels space becomes an instrument for the expression of power relations in class societies.

The production and realization of surplus value is central to the logic of capitalist accumulation. While the competing theories differ in the emphasis placed on the relative primacy of accumulation as an explanatory variable, all recognize its pivotal rôle in the centralization of economic power and thereby in relations of domination and subdomination in urban development. Capital accumulation also occurs at various spatial scales, and urban areas in the Third World play an important rôle. Special emphasis is placed on the rôle of the multinationals and international trade in the dynamics of accumulation at the global level. Petty commodity production is related to the dynamics of accumulation at the local level. Accumulation processes occurring at these various spatial levels are central to the expansion of capitalist relations of production in the urban domain.

The capitalist production system is expansionary in two ways. It is expansionary in a geographic sense, incorporating more and more countries into its system. It is also expansionary in the sense that more and more areas of the activities of a social system come under its influence. A central concept in the political-economic approach is the antagonistic relation between the capitalist mode of production and the pre-capitalist modes it replaces. The disruption of earlier modes of production, such as the domestic and the peasant modes, or the deskilling which occurs through integration of earlier modes with the capitalist (for example, in textiles and metal working), results from impulses which originate in the urban areas of core countries and are then transmitted through the urban system in the Third World. In the contemporary Third World the process of integration is incomplete and is penetrating the original social formation through the urban and regional hierarchy. Theories focusing on the articulation of the modes of production and the development of underdevelopment place primary emphasis on this process.

Third World urban areas play a vital rôle in global restructuring. Rapid urbanization is associated with greater diversification of the economy and increasing class complexity, i.e. a fundamental transformation of Third World societies. Urbanization has permitted the differential incorporation of the capitalist production system into the various sectors of an economy. Urban centres provide the locale for the transmission of capitalist cultural values, i.e. commodification, attitudes towards time, proleterianization of the labourforce, and so on. The repressive rôle of the state, the function of which is to smooth the process of accumulation and manage class struggle, also has an urban expression not only through the location of the agencies of the state, but also by the expression of urban interests through the political apparatus.

This catalogue does not exhaust all the aspects of Third World urban development; these are simply the more central issues elaborated in this chapter. In the discussion which follows it is important to emphasize that the dichotomy between urban and rural is only epiphenomenal. We need to penetrate beyond the superficial expressions of rural–urban differences and look at the interlinkages between them, linkages that allow the penetration of urban-located, but systemically generated impulses causing the fundamental restructuring of the local, national, and international systems.

Third World cities in market societies

The key concepts relating to Third World cities discussed here are: (a) the new international division of labour; (b) dependent urbanization and industrialization; (c) urban characteristics in peripheral capitalism, particularly the informal sector in labour and housing markets; (d) urban conflict and the rôle of the state in collective consumption.

Third World cities and the new international division of labour

The concept of a new international division of labour was first expounded by Marx more than a century ago in *Capital* (vol. 1) while discussing the incorporation of colonies into a global production and consumption system (Walton 1985). Marx saw the Third World countries as primary suppliers of materials and consumers of European manufactured goods, with their pre-capitalist manufacturing systems subject to destruction. Almost a century later the concept has taken a contemporary definition as a powerful descriptor of the modern transnational economy, wherein the Third World is a competitive supplier of manufactured goods and plays a critical rôle in the new restructuring of the global economy. Frobel *et al.* (1980) define the concept of the international division of labour in the context of the shifting world market for labour and production sites. What is the rôle of Third World cities?

Feagin & Smith (1987) note an intimate relationship between the web-like organization of modern capitalism, the world-wide net of multinational corporations, and the global network of cities. Their urban classification takes into account the position a city occupies in the decision chain of the transnational economy: thus, world command cities are those in which the top corporate financial, industrial, and commercial decision makers are concentrated (e.g. New York); specialized command cities are those in which the headquarters of specific industries are located (e.g. Detroit); division command cities those with a high concentration of division headquarters (e.g. Houston); then there are state command cities (e.g. Brasilia); and so on.

Major Third World cities are not at the world command level. Only Seoul, Korea has headquarters of firms belonging to the world's 500 largest transnational corporations. Indeed, cities such as New Delhi, Singapore, Lagos, Cairo, Mexico City, and Sao Paulo do not lend themselves to such a clear classification (a fact which the authors note) because of the diverse mixing of their economic and state functions. Nevertheless, these cities, pivotal for global capital accumulation, are critical elements in the organizational network of the internationally integrated production and consumption system. They occupy a variety of niches in which international capital, particularly multinational corporations, has played a key rôle.

Multinational corporations that have globalized production and marketing strategies have located plants and offices in several Third World cities. Responding to the import substitution and protectionist policies of the Third World in the postwar period that closed off many of their major markets, to increasing domestic labour costs due to gains made by trade unionism, to attractive tax advantages offered by the state and to the expanding costs of meeting environmental protection laws, the multinationals moved their capital and production facilities into Third World cities. These cities offered appropriate infrastructure facilities, low wages, docile labourforces, and attractive urban environments for expatriate management. The 'runaway industries' had characteristic profiles – they were low-wage, labour-intensive industries, using a high proportion of semi-skilled or unskilled labour producing commodities that could be inexpensively transported, e.g. garments, footwear, and electronics (Portes & Walton 1981, p. 156). Third World cities have recently been key agents in the internationalization of capital by acting as gateways and preferred sites for new factories and assembly plants. These cities are predominantly located in countries that either have large domestic markets or have adopted export promotion strategies for their national development – for example, Mexico, Brazil, Colombia, Venezuela, India, South Korea, Malaysia, Singapore, Hong Kong, Taiwan, Phillipines, Indonesia, Thailand, Egypt, Kenya, and Nigeria. When Third World cities are used for offshore production by multinationals and Third World products are exported back to the First World, they can adversely affect the economy of the cities of the advanced economies and their working class, for example, through weakening of bargaining power, and eventual job loss in the competing industries (Sassen 1988, Peet 1987).

Not all Third World cities are incorporated into the global economy to the same degree. The multinationals, with their control over vast sums of international capital and efficient global scanning capacity, are selective in their location policy. They evaluate the characteristics of the urban labour pool, for example its size, discipline, docility, and state policies toward unemployment and social security. Decisions about which cities to choose are guided by a complex set of political and economic issues – for example, subsidies offered by the host country, its profit repatriation policies, and whether it provides shelter from corporate taxes. Cohen (1981) provides a good discussion of multinationals and the spatial ramifications of their decision making, particularly their impact on the urban hierarchy.

Multinational firms have multiplier effects on the urban economy. Secondary and ancillary firms, particularly in the export manufacturing and service sectors, depend on primary multinationals. Clairmonte & Cavanagh (1981) point to a pattern of dependent partnership spreading in underdeveloped countries due to joint ventures between domestic-multinational and state-multinational partnerships. Thus the economies of Third World cities become dependent on decisions made by powerful international firms interested only in rates of surplus appropriation and accumulation. Wilson (1987) using the case of Lima, Peru discusses the vulnerability arising from relegation to a peripheral status in the international division of labour as a result of decisions taken by multinationals and external investors. With their increasing global reach, the multinationals opt to divide the production process of goods such as electronics, textiles, and automobiles, into discrete segments concentrated in a few cities to take advantage of scale economies. Instead of producing for national markets, as they did in earlier phases, they reduce the number of duplicated production facilities in Latin America and export final goods to larger markets. Competition between the Latin American countries, to retain their manufacturing hold, results in even more favourable conditions for the multinationals in terms of tariff reductions and tax incentives. Lima has been a loser while Sao Paulo and Mexico City have been winners. This has had sharp downward effects on employment in Lima and the coastal cities whose growth was initially fostered after World War II by multinational corporations spearheading the country's industrialization process.

Increasing concentrations of economic and political power strengthen the ability of international capital to take advantage of world labour conditions. This increases volatility, further exacerbating the existing uneven spatial development. The net effect on Third World cities is negative, even though a select few cities benefit from the concentration of manufacturing activities. There are many disadvantages of dependent development which are discussed next.

Third World cities and dependent development

The world systems framework and its predecessor, the dependency approach, provide some of the best insights on the interrelationships between Third World urbanization and the global expansion of the mercantilist and capitalist systems. Common to both approaches is the belief that the expansion of the capitalist system on a world scale was only possible through a core–periphery division of labour and unequal exchange between the core/metropole and periphery/satellite (Frank 1979, 1984, Wallerstein 1980). These exchange relationships benefit the core areas while underdeveloping peripheral areas (Wallerstein 1980). Chirot & Hall (1982) provide an excellent review of world systems theory. The dependency approach emphasizes the negative consequences of unequal exchange relations between the core and the periphery within the international economy; Palma (1978) provides an excellent review of the several variants of the Dependency School. Only the urbanization aspects of these approaches are discussed here.

Dependent urbanization developed initially in the colonial period when countries became incorporated into an international division of labour through their rôle as suppliers of primary commodities for production and consumption in the core and as markets for manufactured goods. Ports and administrative centres played pivotal rôles in incorporation and are examples of dependent urban centres. The capitalist imperative was initially weak in import–export oriented colonial cities accelerating, however, during the period of dependent industrialization.

Uneven development is an expression of dependent industrialization and the continuing incorporation of peripheral countries into the capitalist world economy. Uneven development causes dependent urbanization to vary from core urbanization. It aggravates urban primacy and causes increased centralization of activities in the major cities of the national space (Slater 1978). Accelerating rural–urban migration, worker exploitation in urban labour-markets, and reorientation of urban economies toward capital-intensive industrialization all reflect the dependent character of this urban growth (Walton 1982). Using data for 37 peripheral countries, Kentor (1981) shows that several international factors, especially the degree of investment dependency, are significant determinants of the speed of peripheral urban growth. Since the growth of Third World cities has not kept pace with the increase in employment opportunities in the urban labourmarket, the world systems theorists emphasize the 'overurbanization' of Third World countries (Bradshaw 1985, Chase-Dunn 1985).

The concept of overurbanization, first introduced in the development literature in the 1950s, describes an imbalance between the urbanization rate and the level of industrialization of Third World countries relative to that of Europe and North America during similar stages of development. It was quickly discredited since the definition drew from modernization theory and expected the proportion of urban population to total population in all countries to conform to a 'normal' path of development derived from the historical experience of Western countries (Sovani 1964).

In the last few years, world systems theorists have reinterpreted high rates of urban growth in Third World countries relative to their industrialization, particularly the increasing unemployment in the cities, in terms of the global core–periphery division of labour (Timberlake 1987). The contemporary formulation relates to the interplay between industrialization, urbanization, and dependent economic growth (Gugler 1982). It is argued that mechanisms reproducing this global division of labour also produce overurbanization by simultaneously limiting the spread of industrialization and encouraging rural–urban migration to the national cores. Foreign investment in manufacturing is capital–intensive, either export-led or geared to the consumption of the urban middle classes. Such industries have few backward linkages and are unable to stimulate sustained industrial development in the peripheral country. Yet urban industrialization reinforces the illusion of ample employment opportunities which, combined with the displacement of rural labour due to the increasing poverty of the peasantry, fuels the processes of migration and urban expansion. Timberlake & Kentor (1983) estimate the relation between investment dependence and overurbanization, showing that the relative increase in overurbanization has been consistently accompanied by relative declines in per capita economic growth.

Moreover, the major cities are favoured with large public expenditures for infrastructure and subsidized construction due to a need to attract foreign investment and transnational firms (Meyer 1986). These desires interact with domestic class interests and national politics to produce an urban bias in public expenditures. Higher levels of urban amenities and social services produce an illusion of access to those amenities in the minds of migrants, thereby stimulating migration. According to Smith (1984), however, overurbanization is not merely a manifestation of urban bias in expenditures. Criticizing the emphasis on the demographic component of overurbanization, he argues that it must be situated in the context of state policy and the influence of domestic and multinational firms on state decision making. Since public expenditures and labour mobility reduce the costs of urban production and aid the capital accumulation process, pressure is applied through the political process. The rôle of the state in capital accumulation and the management of conflict are discussed in a later section. Dependency and world systems approaches also focus on the employment characteristics of Third World cities, particularly the interconnection between the formal and informal sectors. This too is discussed below.

Dependency and world systems approaches have been criticized for under-emphasizing internal processes in Third World countries (Smith & Tardanico 1987). In focusing on vertically integrated processes extending from the domestic to the global level, they deal inadequately with the manner in which socio-economic and political transformations in different locales are related to local conflicts, reproduction and related gender issues, the intentionality of common people, and adaptive strategies of urban households to economic restructuring. A richer understanding is provided in the literature on peripheral capitalism, which is discussed next.

Third World cities and peripheral capitalism

This section draws primarily on the literature on the articulation of pre-capitalist and capitalist modes of production in peripheral social formations (Foster-Carter 1978). The articulation literature focuses on the structure of production rather than on the relations of exchange central to the world systems and dependency views. This literature illustrates the importance of looking at the conjuncture of different social formations, each described by different organizing principles. During the development or underdevelopment process there is a reciprocal meeting of systems, in which the actual outcome varies.

The Structuralist School draws attention to the segmented articulation of various formations in Third World countries (Amin 1976). The penetrating capitalist and the existing pre-capitalist confront each other and, given the relative strengths of each in the locality, their strategies vary. Although the decisions to adapt to the capitalist formation or retain existing values are taken by households and individuals, this has a systemic representation. In this confrontation some pre-capitalist elements are dissolved, others are retained. Commonly that which is retained, for example the informal economy, is functional for peripheral capitalism. However, the element of human agency has to be considered in explaining the characteristics of Third World cities. This approach can be contrasted to the world systems and dependency perspectives that view capitalism as a unilateral penetration from the core to the periphery. The articulation literature avoids the problems of scale and provides the best insights on how capitalism conserves and dissolves pre-capitalist modes of production in Third World cities. Its explanation of the urban impacts of imperialism and the contemporary urban informal sector is the richest.

The penetration of capitalism into Third World countries began in the colonial period with their dual rôles as suppliers of primary commodities and markets for manufactured products. Ports and internal market centres, places for collection and distribution, played a pivotal rôle; the location of contemporary Third World cities demonstrates this earlier colonial imperative. Most of the presently large cities of Latin America were founded by Spanish and Portuguese imperial interests (Bromley & Bromley 1982). Colonial cities were also added to countries with a strong urban tradition, for example Calcutta, Bombay, and Madras in India and Port Said and Suez in Egypt. Moreover, surplus extraction required the presence of capitalist institutions such as banks, export–import firms, headquarters of extractive industries, and joint stock agencies, which were concentrated in the cities. There was also a need for a repressive state, since surplus extraction could not occur through peaceful means alone, and the new urban centres also became the seats of administration. Contemporary Third World cities retain many of these characteristics from the colonial period. They continue to perform the administrative and repressive functions and the capitalist institutions are still primarily concentrated in the major cities.

The socio-economic organization of colonial cities and their urban landscape differed fundamentally from those of the pre-capitalist world where surplus extraction by élites was carried out through direct means. While clear physical demarcations of the dual city – European and native – in the urban landscape are blurred, their influence is still recognizable in the spatial layout of cities. The Western sections are still preferred by the domestic élite and Western building forms have class connotations. However, there are more complex impacts of colonialism on Third World cities than the urban settlement system, the location of the major cities, or the cultural landscape would suggest.

Relatively rarely did the location of cities crucial for the global expansion of capitalism coincide with that of urban centres used for surplus extraction in pre-capitalist modes of production. Colonialism debilitated and eventually destroyed existing urban settlements unnecessary for the extraction of surplus in the capitalist mode. In Third World countries with a strong urban tradition, such as India and Egypt, the urban settlement pattern reflects this selective retention. However, countries of the Third World are not homogeneous and many did not have an urban tradition. In such countries, particularly in Africa, urban growth had to be induced through labour migration. Imposition of taxes raised revenues for the colonial administration and brought rural households into the money economy. Such administrations also legitimized land appropriation from the peasants through permanent settlement schemes and like instruments; these were deliberately imposed to destabilize the pre-capitalist system. A direct consequence of such interrelated processes of destruction and destabilization is the urban primacy and unbalanced settlement systems typical of many Third World countries. However important these consequences for the urban system, they are only surface manifestations of fundamental structural changes brought about by the integration of pre-capitalist social formations into the global economy.

Central to the argument of peripheral capitalism is the concept of the articulation of the different modes of production in a single world economy (Wolpe 1980). While a number of theorists have provided varying formulations, there is a common central thesis. In peripheral capitalism pre-capitalist structures are retained for functional reasons; they absorb a large element of surplus labour and substantially bear the costs of reproduction of the labourforce (Portes 1985a). There are two aspects to this. Rural–urban migration permits the reproduction of the labourforce in subsistence enclaves with communal modes of production. Some interpret the enclaves of subsistence agriculture, share cropping, and like remnants of earlier communal or feudal modes of production in these terms (Chatterjee 1983).

Portes & Walton (1981), however, point out that rural subsistence economy is not the primary mechanism for the reproduction of urban surplus labour. Reproduction of the urban surplus wage labourforce also occurs in the urban economy. The presence of the urban informal sector through casual wage labour household subsistence, self-built housing, and similar self-provisioning strategies performs this function (Gerry 1978, Bromley & Gerry 1979, Bromley 1985, Shaw 1985, Kalpagam 1985).

A principal characteristic of peripheral capitalism is the articulation of various modes of production, since the capitalist mode of production confronts and partially dissolves earlier modes of production. Consequently, significant sections of the labourforce in the cities are either marginally related to the capitalist mode of production or are totally excluded from it (Sandbrook 1982, p. 37). For example, in the textile sector previous artisanal production is not eliminated with the development of industrial production; in the housing sector craft methods of construction are only partially replaced by modern construction technologies; in the footwear industries cobblers exist side by side with manufacturers of shoes. However, the capitalist sector is hegemonic and the earlier relations of production serve the interest of the capitalist sector through lowering the cost of reproduction of a significant portion of the labourforce, sustaining the industrial reserve army during their periods of irregular employment, and depressing wages in the non-unionized sector. Thus, this co-existence of various forms is not dysfunctional to the capitalist sector; indeed, they aid the accumulation process by permitting higher rates of surplus value appropriation. Obregon (1980) suggests that the marginal pole of the economy and the marginalized labourforce perform this rôle in Latin American cities. Santos (1979), describing the two sectors as linked and dependent, two circuits of capital, notes that the lower circuit produces goods and services through petty commodity production, using low technology, petty trading, and personal services affordable by the poor.

Dependency theorists interpret the relationship between the formal and informal sectors in dualistic terms (Evans & Timberlake 1980). The formal or organized sector is integrated into the networks of global capitalism. The informal sector constituting a vast pool of unemployed and subemployed labour, a marginalized mass seeking to survive in the informal economy through various subsistence and petty commodity production strategies, is excluded from participation. The dualistic model has been challenged as greater knowledge about the informal sector documents the interdependence between the two. Portes & Walton (1981), in challenging the dualistic interpretation, provide empirical evidence showing articulation between the formal and informal sectors in which informal enterprise is a stable component of the national and international economies. They refer to several studies documenting the use of second hand, but modern equipment in transportation, garment manufacturing, footwear production, and the metalworking industries.

The international linkages are clearly seen in the home-based market production of women in Third World cities which spans a wide range of activities, with traditional and non-traditional (Singh & Kelles-Vitanen 1987). Much of the activity carried out by women in the informal sector of the large cities is in industrial subcontracting which produces commodities for the export market as in food processing, textiles, and clothing. The labour conditions are marked by exploitation through low wages, long hours, poor working conditions, non-payment of overhead costs, and the use of child labour. Moreover, low cash wages limit the purchasing power of the household with respect to food and other household necessities that require payment in money. Due to gender inequalities in intra-household allocation of consumption items, urban women suffer from higher rates of malnutrition and morbidity (Steady 1982). Women in the informal sector suffer not only from poverty through belonging to poor households but, in addition, they suffer from the gender subordination and sexual discrimination characteristic of the patriarchial values of the majority of Third World countries.

The degree of hegemony of the capitalist sector varies between cities in the Third World and is dependent on the level of capitalist development in the nation and the region. Cities perform a hegemonic rôle with respect to small towns and rural areas since the capitalist industrial sector has greater urban concentration. As the urban interests are more articulate and powerful, they are able to bias resource allocations, particularly infrastructure investments, in their favour. This urban bias has on occasion been interpreted incorrectly in terms of simple rural–urban antagonisms (Lipton 1977).

The rôle of the state in Third World cities

The state is very influential in determining the characteristics of Third World urbanization. Some of its rôle *vis-à-vis* transnationals and urban infrastructure has been discussed. In this section its rôle in collective consumption by the poor and conflict management is highlighted.

Low-income housing is largely supplied through the informal or popular sector and a substantial portion of the urban poor live in illegal settlements. For example, 70 per cent of all new housing supplied in Cairo between 1970 and 1980 was in the informal sector. Twenty-five per cent of the urban working class in Rio de Janeiro and Lima, and 40 per cent in Caracas live in pirate settlements (Angel *et al.* 1983, Portes & Walton 1981). Popular settlements encompass a wide variety of housing types as, with time, the initial squatter settlements become consolidated (Gilbert & Ward 1982, Conway 1982). Baross (1983) provides an excellent discussion of the mechanisms of land supply for low-income house builders, particularly the forms of articulation through which land takes on a commodity form in a wide range of countries.

Housing the urban poor is viewed as a major problem in Third World cities, particularly as the state ignores these areas in terms of infrastructure and service provision. Perspectives on informal housing have varied from the early interpretations of marginality in which the urban poor were viewed as an undifferentiated mass, through self-help housing as a vehicle for upward mobility, to the notion of housing as petty commodity production (Burgess 1978, Conway 1982). The emergence of these settlements has been seen as a failure of the state; in reality informal housing is a stable component of the structure of peripheral urban economies. Such settlements do not make claims on national resources, allow low-cost reproduction of the labourforce, incorporate the poor into the values of property ownership, and act as stimuli to overwork by the poor who often work in extra jobs to raise funds for their housing.

Political parties and populist movements have used the consolidation and delivery of urban services as a means for gathering political support (Castells 1982). Urban issues surrounding the built environment and collective consumption of goods and services are some of the most visible forms of social protest (Lubeck & Walton 1979). Land speculation and accumulation through the circulation process are issues in which the state allies itself with the landowning and rentier classes (Harvey 1982). There is a rich literature in this area and the reader is directed to the several volumes on Third World cities referred to in this chapter.

Relations derived from pre-capitalist, commercial, colonial, competitive, and monopolistic phases of capitalism combine in each economic sector of production in Third World cities. The incorporation of newer forms is mediated either by external agents, as in the colonial phase, or by domestic agents in alliance with external agents, as in the monopolistic phase.

The socialist city in the Third World

The socialist city in the Third World does not belong to an homogeneous category; attributes vary with the history, resource base, civil society, brand of socialism, and the spatial planning policies of the various countries. In this section, however, some common elements are discussed.

The abolition of the antitheses between the urban and the rural has been a classical objective of socialism since Marx (Merrington 1978). A predominant objective of socialism in the Third World was to reduce the rate of urban growth, particularly the degree of urban primacy. Some countries sought to decrease the size of their large cities through de-urbanization policies, for example Tanzania, Vietnam, and Kampuchea; others such as China attempted to arrest their large city growth through strict control of labour migration (Forbes 1984). Cuba adopted the policy of promoting the growth of the intermediate sized towns while suspending developmental and industrial investment in Havana (Barkin 1978, Slater 1982). The provincial capitals were selected for industrial investment with a specific bias in favour of agro-industries.

The relative success of the socialist countries in meeting the objectives of polarization reversal depends on the length of their record of socialism and their ability to control population movements through various regulatory mechanisms (Forbes & Thrift 1987). Tanzania and Mozambique in the period between 1970 and 1982 had average urban growth rates in excess of 8 per cent per annum, in contrast to Cuba and Vietnam with rates between 2 per cent and 3 per cent respectively. Whereas China and North Korea have less than 20 per cent of their urban population in their largest cities, in Mozambique and Guinea the comparable figure is around 80 per cent. Fuchs & Demko (1981) describe the various instruments used for urban containment such as residence registration, the issuing of work and food permits, access to housing and schooling, and forced resettlement schemes.

Socialist countries make a distinction between consumer and producer cities, the former being considered parasitic. Most large cities, particularly if they were important to the colonial economy, were considered to be consumer cities. Socialist countries use capital investments in infrastructure and industry to convert consumer cities into producer cities. For example, in Hanoi the output of large, small, and handicraft industry increased several fold in the period between 1954 and 1960. Other measures include the limiting of the labourforce in service and trade occupations and efforts to make the city self-sufficient in food. For example, suburban and peri-urban agriculture using the urban labourforce has been intensified in Cuba and China.

Expenditures for consumption activities, such as housing and retail establishments, were curtailed in favour of job-creating expenditures in the industrial and agricultural sectors. For example, in Hanoi almost no new inner-city housing has been built since 1955, and one third of the urban labourforce lives in rural areas. This is also the pattern in Chinese cities (Xu Xue 1984). Restrictions on investment in social overhead capital result in an undersupply of urban consumption items in the majority of socialist cities. On the positive side, in China this policy discouraged migration and the creation of impoverished urban masses during the period of reconstruction (Kirkby 1985). Rents are inexpensive and linked to wages. In Cuba a Rent Reduction Law gave rent reprieves to those who had a long record of tenancy. Homelessness has been all but eliminated in China, Vietnam, and Cuba, even though space and amenity standards are kept low. On the negative side, the high costs of construction, combined with low rents, cause severe housing shortages. Serious attention is placed by planners on increasing housing consumption standards and supply through construction of apartment blocks in suburban areas, new towns, and satellite cities.

As noted earlier, there is considerable diversity among Third World socialist countries. In several African countries, such as Mozambique, Tanzania, Angola, and Zimbabwe, the proportion of urban population in the largest cities rapidly increased. For example, in Zimbabwe the influx of black people to the cities continues to accelerate, giving rise to squatter settlements, self-employment in petty commodity trading and other attributes common to market-based Third World cities (Drakakis-Smith 1987). Unable to control the influx of migrants, Mozambique and Tanzania have adopted self-help sites and services, and slum upgrading programmes, together with co-operative housing schemes similar to those of the rest of the non-socialist Third World (Skinner & Rodell 1983). While the stated goal is to reduce primacy, the growth of Dar es Saalam and Harare demonstrates that implementation is unsuccessful. There are problems with urban housing supply in Cuba as well. Self-built, single-family detached dwellings in the urban areas of Cuba averaged 30 000 units annually in the early 1980s. While Cuba's urban population has doubled since 1959, its urban land use has tripled due to the construction of these units without planning permission (Susman 1987).

According to neoclassical economics, cities provide external economies of scale – that is, the costs of production decline as cities increase their scale of economic activities. China, Cuba, and Vietnam, countries which have successfully managed to contain the growth of their large cities, have discovered the operation of this economic principle with respect to the development of the forces of production. China and Vietnam, after partially achieving their equity goals, realize the importance of increasing the productive base. China is seeking to harness scale efficiencies through its policy of core cities and New Economic Zones: the former places investment emphasis on existing district towns with high levels of industrialization; the latter are situated around the major metropolises. The Beijing–Tianjin region, the Shanghai–Changjiang Delta region, and the Guangzhou region are selected for priority development in China. As these are the existing industrialized areas, the backwash effects are likely to widen regional disparities in income and welfare (Lakshmanan & Hua 1987). Economic liberalization policies, introducing free market principles, increase urban inter-household income inequalities. These are expressed in differences in standards of housing consumption as urban property reforms permit homeownership. However, strict regulations regarding uniformity of infrastructure and building construction standards and avoidance of income-based neighbourhood segregation are expected to retain the basic principles of socialist development. Vietnam is currently emphasizing growth in Hanoi and Ho Chi Minh City, its two largest cities; the latter will have satellite cities to divert population growth from the centre.

Conclusions

This chapter has briefly reviewed some of the useful insights provided by recent scholarship on Third World cities. It has drawn on several variants of the political-economy theoretical tradition in the belief that all contribute to our understanding of this complex phenomenon. All approaches attempt to explain empirically observable phenomena such as primacy, choice of capital-intensive technology, labourmarket characteristics such as the informal sector, worsening income distribution, speculation in the land and housing markets, political processes, and the rôle of the state. However, their explanatory power with respect to a specific phenomenon varies with the match between the emphasis placed in their theoretical framework on the underlying forces and the urban processes they seek to explain. They look at the reality of Third World urbanization through different lenses.

References

Abu-Lughod, J. 1984. Culture, modes of production and the changing nature of cities in the Arab world. In *The city in cultural context*, A. Agnew, J. Mercer & D. Sopher (eds), 94–119. Boston: Allen & Unwin.

Abu-Lughod, J. & R. Hay (eds) 1977. *Third World urbanization*. Chicago: Maaroufa Press.

Amin, S. 1976. *Unequal development: an essay on the social formations of peripheral capitalism*. New York: Monthly Review Press.

Angel, S. 1983. Upgrading slum infrastructure: divergent objectives in search of a consensus. *Third World Planning Review* **5**, 5–22.

Angel, S., R. Archer, S. Tanphiphat & E. Wegelin (eds) 1983. *Land for housing the poor*. Singapore: Select Books.

Barkin, D. 1978. Confronting the separation of town and country in Cuba. In *Marxism and the metropolis*, W. Tabb & L. Sawyers (eds), 317–37. New York: Oxford University Press.

Baross, P. 1983. The articulation of land supply for popular settlements in Third World cities. In *Land for housing the poor*, S. Angel, R. Archer, S. Tanphiphat & E. Wegelin (eds). Singapore: Select Books.

Bradshaw, Y. 1985. Overurbanization and underdevelopment in Black Africa: a cross national study. *Studies in Comparative International Development* **1**, 43–56.

Bromley, R. 1985. Cities, poverty and development. *Labor, Capital and Society* **18**, 211–19.

Bromley, R. & R. Bromley 1982. *South American development: a geographical introduction*. Cambridge: Cambridge University Press.

Bromley, R. & C. Gerry (eds) 1979. *Casual work and poverty in Third World cities*. New York: Wiley.

Brunn, S. & J. Williams 1983. *Cities of the world: world regional development*. New York: Harper & Row.

Burgess, R. 1978. Petty commodity housing or dweller control? A critique of John Turner's views on housing policy. *World Development* **6**, 1105–33.

Cardoso, F. & E. Falletto 1978. *Dependency and development in Latin America*. Berkeley: University of California Press.

Castells, M. 1982. Squatters and politics in Latin America: a comparative analysis of urban social movements in Chile, Peru and Mexico. In *Toward a political economy of urbanization in Third World countries*, H. Safa (ed.), 249–82. New Delhi: Oxford University Press.

Chase-Dunn, C. K. 1985. The coming of urban primacy in Latin America. *Comparative Urban Research* **11**, 14–31.

Chatterjee, L. 1983. Migration, housing and equity in developing countries. *Regional Development Dialogue* **4**, 105–31.

Chatterjee, L. 1985. The construction sector in developing countries. In *Spatial, environmental, and resource policy in the developing countries*, M. Chatterji, P. Nijkamp, T. Lakshmanan & C. Pathak (eds). Aldershot: Gower.

Chirot, D. & T. Hall 1982. World systems theory. *Annual Review of Sociology* **8**, 93–105.

Clairmonte, F. & J. Cavanagh 1981. *The world in their web*. London: Zed Press.

Cohen, R. 1981. The new international division of labor, multinational corporations and urban hierarchy. In *Urbanization and urban planning in capitalist society*, M. Dear & A. Scott (eds), 287–315. London: Methuen.

Conway, D. 1982. Self help housing, the commodity nature of housing and amelioration of the housing deficit continuing the Turner–Burgess debate. *Antipode* **14**, 40–6.

Cooper, F. (ed.) 1983. *Struggle for the city: migrant labor, capital and the state in urban Africa*, Ch. 1. Beverly Hills: Sage Publications.

Dear, M. & A. Scott (eds) 1981. *Urbanization and urban planning in capitalist society*. London: Methuen.

Demko, G. & J. Regulska, 1987. Socialism and its impact on urban process and the city. *Urban Geography* **8**, 289–92.

Drakakis-Smith, D. 1987. Urban and regional development in Zimbabwe. In *The socialist Third World*, D. Forbes & N. Thrift (eds). Oxford: Basil Blackwell.

Durand-Lasserve, A. 1983. The land conversion process in Bangkok and the predominance of the private over the public sector. In *Land for housing the poor*, S. Angel, R. Archer, S. Tanphiphat & E. Wegelin (eds), 284–309. Singapore: Select Books.

Eckstein, S. 1977. *The poverty of revolution, the state and the urban poor in Mexico*. Princeton, NJ: Princeton University Press.

Eisenstadt, S. & M. Shachar 1987. *Society, culture and urbanization*. Beverly Hills: Sage Publications.

Evans, P. & M. Timberlake 1980. Dependence, inequality and the growth of the tertiary: a comparative analysis of less developed countries. *American Sociological Review* **45**, 531–52.

Feagin, J. & M. Smith 1987. Cities and the new international division of labor: an overview. In *The capitalist city: global restructuring and community politics*, M. Smith & J. Feagin (eds). Oxford: Basil Blackwell.
Forbes, D. 1984. *The geography of underdevelopment: a critical survey*. Baltimore: Johns Hopkins University Press.
Forbes, D. & N. Thrift 1987. *The socialist Third World: urban development and territorial planning*. Oxford: Basil Blackwell.
Foster-Carter, A. 1978. The modes of production controversy. *New Left Review* **107**, 47–77.
Frank, A. G. 1979. *Dependent accumulation and underdevelopment*. New York: Monthly Review Press.
Frank, A. G. 1984. *Critique and anti-critique: essays on dependence and reformism*. New York: Praeger.
Frobel, F., J. Heinrichs & O. Kreye 1980. *The new international division of labor: structural unemployment in industrialised countries and industrialization in developing countries*. Cambridge: Cambridge University Press.
Fuchs, R. & G. Demko 1981. Population distribution measures and the redistribution mechanism. *Population Studies* no. 75, 78–84.
Gerry, C. 1978. Petty production and capitalist production in Dakar: the crisis of the self employed. *World Development* **6**, 1147–60.
Gilbert, A. & J. Gugler 1983. *Cities, poverty and dependent urbanization in the Third World*. Oxford: Oxford University Press.
Gilbert, A., J. Hardoy & R. Ramirez 1982. *Urbanization in contemporary Latin America*. New York: Wiley.
Gilbert, A. & P. Ward 1982. Residential movements among the poor: the constraints on housing choice in Latin American cities. *Transactions of the Institute of British Geographers* **7**, 129–49.
Griffen, K. & J. Gurley 1985. Radical analysis of imperialism, the Third World and the transition to socialism: a survey article. *Journal of Economic Literature* **23**, 1089–143.
Gugler, J. 1982. Overurbanization reconsidered. *Economic development and cultural change* **31**, 173–89.
Harvey, D. 1982. *The limits to capital*. Oxford: Basil Blackwell.
Hauser, P. M. & R. W. Gardner 1982. World urbanization trends and prospects. In *Population and the urban future*. New York: State University of New York Press.
Kalpagam, U. 1985. Coping with urban poverty in India. *Bulletin of Concerned Asian Scholars* **17**, 2–8.
Kentor, J. 1981. Structural determinants of peripheral urbanization: the effects of international dependence. *American Sociological Review* **46**, April, 201–11.
Kirkby, R. 1985. *Urbanization in China: town and country in a developing economy 1949–2000 A.D.* London: Croom Helm.
Lakshmanan, T. & C. Hua 1987. Regional disparities in China. *International Regional Science Review* **11**, 97–104.
Lipton, M. 1977. *Why poor people stay poor*. Cambridge, Mass.: Harvard University Press.
Lowder, S. 1986. *The geography of Third World cities*. Totowa, NJ: Barnes & Noble.
Lubeck, P. & J. Walton 1979. Urban class conflict in Africa and Latin America: comparative analysis from a world systems perspective. *International Journal of Urban and Regional Research* **3**, 3–28.
McGee, T. 1979. The poverty syndrome: making out in the South East Asian city. In *Casual work and poverty in Third World cities*, R. Bromley & C. Gerry (eds). New York: Wiley.
Merrington, J. 1978. Town and country in the transition to capitalism. In *The Transition from Feudalism to capitalism*, R. Hilton (ed.), 170–95. New York: Schocken.
Meyer, D. 1986. The world system of cities: relations between international financial metropolises and South American cities. *Social Forces* **64**, 553–81.

Obergon, A. Q. 1980. The marginal pole of the economy and the marginalised labor force. In *The articulation of modes of production*, H. Wolpe (ed.). London: Routledge & Kegan Paul.
Pacione, M. (ed.) 1981. *Problems and planning in Third World cities.* New York: St Martin's Press.
Palma, G. 1978. Dependency: a formal theory of underdevelopment or a methodology for the analysis of concrete situations of underdevelopment? *World Development* **6**, 881–924.
Peet, R. (ed.) 1987. *International capitalism and industrial restructuring*. Boston: Allen & Unwin.
Portes, A. 1985a. The informal sector and the world economy: notes on the structure of subsidised labor. In *Urbanization in the world economy*, M. Timberlake (ed.). Orlando: Academic Press.
Portes, A. 1985b. Urbanization, migration and models of development in Latin America. In *Capital and labor in the urbanized world*, J. Walton (ed.). Beverly Hills: Sage Publications.
Portes, A. & J. Walton 1981. *Labor, class and the international system.* New York: Academic Press.
Pryor, E. 1983. Housing needs and related urban-development programs and processes in Hong Kong. In *A place to live: more effective low cost housing in Asia*, Y. Yeung (ed.). Canada, IDRC.
Ruccio, D. & L. Simon 1988. Radical theories of development: Frank, the Modes of Production School and Amin. In *The political economy of development and underdevelopment*, C. Wilbur (ed.). New York: Random House.
Safa, H. (ed.) 1982. *Towards a political economy of urbanization in Third World countries.* Delhi: Oxford University Press.
Safa, H. 1987. Urbanization, the informal economy and state policy in Latin America. In *The capitalist city*, M. Smith & J. Feagin (eds). Oxford: Basil Blackwell.
Sandbrook, R. 1982. *The politics of basic needs: urban aspects of assaulting poverty in Africa.* Toronto: University of Toronto Press.
Santos, M. 1979. *The shared space*. London: Methuen.
Sarin, M. 1983. The rich, the poor, and the land question. In *Land for housing the poor*, S. Angel, R. Archer, S. Tanphiphat & E. Wegelin (eds), 237–53. Singapore: Select Books.
Sassen, S. 1988. *The mobility of labor and capital.* New York: Cambridge University Press.
Singh, A. & A. Kelles-Vitanen (eds) 1987. *Invisible hands: women in home-based production.* New Delhi: Sage Publications.
Shaw, A. 1985. The informal sector in Third World urban economy. *Bulletin of Concerned Asian Scholars* **17**, 42–53.
Skinner, R. J. & M. J. Rodell 1983. *People, poverty and shelter: problems of self help housing in the Third World.* New York: Methuen.
Slater, D. 1978. Towards a political economy of urbanization in peripheral capitalist societies: problems of theory and method with illustrations from Latin America. *International Journal of Urban and Regional Research* **5**, 26–52.
Slater, D. 1982. State and territory in post revolutionary Cuba: some critical reflections on the development of spatial policy. *International Journal of Urban and Regional Research* **9**, 1–34.
Smith, M. P. (ed.) 1985. *Cities in transformation: class, capital and the state*. Beverly Hills: Sage Publications.
Smith, M. & J. Feagin (eds) 1987. *The capitalist city*. Oxford: Basil Blackwell.
Smith, M. & R. Tardanico 1987. Urban theory reconsidered: production, reproduction and collective action. In *The capitalist city*, M. Smith & J. Feagin (eds), 87–110. Oxford: Basil Blackwell.

Smith, N. 1984. *Uneven development*. Oxford: Basil Blackwell.
Sovani, N. 1964. The analysis of over-urbanization. *Economic Development and Cultural Change* **12**, 113–22.
Steady, F. 1982. Urban malnutrition in West Africa: a consequence of urbanization and underdevelopment. In *Towards a political economy of urbanization in Third World countries*, H. Safa (ed.). Delhi: Oxford University Press.
Susman, P. 1987. Spatial equality and socialist transformation in Cuba. In *The socialist Third World*, D. Forbes & N. Thrift (eds), 250–81. Oxford: Basil Blackwell.
Timberlake, M. (ed.) 1985. *Urbanization in the world economy*. Orlando: Academic Press.
Timberlake, M. 1987. World systems theory and comparative urbanization. In *The capitalist city*, M. Smith & J. Feagin (eds), 37–65. Oxford: Basil Blackwell.
Timberlake, M. & J. Kentor 1983. Economic dependence, overurbanization and economic growth: a study of less developed countries. *Sociological Quarterly* **24**, 489–508.
Wallerstein, I. 1980. *The modern world system II. Mercantilism and the consolidation of the European world economy, 1600–1750*. New York: Academic Press.
Walton, J. 1982. The international economy and peripheral urbanization. In *Urban policy under capitalism*, N. Fainstein & S. Fainstein (eds). Beverly Hills: Sage Publications.
Wilson, J. 1987. Lima and the international division of labor. In *The capitalist city*, M. Smith & J. Feagin (eds), 199–211. Oxford: Basil Blackwell.
Wolpe, H. (ed.) 1980. *The articulation of modes of production: essays from economy and society*. London: Routledge & Kegan Paul.
Xu Xue, Q. 1984. Character of urbanization of China – changes and cause of urban population growth and distribution. *Asian Geographer* **3**, 15–29.
Yeh, S. (ed.) 1975. *Public housing in Singapore: a multidisciplinary study*. Singapore: Singapore University Press.

Part III
NEW MODELS OF CIVIL SOCIETY

Introduction

Nigel Thrift

Since the mid-1970s, civil society has been the subject of more and more research in political economy. As researchers have become interested in the analysis of civil society, so their research has had increasingly important effects on the direction that the political-economy approach has taken: it has lead to some important theoretical adjustments; it has meant a widening of what is included within the approach; and it has meant some quite substantial changes in methodology.

Each of the following four chapters illustrates the shift from a purely economistic mode of analysis of four of the various dimensions of civil society – in which gender, race, culture, and locality are completely determined by capital – to a mode of analysis which is much looser, in which each of these dimensions of civil society takes on a life of its own; the unproblematic is problematized. It is, therefore, no surprise that in each of the dimensions of civil society analyzed in these chapters positions are taken up which are quite similar to one another.

Of these four dimensions, gender has had to wage the hardest battle to win acceptance within the political-economy approach. Feminists have made fewer inroads into the approach in human geography than in other social science subjects. Much of the writing in political economy still adopts a fairly patronizing/patriarchal tone, seeing the matter of gender as something to be cleared up after capitalism has been analyzed. Gender is eternally next on the agenda. This kind of attitude is no longer acceptable, as the chapter by Bowlby, Foord, Lewis and McDowell makes clear, for gender raises all kind of issues which are central to social science now about how people become 'men' and 'women'.

As Bowlby and her colleagues point out, much of the early work on gender tended to fit it unproblematically into the problematic of capitalism. The oppression of women was viewed, in one way or another, as an artefact of the prevailing system of economic exploitation. Thus, the liberation of women was part and parcel of the liberation of the working class. The assumption that women's oppression was bound up in the system of economic exploitation even underlay the 'domestic labour' perspective, where it was argued that women's labour in the home, reproducing the labourforce, reduced the overall cost of labour to capital (see Dalla Costa & James 1975, Himmelweit 1983). This unidirectional analysis was subsequently followed by another, in which capitalism and patriarchy form two separate, but interrelated structures, the so-called 'dual systems' approach (Hartmann 1979, Cockburn 1986). Finally, another position has emerged in which it is argued that

> the oppression of women is not to be viewed as 'secondary to' – and therefore, by implication, less important than – class oppression as a whole. Women . . . are oppressed as a class by *men*, and patriarchal structures are geographically and historically almost universal, predating capitalism and persisting in the so-called 'socialist' societies. In short, the major axis of differentiation is not class as such, but gender, and it is women who face the 'longest revolution' (Crompton & Mann 1986, p. 5).

The study of race and racism has had fewer problems in gaining acceptance in the political-economy approach to human geography than the study of gender. Problems of combating racial discrimination were bound up with the beginnings of radical geography in the 1960s and the study of race has formed a continuing strand of work since then. What is noticeable in studies of race and racism is the way that they too have followed a progression from ideas of strong economic determination to more elliptical formulations. Fewer researchers have taken up the extreme position that race can be read off from class. Rather, in the earlier work, racial groups were identified with particular aspects of the structure of capital and labour. Thus, such groups formed an industrial reserve army (Castles & Kozack 1985), or they were racialized class fractions (Phizacklea & Miles 1980). These kinds of influence were clearly quite abstract, since racial groups are very diverse and not all can be assigned easily to the reserve army or specific class fractions. So it is no surprise to find that these kinds of direct determination have either given way to or been supplemented by other less deterministic formulations. These formulations have ranged from neo-Weberian to neo-Marxist analyses. But, most particularly, new work on race and racism tends to stress the dynamics of social mobilization, eschewing simple links to capital–labour relations in favour of a notion of social relations structured in dominance (Hall 1980). It is particularly concerned with the racialization of politics and the politics of racial categorization. Some radical commentators are now willing to break the links with economic determinism in the same way as has happened in the study of gender:

> New types of class relations are being shaped and reproduced in the novel economic conditions we inhabit. The scale of these changes, which can be glimpsed through the pertinence of the populist politics of 'race' and nation, is such that it calls the vocabulary and analytic frameworks of class analysis into question. It emphasises the fact that class is not something given in economic antagonisms which can be expressed straightforwardly in political formations. It no longer has a monooply of the political stage, if indeed it ever had one (Gilroy 1987, p. 34).

In the analysis of culture, the same trends can be found as in gender and race. But here the trajectory has been from simple and simplistic notions of ideology, of policing civil society on behalf of capital, to more specific notions involving contestation in the symbolic realm between different social groups. Within human geography, the analysis of culture has tended to fix on notions of hegemony, taken originally from Gramsci and developed by writers such as Williams (1979), Hall (1980), and Willis (1977), and has involved investigations of the processes by which consent is induced in populations by dominant social groups via what are called variously, 'symbolic fields' (Bourdieu 1984), 'maps of meaning' (Hall 1980) or 'ways of seeing' (Berger 1972). Each of these three concepts is meant to convey the idea that social groups *actively* create meaning, but do so in ways which can pinch out emancipatory impulses. It is perhaps the sheer complexity of the processes of production and distribution of meaning that has taxed human geographers most. Take the special case of landscape:

> A landscape is a cultural image, a pictorial way of representing, studying or symbolising surroundings. This is not to say that landscapes are immaterial. They may be represented in a variety of materials and on many surfaces – in paint on canvas, in writing on paper, in earth, stone, water and vegetation on the ground. A landscape park is more palpable but no more real, nor less imaginary, than a landscape painting or poem. Indeed the meanings of verbal, visual and built landscapes have a complex interwoven history. To understand a landscape, say an eighteenth century English park, it is usually necessary to understand written and verbal illustrations of it, not as 'illustrations', images standing outside it, but as constituent elements of its meaning or meanings. And, of course, every study of a landscape further transforms its meaning, depositing yet another layer of cultural representation (Cosgrove & Daniels 1988, p. 1).

This complexity has lead some of those drawing on postmodern ideas to opt for ever more fragmentary ideas of culture, in which the only dominant theme is difference. But for most human geographers involved in the political-economy approach the problem is one of understanding *and* representing the production and distribution of meaning as determinate but not wholly determined.

The final dimension of civil society considered in this group of chapters is that of locality. The study of locality is a relatively recent phenomenon in human geography (Urry 1981, Massey 1984) although, as Duncan shows, it has lengthy intellectual roots. In its original conception, it was meant to capture the outcome of the complex dialectic of international and national restructuring and local responses: 'the social and economic roles of any given local area will be a complex result of the combination of that area's succession of roles within the series of wider, national and international, spatial divisions of labour' (Massey 1978, p. 116). But, more recently, those involved in the consideration of locality have tended to split into two camps. In one camp are those, like Duncan in this volume, who retain a fairly strong notion of economic and social determination and who see the spatial aspects of social life as following on from such determination. Whilst they see local specificity as important, they see it as having causal powers only in quite rare instances when a locality can be unambiguously identified as a *community*, as a distinctive local culture with a historical geography of quite special social relations. By contrast, Urry (1987) and others argue that localities must be seen as economically and socially determined, but because the combination of determinants that can be found at a local scale are so complex and variable, locality can have important effects: 'there are a wide variety of local effects which are in part dependent upon the interconnection between the national and the international and the local. To focus in on the local as in effect the "community" is to concentrate upon a traditional, premodern concept' (Urry 1987, p. 443). In other words, contingency can have causal power.

Integrative shifts in the study of civil society

In each of the four dimensions of civil society, it is therefore possible to see the same themes cropping up at much the same point in time. It is also important to note how the study of each of these four dimensions is becoming intertwined, moving towards the same intellectual targets. This integration is coming about in four main ways.

The first of these integrative shifts consists of a similar redefinition of the problematic of each dimension. In the case of feminism and feminist geography, there has tended to be a move away from considering womens' problems towards the study of men and masculinity in the belief that it is the dominant power group of heterosexual men who are the chief problems for women (Connell 1987, Hearn 1987). In the case of race and racism there has been a move away from studying racial groups as victims or problems, to the study of racism in all its forms, in the belief that it is dominant social groups of white males who are the main problem (Jackson 1987). In the case of culture and cultural studies, the move has been away from the study of working-class culture (with the attendant charges of academic exploitation and stigmatization) towards the study of dominant upper- and upper-middle class social groups, in the belief that it is they who are causing many working-class problems. Finally, in the case of locality studies, middle- and upper-class localities are now receiving at least as much attention as working-class localities, in the belief that the study of the unequal geography of social power requires research into the powerful. Certain studies have already managed to combine all these different moves. Davidoff & Hall's (1987) remarkable locality study is a case in point.

The second integrative shift in the study of the different dimensions of civil society consists of a shift in subject matter, within the general redefinition of problematics, towards a common concern with how diverse ideological forms and practices create, sustain, or resist material inequality. Four chief areas of investigation now predominate in the literature, each one closely linked to the others. The first of these is simply the study of symbolic production as a contested process, whether in the form of the printed word, various images, or actual practices like carnivals (Jackson 1988). Daniels presents one aspect of this process in his chapter, where he considers landscape as a concept 'in high tension', in which different meanings are continually being inscribed and contested. Another major area of investigation concerns the processes of subjectification, the means by which people are made into subjects in a process which begins at birth and ends only with death. How are people made into men and women, children and adults, workers and bosses, black or white men and women, English and Americans, and so on (Benhabib & Cornell 1987)? All manner of problems present themselves – the explanation of practical reason, how people fashion accounts, the nature of the self. These problems are all intensely geographical. People are socialized in localized contexts (although the institutions of socialization are now rarely local) and the exigencies of these contexts produce different people with different capacities to think, to cooperate, to dominate, and to resist (Thrift 1983, 1985, 1986, 1989). A third area of investigation with intimate links to the previous area concerns the study of the state. From Althusser's ideological state apparatuses onwards, the study of the rôle of the state and other disciplinary institutions in the process of producing 'a regulated formation of identities and subjectivities' (Corrigan & Sayer 1985, p. 141) has become a vital area of research in social science. In human geography the main stimulus to this work has probably been the work of Foucault (see Driver 1985, Philo 1984; and, later, Corrigan & Sayer (1985)), on the micrologics of power. As a result, geographers have become more and more aware of the need to study

> the way in which the state organises and legitimises itself not simply by direct force nor by promulgating integrative ideologies but by the massive and disseminated operation of myriad forms of regulation and machining that classify, designate, identify, gender, homogenise and normalise, so that those material differences and practices and behaviours that resist these qualities are relegated to the status of the unthinkable, the aberrant, the unimaginable (Musselwhite 1987, p. 5).

Space is continually involved in this great cultural revolution, whether as part of the structure of discipline or as the gap in which resistance can still blossom.

A final area of investigation is the subject of global culture. All around the world the rise of commodity aesthetics, especially advertising (Peet 1982), a global communications industry, mass tourism, and international labour migration has made for something that begins to look like a global culture with similar images, motifs, and styles recurring in dissimilar locations. But, paradoxically, this growth of a global culture may also have served to reinforce the power of local cultures (Lash & Urry 1987, Cooke 1988). As Clifford (1988) puts it:

> This century has seen a drastic expansion of mobility . . . . More and more people 'dwell' with the help of mass transit, automobiles, airplanes. In cities on six continents foreign populations have come to stay – mixing in but often in partial, specific fashions. The 'exotic' is uncannily close. Conversely, there seem to be no distant places left on the planet where the presence of 'modern' products, medicine and power cannot be felt. An older topography and experience of travel is exploded. One no longer leaves home confident of finding something radically new, another time or space. Difference is encountered in the adjoining neighbourhood, the familiar turns up at the end of the earth (Clifford 1988, pp. 13–14).

The interrogation of this postmodern global culture has only just begun, but already it points to some significant developments around the world, including the increasing production/design of consumption via various consumer spectacles (Debord 1987, Harvey 1987, Ley & Olds 1988, Zukin 1988b), the burgeoning landscapes of gentrification (Jager 1986, Wright 1985, Mills 1988) and the rise of heritage motifs and the heritage industry (Hewison 1987, Wright 1985, Samuel 1988, Thrift 1989). Sometimes it does seem that all the world is becoming a stage.

The third integrative shift in the study of the different dimensions of civil society consists of a common focus on the problems of methodology. The study of the development of civil society has depended upon the use of specific kinds of method, hermeneutical in character, that allow meanings to be illuminated and understood. Various kinds of method have been applied. Perhaps the most common of these methods has been ethnography in its various forms, whether drawn from social anthropology or the nexus between the Chicago School and symbolic interactionism (Western 1981, Jackson & Smith 1984). Other methods have included oral history (Rose 1988), group analytical techniques (Burgess & Limb 1988a), discourse analysis (Potter & Wetherell 1987), even iconography (Cosgrove & Daniels 1988). All of these methods have in common the requirements of rigour in application and in analysis. Thus, many commentators now remark upon the rise of a qualitative geography, paralleling quantitative geography (Eyles & Smith 1988).

The final integrative shift in the study of the different dimensions of civil society consists of a common problem of representation. What strategies can be used to represent the otherness of social groups, at least partly on their own terms? Clearly, new modes of writing and other forms of expression than writing are required. In their analysis of cultural representation, Clifford & Marcus (1986, p. 6) point to some of the difficulties inherent in representing this otherness:

> Cultures do not hold still for their portraits. Attempts to make them do so always involve simplification and exclusion, selection of a temporal focus, the constitution of a self–other relationship, and the imposition or negotiation of a power relationship.

Quite clearly, the way that theoretical and empirical material is turned into text and image by an 'author' critically influences the interpretations the 'reader' will put on it. But, until recently, Olsson (1982) trod a lonely path in his attention to the problem of representation. He was one of the very few geographers to display a self-conscious understanding of textuality and his writing still remains a source of inspiration (Philo 1984). However, it is true to say that geographers have also turned to other exemplars. Most particularly, they have been influenced by Geertz's (1973, 1988) conceptualization of culture as 'text', and his dual method of 'thick description' – setting down meanings as they are for their actors, and 'diagnosis' – stating what the knowledge thus attained demonstrates about the society in which it is found. Geographers have also turned to experiments in postmodern ethnography that postdate Geertz's original work (e.g. Clifford & Marcus 1986, Gregory 1989), to literary theory, especially in its deconstructionist guise (Eagleton 1983, Eagleton 1986), and to social history (e.g. Darnton 1984). As yet, it is too early to tell the degree to which innovations in these areas of knowledge will be adopted by geographers, or the degree to which they will fend off the crisis of representation which is now general to the social sciences. Currently, calls for better writing (Gregory 1989) do not seem to have met with a particularly enthusiastic response. Yet it is hard to deny that there is a need for greater attention to aesthetics. The problem is how to answer this call without becoming merely an aesthete.

In summary, geographical research into civil society within the orbit of the political-economy approach is in an upswing, as each of the following four chapters demonstrate. Perhaps the only surprise is that this upswing has taken so long to occur, for the theme of the production of difference which typifies the modern analysis of civil society fits geography like a glove.

References

Benhabib, S. & D. Cornell (eds) 1987. *Feminism as critique*. Cambridge: Polity Press.

Berger, J. 1972. *Ways of seeing*. Penguin: Harmondsworth.

Bourdieu, P. 1984. *Distinction: a social critique of the judgement of taste*. London: Routledge & Kegan Paul.

Burgess, J., M. Limb & C. M. Harrison 1988. Exploring environmental values through the medium of small groups. 1 Theory and practice. 2 Illustrations of a group at work. *Environment and Planning A*, **20**, 309–26, 457–76.

Castles, S. & G. Kozack 1985. *Immigrant workers and class structure in Western Europe*, 2nd edn. Oxford: Oxford University Press.

Clifford, J. 1988. *The predicament of culture: Twentieth century ethnography, literature and art*. Harvard, Mass.: Harvard University Press.

Clifford, J. & B. Marcus (eds) 1986. *Writing cultures: the poetics and politics of ethnography*. Berkeley: University of California Press.

Cockburn, C. 1986. *Machinery of dominance. Women, men and technical know-how*. London: Pluto Press.

Connell, R. W. 1987. *Gender and power*. Cambridge: Polity Press.

Cooke, P. 1988. Modernity, post modernity and the city. *Theory, Culture and Society* **5**, 475–92.

Corrigan, P. & D. Sayer 1985. *The great arch*. Oxford: Basil Blackwell.

Cosgrove, D. & S. Daniels (eds) 1988. *The iconography of landscape*. Cambridge: Cambridge University Press.

Crompton, R. & M. Mann (eds) 1986. *Gender and stratification*. Cambridge: Polity Press.

Dalla Costa, M. & S. James 1975. *The power of women and the subversion of the community*. Bristol: Falling Wall Press.

Darnton, R. 1984. *The Great Cat Massacre and other episodes in French cultural history*. London: Allen Lane.

Davidoff, L. & C. Hall 1987. *Family fortunes: men and women of the English middle class 1780–1850*. London: Hutchinson.

Debord, G. 1987. *Society of the spectacle*. Boston: Rebel Press. (Originally published 1973.)

Driver, F. 1985. Power, space and the body: a critical assessment of Foucault's *Discipline and Punish*. *Environment and Planning D, Society and Space* **3**, 425–46.

Dyer, G. 1986. *Ways of telling: the work of John Berger*. London: Pluto Press.

Eagleton, M. (ed.) 1986. *Feminist literary theory*. Oxford: Basil Blackwell.

Eagleton, T. 1983. *Literary theory: an introduction*. Oxford: Basil Blackwell.

Eyles, J. & D. Smith 1988. *Qualitative geography*. Cambridge: Polity Press.

Geertz, C. 1973. *The interpretation of cultures*. New York: Basic Books.

Geertz, C. 1988. *Works and lives. The anthropologist as author*. Cambridge: Polity Press.

Gilroy, P. 1987. *There ain't no black in the Union Jack*. London: Hutchinson.

Gottdiener, M. 1985. *The social production of space*. Austin: University of Texas Press.

Gregory, D. 1989. Areal differentiation and postmodern human geography. In *New horizons in human geography*, D. Gregory & R. Walford (eds). London: Macmillan.

Hall, S. 1980. *Culture, media, language*. London: Hutchinson.

Hartmann, H. 1979. The unhappy marriage of Marxism and feminism: towards a more progressive union. *Capital and Class* **8**, 1–33.

Harvey, D. 1987. Flexible accumulation through urbanisation: reflections on post-modernism in the American city. *Antipode* **19**, 260–86.
Hearn, J. 1987. *The gender of oppression*. Brighton: Wheatsheaf.
Hewison, R. 1987. *The heritage industry*. London: Methuen.
Himmelweit, S. 1983. *Development of family and work in capitalist society*. Milton Keynes: Open University Press.
Jackson, P. 1986. Urban ethnography. *Progress in Human Geography* **9**, 157–76.
Jackson, P. (ed.) 1987. *Race and racism*. London: Allen & Unwin.
Jackson, P. 1988. Street life. The politics of Carnival. *Environment and Planning D, Society and Space* **6**, 213–27.
Jackson, P. & S. J. Smith 1984. *Exploring social geography*. London: Allen & Unwin.
Jager, M. 1986. Class definition and the aesthetics of gentrification. Victoria in Melbourne. In *Gentrification of the city*, N. Smith & P. Williams (eds), 78–91. Hemel Hempstead: Allen & Unwin.
Lash, S. & J. Urry 1987. *The end of organised capitalism*. Cambridge: Polity Press.
Ley, D. & K. Olds 1988. Landscape as spectacle: world's fair and the culture of consumption. *Environment and Planning D, Society and Space* **6**, 191–212.
Massey, D. 1978. Regionalism. Some current issues. *Capital and Class*, no. 6, 106–25.
Massey, D. 1984. *Spatial divisions of labour. Social structures and the geography of production*. London: Macmillan.
Mills, C. A. 1988. Life on the upslope: the postmodern landscape of gentrification. *Environment and Planning D, Society and Space* **6**, 169–89.
Musselwhite, M. 1987. *Partings welded together. Politics and desire in the nineteenth century English novel*. London: Methuen.
Olsson, G. 1982. *Birds in egg/eggs in bird*. London: Pion.
Peet, R. J. 1982. International capital, international culture. In *The geography of multinationals*, M. J. Taylor & N. J. Thrift (eds), 275–302. London: Croom Helm.
Philo, C. 1984. Reflections on Gunnar Olsson's contribution to the discourse of contemporary human geography. *Environment and Planning D, Society and Space* **2**, 217–40.
Philo, C. 1987. Fit localities for an asylum: the historical geography of England's nineteenth century idiot asylums. *Journal of Historical Geography* **13**, 348–415.
Phizacklea, A. & R. Miles 1980. *Labour and racism*. London: Routledge & Kegan Paul.
Potter, J. & L. Wetherell 1987. *Discourse and social psychology*. London: Sage Publications.
Samuel, R. 1988. *Theatres of memory*. London: Verso.
Thrift, N. J. 1983. On the determination of social action in space and time. *Environment and Planning D, Society and Space* **1**, 23–57.
Thrift, N. J. 1985. Flies and germs: a geography of knowledge. In *Social relations and spatial structures*, D. Gregory & J. Urry (eds), 366–403. London: Macmillan.
Thrift, N. J. 1986. Little games and big stories. In *Politics, geography and social stratification*, K. Hoggart & E. Kofman (eds). Beckenham: Croom Helm.
Thrift, N. J. 1989. Images of social change. In *The changing social structure*, L. McDowell, P. Sarre & C. Hamnett (eds). London: Sage Publications.
Urry, J. 1981. Localities, regions and social class. *International Journal of Urban and Regional Research* **5**, 455–74.
Urry, J. 1987. Society, space and locality. *Environment and Planning D, Society and Space* **5**, 435–44.
Williams, R. 1979. *Politics and letters*. London: New Left Books.
Willis, P. 1977. *Learning to labour*. Farnborough: Saxon House.
Wright, P. 1985. *On living in an old country*. London: Verso.
Zukin, S. 1988a. *Loft living. Culture and capital in urban change*, 2nd edn. London: Hutchinson.
Zukin, S. 1988b. The postmodern debate over urban form. *Theory, Culture and Society* **5**, 431–46.

7 *The geography of gender*

Sophie Bowlby, Jane Lewis, Linda McDowell, & Jo Foord

Introduction: a brief history of feminist geography

In writing a chapter about feminist geography, the title of this collection, *New models in geography*, loses its appropriateness. For there are no new models in feminist geography nor, indeed, any established set of models at all as this area of our discipline is a very recent arrival on the geographical agenda. It was less than ten years ago that the issue of the invisibility of women, both as the subjects of geographical study and as practitioners of the discipline, began to be raised. In the intervening years we have seen a remarkable burst of energy by geographers interested in feminist theory, and in the documenting of women's inequality and oppression in all areas of social and economic life, in all parts of the globe. This period has been one marked by intellectual excitement as well as by the variety of methods and approaches adopted by all the teachers and researchers who have become participants in the geographical debate about women's position and about changes in gender divisions over space and time.

In order to explain this diversity we have chosen, in this chapter, to begin by outlining briefly some of the ways in which the issue of gender divisions has been tackled by geographers over the last decade or so. This is followed by two more detailed discussions of the type of work undertaken in two areas: economic and social geography. However, as we shall explain, the division into economic and social itself mirrors a particular patriarchal view of social divisions. Finally, we shall conclude with a section that outlines some of the current theoretical issues that preoccupy feminist geographers.

Like all areas of geography, feminist geography has developed in a social and political context that has had an influence over the type of issues selected for analysis and the methods adopted to investigate them. Feminist geography spans two areas of discourse – geography and feminism – and both areas in themselves encompass a significant diversity of approaches. From geography, different feminist geographers adopted, particularly in the early years of study, one of the several approaches to their subject matter that were current at the time. Thus we have seen feminist geography written within conventional geographic paradigms. There has been work influenced by humanist geography, Weberian or Marxist analyses, and by welfare geography. A range of analyses of gender inequalities was undertaken within these perspectives. In the more specifically feminist literature over the same period there has been an intense theoretical debate between radical and socialist feminists, each committed to a different view of the origins of women's subordination and the relationship between gender and class oppression. In more recent years, there has also been the development of writing about ideology, subjectivity, and interpersonal relations, and about the construction of gender identities in different spheres of life – at home, at school, in the community, and in the workplace. In many ways this writing has brought radical and socialist feminists together, has produced changes in feminist politics, and has begun to have an influence on feminist geography.

How should someone arriving new to the area begin to understand the key concerns of feminist geographers? One way to start is with an historical survey of the area – after all, it will only be a brief survey. We can characterize most of the feminist geography undertaken during the mid- to late 1970s as trying to create 'a geography of women'. The key concern of feminist geographers at this time was to document the extent to which women were systematically disadvantaged in many areas of life by the sets of assumptions made about 'women's place' and by the resulting material constraints on their activities. At the time work in this area was strongly influenced by managerial and welfare approaches in geography. Tivers's (1986) book on the constraints on the activities of women with young children in a London suburban area is an excellent example of this approach, and a good review of research documenting differences between men and women is to be found in the essay by Zelinsky *et al.* (1982). Other researchers were more strongly influenced by Marxism, both by the structuralist developments within geography and by socialist feminists' critique of the type of orthodox Marxism that ignored gender divisions or attributed them solely to capitalist social relations. The work by Mackenzie & Rose (1983), for example, on the structure of urban areas and the significance of the spatial divisions between work and home and between areas of production and reproduction, falls into this second main strand of work in feminist geography.

As well as documenting gender divisions and the inequalities in women's position in both the advanced and less developed economies (see Women and Geography Study Group (WGSG) 1984), feminist geographers became increasingly concerned to *explain* the patterns of inequalities. From their initial interest in describing, rather than explaining, *gender rôles* and the ways in which the acceptance of gender rôles disadvantaged women, attention began to be focused during the first half of the 1980s on *gender relations* and on trying to identify the reasons for male dominance over women, to document the consequences, and suggest alternatives. Here literature from feminism rather than from geography was the main source for feminist geographers. In a working paper published in 1985, Foord, Bowlby, and McDowell, for example, looked to a literature ranging from theories about the construction of gender identities, sexuality, and femininity, to work by feminist economists and sociologists about the uses of male power in the workplace, to begin to explain regional and local variations in women's and men's lives in contemporary Britain. In many ways, this work is only just beginning. We thus are moving towards a feminist geography rather than a geography of women. It is a geography that will not be comfortable for many as it challenges conventional assumptions about the theory and practice of our discipline. It is also a geography that requires areas of everyday life and of the academic literature that we do not usually regard as geographical to be included within the subject matter of our discipline. In the final section of this chapter we shall spell out some of the implications of a truly feminist geography in greater detail. First, however, we turn to a more detailed survey of the existing studies that fall into one or other of our two categories: the geography of women and feminist geography.

Analyses of gender in urban and social geography

A great deal of feminist geographical work has been concerned with analyses of women's lives and of gender relations outside the workplace, in the home, and in the community. This work has thus fallen within the ambit of urban and social geography. However, unlike most work in these fields, one of the principal themes running through these feminist studies has been the importance of the *interconnections* between workplace- and non-workplace-based activities and social relations.

Perhaps because of the general lack of interest in the interconnections between work and home in social and urban geography (Duncan 1979), the changes and tensions in women's domestic and community lives that inevitably were part of their postwar entry into the waged labourforce have until recently been researched almost exclusively by feminist geographers. As we shall argue later, industrial geographers were forced to notice women because they became an important category of workers, but there were no comparable, striking empirical changes in women's or men's domestic and political lives – in large measure because those women who did enter the labourmarket continued to carry out the bulk of domestic tasks. Thus during the late 1970s and early 1980s work on women and gender was treated as a specialized topic that could be ignored by those studying other areas of the subject. In particular, and somewhat ironically given the common identification of women with 'the home', the extensive radical literature on housing markets which was developed during the 1970s ignored issues of gender.

More recently, however, as male unemployment has increased and the nature of and need for paid work has been questioned popularly, there has been greater recognition of the interconnections between the worlds of home and paid work, and increased interest in household work strategies, and in the impacts of the restructuring of employment on political and social relations outside the workplace (Pahl 1984, Redclift & Mingione 1984, Cooke 1986). Feminist work, both within and outside geography, has played an important rôle in this shift of emphasis.

In this section, we trace the history of feminist work on gender rôles and gender relations in the home and community and show how, recently, the theoretical concerns of feminist geographers in the urban and social field have converged with those in the field of industrial and regional geography.

Making women's inequality visible

In the mid- to late 1970s feminist geographers were concerned with two issues. First, they argued that the study of women's lives and activities was of equal importance to the study of men's lives and that the distinctiveness and significance of women's experiences and activities should be recognized (Zelinsky *et al.* 1982, Monk & Hanson 1982). Second, they sought to demonstrate that women's access to opportunities was not equal to that of men. In challenging the gender-blindness of much geography and its implicit assumptions that men's lives were 'the norm', feminist geographers were reflecting both the political demands of the women's movement and feminist work in other academic disciplines (Rowbotham 1973). In their work on women's access, however, the influence of existing geographical theoretical approaches on the questions addressed was as strong as the influence on the debates within feminism.

During the late 1970s, in both British and American feminist geography, the dominant approach to the study of gender issues in geography was via the concepts of constraint and social rôle. In this, feminist geographers were reflecting the concerns of welfare geographers and neo-Weberian analysts (Smith 1973, 1977, Pahl 1969, 1975) which were so influential in geography at this period.

An important example is the paper by Tivers (1977). She suggested that women's problems of access could be seen as resulting from their 'gender role constraint' – in other words, the social expectation that women's main activities should be those of family care. She further suggested that this gender rôle constraint should be seen as structuring the impact of other time–space constraints on women's opportunities. Like Palm & Pred (1976) in the United States, she drew on the broad framework of Hägerstrand's time–space approach to conceptualize choice and constraint. This type of approach was adopted in Britain by Pickup who studied the ways in which the gender rôle constraint limits women's access to paid work (Pickup 1984). He used Hägerstrand's approach to model women's access under different childcare and mobility constraints as well as drawing on intensive interview material to explore women's attitudes and feelings about their domestic obligations. Other work drew attention to the inequalities in men's and women's domestic work and its implications for their daily lives through a comparison of women's and men's travel behaviour (Hanson & Hanson 1980, 1981).

These studies demonstrated the significance of both material and ideological constraints, deriving from women's domestic rôle, on women's access to a variety of activities, including paid work. In particular, they demonstrated the crucial significance of childcare constraints in bounding the extent of women's activities both spatially and temporally. They also drew attention to the links between the assignment of domestic, particularly childcare tasks to women and their poor position in the job market.

There were other studies of women's access which fell within the welfare geography approach of considering groups' differential access to opportunities. These studied women as a group with either particular problems of mobility (resulting from low access to cars) (e.g. Hillman *et al.* 1974, Bowlby 1979), or with particular demands for access that other groups did not have – such as access to childcare or particular medical facilities (Brown *et al.* 1972). While interest in women as a group may have been stimulated by feminist demands that women's lives be made visible, much of this work was not carried out by feminists and was concerned neither with gender rôles nor with the interconnections between home and work. However, while limited in their theoretical concerns, these studies did provide further evidence of the material limitations and problems faced by many women.

There were two other important feminist geographical approaches that were emerging during the 1970s. One, principally found in North America, argued for a humanistic approach to exploring the significance of women's experiences of the physical and social landscape (Buchanan 1970, Burton 1976, Buttimer 1976). Those sympathetic to this approach also tended to suggest that women's ways of knowing and understanding are qualitatively different from those of men and are more sensitive and perceptive (Libbee & Libbee 1978, Larimore 1978). This and the liberal welfare geography perspective have tended to be favoured subsequently by feminist researchers within the United States, but it remains an underdeveloped approach in feminist geography.

The other important approach that was being developed used Marxist ideas and methodology to explore women's social and economic position within capitalist society. Perhaps the first paper in geography advocating the development of Marxist categories and theories to account for women's subordination was by Burnett and appeared in *Antipode* in 1973. Other papers by Hayford (1974) and Mackenzie & Seymour (1976) developed the theme.

Like socialist feminist work in other disciplines, the geographical research on women that has used a Marxist approach has been concerned with analyzing the relationship between the type and socio-economic significance of the work that women carry out within the home and with the organization of a capitalist economy (Malos 1980). The geographical research explored this theme through examining those historical changes in the relationship between home and work and their expression in the physical form of the city that took place with the development of capitalism. Work by Mackenzie (1980) and McDowell (1980, 1983) examined the development of separate residential, industrial, and commercial areas and the exclusion of women from formal paid work in the 19th-century British and American city. This work and that of others also explored the ways in which, in the 20th century, ideologies which saw women's place to be in the home informed the development of planning practices which reinforced the separation of home and work, reproduction and production, private and public, in the urban form (McDowell 1983, Mackenzie & Rose 1983, Women and Geography Study Group 1984, Bowlby 1984, Boys 1984).

The contradictory nature of the home and its separation from work was discussed by Rose (1981) in a comparative historical study of an industrializing city and a rural mining area. She showed how the home could be both a place within which production could occur in opposition to capitalist wage relations as well as a tie binding people into capitalist work relations.

A further development of this work on the social and spatial separation of production and reproduction was to examine its implications for women's involvement in political action (Mackenzie 1980, 1986). Bondi and Peake criticize the restricted definitions of political action common amongst male political scientists and political geographers who, they suggest, ignore the important activities of women in community politics.

This work has suggested the need for a reappraisal of the potential political importance of domestic labour and childcare activities carried out in the home – as against the concentration on collective consumption in the influential work of

Castells (1977) and Dunleavy (1980). It also draws attention to the political and social significance of the movement of activities between home-based consumption and reproduction, and the public sector and the private sector (McDowell 1983).

A significant aspect of these socialist feminist analyses was their development of a focus on gender *relations* rather than gender *rôles* and recognition of the significance of male power in these relations (Boys 1984). However, the nature of gender rôles and of male power was not explored in depth. It is also worth noting that in all the work reviewed so far, little attention was paid either to describing or to analyzing the variation in women's position in society either from country to country or region to region. It is only more recently that feminist geographers have begun to contribute to the mapping of women's lives (Seager & Olson 1986) or to research on gender relations and variations in patterns of development in the Third World (Momsen & Townsend 1987). Moreover, few feminist geographers at this time concerned themselves with issues such as variations from place to place in the processes by which people are socialized into feminine or masculine patterns of behaviour, although the issue of such socialization was – and remains – an important area of feminist research and policy discussion. It is perhaps somewhat ironic that the geographic tradition of areal comparison was ignored so that such research was done by sociologists or economists (see, for example on spatial variations in school achievement, Synge (1973, 1974)). As we shall see, interest in this tradition amongst feminist geographers has revived as a result both of new theoretical approaches to understanding women's inequality and of changes within geography.

Explaining women's inequality

As we have seen, much of the early work of feminist geographers was concerned to demonstrate the existence of women's inequality using the approaches of welfare geography and neo-Weberian analysis. It also sought to examine how that inequality was maintained through material and ideological pressures, but it paid little attention to explanations of the origins of gender inequality. The socialist feminist work moved away from the earlier use of the concept of gender rôles to use a terminology that recognized more explicitly the need to explain gender inequality and the existence of male power. As Foord & Gregson (1986) have argued, this shift was very important because a continued focus on gender rôles would have resulted in a restricted understanding of how changes in social structures and practices come about and how the social meanings of particular rôles are altered through processes of negotiation and conflict. However, although the words 'gender relations', 'male power' and 'patriarchy' were introduced into these analyses, there was little analysis or questioning of the basis of male power or of the nature of patriarchal relations or gender relations. In so far as these issues were considered, either implicitly or explicitly, the position adopted was either to analyze gender inequality as a product of capitalist economic relations or as resulting from the incorporation of earlier forms of gender inequality within capitalist social relations. If we use Walby's five-fold categorization of analyses of gender relations, the positions adopted were approaches 2 and 4 (see Table 7.1). The main focus of attention was therefore on the changing ways in which gender inequality was structured by capitalism.

Table 7.1 Theoretical approaches to the analysis of gender inequality.

	Approach
1	Gender inequality as theoretically insignificant or non-existent
2	Gender inequality as derivative from capitalist relations
3	Gender inequality as a result of an autonomous system of patriarchy which is the primary form of social inequality
4	Gender inequality as resulting from patriarchal relations so intertwined with capitalist relations that they form one system of capitalist patriarchy
5	Gender inequality as a consequence of the interaction of autonomous systems of patriarchy and capitalism

Source: Walby, 1986.

Thus the main questions addressed in most of the work reviewed so far have not been concerned with *explaining* women's inequality as a problem of gender, but either with establishing its empirical significance or with demonstrating that a full understanding of geographical issues – such as journey-to-work patterns or urban development – required analysis of gender rôles or gender relations. However, exploration of the latter argument has resulted in recognition of the need to develop a more adequate theoretical analysis of gender relations and male power. This new theoretical focus has encouraged and been encouraged by a convergence of the interests of feminist geographers working in both social and economic geography. As we will explain, this convergence of interest is itself linked both to the recent focus on locality and to recent developments in feminist empirical work and theory in other disciplines.

The recent upsurge of work in Britain on localities signals a renewed awareness of the importance of variation over space in the form and nature of social relations, and renewed recognition 'that a number of important social and economic changes cannot be investigated satisfactorily without analysing how these processes are embedded within different distinct localities' (Urry & Warde 1985, p. 1). A focus on locality has been urged as a demonstration that 'geography matters' or as 'a method for obtaining data on those processes of social change for which the locality is the appropriate level of analysis' (Newby 1986, p. 213).

As the quotation from Urry & Warde (1985) cited above indicates, however, interest in localities has been primarily an interest in how social relations within them respond to change, in particular, to the restructuring of capital that has been taking place rapidly – and for many people disastrously – in recent years.

Those involved in research on localities and restructuring have recognized the importance of gender relations in influencing the course of economic change and their importance in the politics of different areas (see, in particular, Murgatroyd *et al.* 1985). But despite this welcome extension of concern with gender relations, and the substantial theoretical and empirical contributions of much of this new work on restructuring, social class, and gender, its primary focus remains the economic. Local culture or relations in the home are seen as important only in so far as they impinge upon or are affected by economic change. In general, therefore, in existing locality research gender relations and changes in them within a particular locality have not been treated as important in their own right (Foord *et al.* 1986).

However, the focus on locality has provided an opportunity both to draw together existing feminist geographic work on gender relations outside the workplace and existing geographic work on women's position in the labour-market and thereby to attempt to develop an improved understanding of the nature of patriarchal gender relations. These developments have been encouraged by recent feminist writing outside geography on sexuality and patriarchy (e.g. Cockburn 1986, Coveney *et al.* 1984, Weeks 1985, 1986, Coward 1983, Walby 1986).

Foord & Gregson (1986) have recently initiated discussion in geography of these issues in a paper in which they attempt to define the necessary and contingent conditions for the existence of patriarchal gender relations. They locate these relations in the beliefs and practices which produce male domination of the social forms of biological reproduction and sexuality. They argue that patriarchy is conceptually distinct from capitalism although in social practices the two are linked (i.e. they adopt Approach 5 in Table 7.1). They also argue that, in the future, analysis of the spatially and historically specific forms of patriarchal relations should be a major focus for feminist geographical work. McDowell (1986) commenting on Foord and Gregson's paper, while agreeing with their stress on the importance of patriarchal relations as a focus for future research, disagrees with their analysis of patriarchy. She adopts Approach 4 in Table 7.1, arguing that patriarchal and capitalist relations should not be seen as distinct. She reasons that the particular form of women's oppression under capitalism lies in the ways in which capital resolves the essential contradiction that arises from the daily and generational reproduction of the labourforce. Male control of women's sexuality and particular forms of familial and kinship relations are a common resolution in capitalist societies of the need both to appropriate women's surplus labour in the short run and to provide for their maintenance during childbirth and childrearing in order to ensure the long-term reproduction of the working class. Despite their differences, however, McDowell, Foord and Gregson all argue for a focus on spatially and historically specific forms of sexuality and familial gender relations.

Clearly, the debate on these issues is only just beginning in geography, but it suggests many exciting new theoretical and empirical directions. However, before we can discuss these we need to examine the work on gender issues that has been carried out in industrial and regional geography.

Analysis of gender in industrial and regional geography

We have noted already that within most geographical research, analysis of gender has been incorporated into existing theoretical frameworks within geography. This is nowhere more true than of recent analyses of gender in industrial geography and regional studies. Here, analysis of gender has been incorporated within Marxist and neo-Marxist frameworks where the primary focus is on capital and class relations and on industrial/economic restructuring. While the significance of gender has increasingly been recognized within recent Marxist and neo-Marxist informed analysis of industrial and spatial restructuring, the gender division of labour has been discussed only in so far as it affects class relations and class recomposition within particular localities (for example, see Cooke 1981a, 1981b, Massey 1983). This section looks briefly at the treatment of gender in recent radical industrial geography and concludes by examining recent feminist analyses of industrial and labour process restructuring.

Gender in recent Marxist and neo-Marxist approaches in industrial and regional geography

Reference to gender in Marxist and neo-Marxist analysis of uneven spatial development and industrial location change has become increasingly commonplace over recent years from the perspectives of both feminist and non-feminist writers. To a very large extent, of course, this increasing reference to gender reflects empirical trends, particularly the changing structure and gender composition of the workforce. The growth of women's employment (of low-skilled, poorly paid, part-time jobs) and the growth of men's unemployment taken with the regional variations in these trends and with the influence pools of women's labour in assisted and suburban areas had on the locational strategies of many firms and industries during the 1960s and early 1970s, together made it increasingly difficult to ignore the significance of women and of the gender division of labour in contemporary processes of urban and regional restructuring. In particular, the feminization and associated spatial decentralization of manufacturing mass-production work to the Development Areas has received much attention in recent research concerned with uneven spatial development in the UK (Firn 1975, Hudson 1980, Cooke 1981b, Massey 1983, Lewis 1984).

To digress slightly, it should be pointed out that by the time the processes of feminization and the spatial decentralization of manufacturing production work in search of pools of female labour were being discussed in the literature, such developments, if they had not been reversed, had at least been substantially overturned. Evidence suggests, in fact, that there has been a considerable *de*-feminization of manufacturing production work in the UK which goes back to the mid-1970s at least, much of it associated with recent waves of manufacturing disinvestment involving the closure of plants opened in the assisted areas during the earlier postwar period, or with the automation of previously labour-intensive production processes (Lewis 1984, 1986, Sayer 1985). During the 1970s, for example, more than one in four of women's manufacturing jobs was lost in the UK and although the bulk of such job loss was concentrated in the traditional centres of the textile and clothing industries, substantial losses of women's jobs were also taking place in electrical engineering and in food and drink processing – in other words, in those very industries typically characterized as having become increasingly feminized during the period (Table 7.2). In fact, in the electronics industries, the 1970s saw the loss of almost 25 per cent of women's jobs as compared with a loss of 3 per cent of men's (Table 7.3). Moreover, between 1978 and 1981, between 20 per cent and 25 per cent of women's manufacturing jobs were lost in the North, Wales, and Scotland as well as in the North West, Yorkshire and Humberside and the West Midlands (Lewis 1986). This is closely associated with Sayer's (1985) comments that much radical research concerned with industry and space had tended to rework stereotypes which tend to 'freeze, and then present as universal, relationships

Table 7.2 Women's employment change, 1971–81, Great Britain.

	Total employment 1971–81 No.	%	Full-time 1971–81 No.	%	Part-time 1971–81 No.	%
Manufacturing	−632,018	−27%	−536,442	−29%	−536,442	−20%
Services	+1,341,176	+25%	+362,251	+25%	+1,038,925	+48%
Totals	+764,668	+9%	−209,312	−4%	+973,980	+35%

Source: DOE *Employment Gazette*.

Table 7.3 Women's regional employment change in electronics industries. (MLH 363–7) 1971–81.

	No.	%
South East	+504	+1%
Wales	−380	−4%
Scotland	−628	−5%
West Midlands	−1,126	−8%
East Midlands	−1,541	−22%
Yorkshire & Humberside	−1,665	−46%
East Anglia	−3,883	−54%
South West	−3,994	−38%
North	−5,520	−41%
North West	−5,543	−35%
London	−13,592	−37%
Great Britain	−37,368	−22%

Source: DOE *Employment Gazette*.

which are contingent and historically specific' (Sayer 1985, p. 17). This has become particularly true with respect to analyses of the gender division of labour and of regional variation in women's employment where, in terms of empirical trends at least, analysis appears to have become somewhat stuck in a time-warp more concerned with the changes which took place in the 1960s than with the changes currently taking place in the rather different economic conditions of the 1980s.

Much of the initial work in industrial and regional geography on women's employment and gender divisions in the UK emerged from research concerned primarily with the problem of the rapidly increasing loss of men's jobs from the coalmining, shipbuilding, and mechanical engineering industries in the Development Areas and with the fact that most new manufacturing jobs created were low-skilled and low-paid women's jobs in light engineering (Firn 1975, Massey 1979, Hudson 1980, Cooke 1981b). The processes of production and location change underlying these shifts in the industrial and employment structures of the assisted areas are now familiar enough. Rarely, however, does this work challenge the sexist bias of regional policy. Instead, it actively reinforces a view of gender rôle in which women are primarily housewives and mothers and men are breadwinners. Indeed, women with little tradition of paid employment may well not have had the political force that traditionally well-organized redundant coalminers and shipyard workers could command. Critical analysis of regional policy continued to marginalize women's unemployment and women's right to earn a wage with its frequent gripes that new jobs were 'only for women' (Hudson 1980, p. 361) and their equally frequent references to these women in relation only to their position as 'the wives and daughters of ex-coalminers'. What was uncritically accepted was the bias of regional policy towards areas of high male unemployment in the first place. As McDowell & Massey (1984) remind us, despite losses of women's jobs in the textile areas of the North West comparable to the losses of men's jobs in coalmining and shipbuilding in the North East during the 1950s and 1960s (approximately 150 000 in each case), the textile towns of the North West were never given assisted area status.

Along with major empirical changes which have made it increasingly difficult to ignore women if not gender, changes in theoretical approach have also led to this increasing recognition of the significance of gender in recent neo-Marxist research in industrial geography. Over the past ten years, conventional analysis has been heavily criticized from within geography and elsewhere for both prioritizing the rôle of capital as the major determinant of processes of economic restructuring (and thereby ignoring the active rôle of labour), and for ignoring the significance of divisions within labour in determining change. In particular, of course, the rôle of spatial divisions of labour has been emphasized as a major factor underlying changes in the organization and location of production, but the rôle of the gender division of labour has also increasingly been recognized (Massey 1984, Lewis 1984). The gender division of labour is now seen as a major factor in determining postwar changes in the geography of industry and employment in the UK. In addition, the recent focus on localities research is based on a recognition that the uniqueness of particular areas is not merely the product of general processes of economic change but that local differences are themselves a major determinant of the ways in which these general processes actually function (Massey 1984, Murgatroyd *et al.* 1985). In other words, this approach recognizes that local social relations are both outcomes and determinants of processes of economic and spatial restructuring. Recognition of the fact and importance of local variations in gender relations has itself played a major rôle in this theoretical shift (Massey 1984, McDowell & Massey 1984, Murgatroyd *et al.* 1985).

While reference to gender is certainly more commonplace in Marxist and neo-Marxist informed research in industrial geography, unease has recently been voiced about the specific way in which gender is treated (Bowlby *et al.* 1986). One point made is that within much of the literature discussion of gender is simply discussion of women, rather than being – as it should be – discussion of the social divisions between both women and men. Also, much of the existing work which refers to gender merely describes changes in the gender composition of employment and unemployment within particular local and regional labourmarkets. These points are, however, symptomatic of a more general problem of Marxist analysis – its primary concern with class relations and restructuring. Even within those studies in which analysis of gender has been central, the primary concern has been with the ways in which changes in the gender composition of employment and unemployment affect processes of class restructuring and class recomposition in particular localities (Cooke 1981a, 1981b, Massey 1983, 1984). It is not the gender division of labour which is being explained but the way in which changes in the gender composition of the paid workforce and of the unemployed affect local class relations. Much of this work, for example, has focused on the ways in which the increasing feminization of the workforce in the traditional industrial regions of the North East and south Wales has fragmented the former cohesion and homogeneity of the working class – indeed may even be leading to a process of 'declassement' in these areas (Cooke 1981b, Massey 1983).

Thus, in recent Marxist and neo-Marxist informed research on industrial and spatial restructuring, analysis of gender falls into the following three interrelated categories. First, and most obviously, gender divisions are seen as empirically important in that changes in the gender composition of employment and unemployment continue to be a major aspect of recent changes in particular local and regional labourmarkets and are a major outcome of the processes of economic restructuring being studied. Second, it is now increasingly recognized that gender divisions have played a major rôle in influencing the types of change which have taken place in the geography of industry and employment over the postwar period – that gender divisions in the labourmarket are a major determinant of production and location change. It is widely recognized that capital has made use of gender divisions in restructuring for profitable production. Third, the changing gender composition of employment and unemployment has been examined in terms of its impact on local class relations and recomposition. What this research has not sought to explain is how the gender division of labour is itself produced and reproduced in particular places. This, of course, reflects more general criticisms, particularly feminist criticisms, of Marxist theory, whereby the primary focus on class relations has left other forms of social relations – such as gender and race – largely untheorized. To use again Walby's categorization of analysis of gender relations, the treatment of gender in much existing Marxist and neo-Marxist analysis of economic and spatial restructuring falls into the first group (Table 7.1).

With respect to the analysis of gender in this literature there is little analysis of causation – of why women obtained the new jobs in the assisted areas, of why pools of women's and not men's labour were available, of why few jobs were previously available for women in these areas, and why men dominated the coalmining and shipbuilding industries, and of why jobs became labelled as 'men's jobs' and 'women's jobs' and characteristics such as dexterity and patience became associated with women and those of strength with men. It is generally agreed that capital has made use of a gender division of labour to restructure for more profitable production but, as we have said, little attempt has been made to explain how and why such an unequal gender division of labour exists. Implicitly, the assumption is made that capital has utilized a gender division of labour which is constructed elsewhere, outside of the realm of production, in the family and the home. As Walby suggests, it is assumed that 'women's disadvantaged labour market position is sited in the family and outside of the field of vision of regionalist analysis' (Walby 1986, p. 86). As such, gender relations are left unexplained and untheorized.

This view that women's disadvantaged position in the labourmarket is but a reflection of an unequal gender division of labour created elsewhere in the family and the home is not peculiar to regional analysis. In much feminist theory throughout the 1970s the assumption was made that women's position in the labourmarket reflected the unequal gender division of unpaid work in the family. However, this assumption has now been overturned as feminist research has begun to show that patriarchal gender relations are constructed and reconstructed in every sphere, in paid work – in the labourmarket and through the organization of the labour process – as well as in the home and community.

This research has begun to identify the variety of ways in which patriarchal gender relations are produced and reproduced within the wage workplace (Birnbaum n.d., Phillips & Taylor 1980, Rubery 1980, Coyle 1982, Cockburn 1983, 1985, Beechey 1983, Lewis 1984, Game & Pringle 1983, Walby 1986). In particular, this research has identified the rôle of male-dominated trade unions in controlling women's access to particular occupations and skills and to paid work in general, drawing on a wealth of both historical (Birnbaum n.d., Walby 1986) and contemporary (Cockburn 1983, 1985) examples. To date, however, much of this research has been aspatial and has not examined local variations in the construction of gender relations within the wage workplace.

Industrial and regional geography therefore needs to address those processes through which gender is constructed in the sphere of production as well as in the family and community if it is to understand how gender divisions of labour are produced and reproduced in particular places.

Future directions

Feminist geography is now entering a new phase. It is one marked by two features. First, approaches to the subject matter are becoming far more self-consciously theoretical and there are now attempts to reconceptualize the key terms and ideas of the developing feminist geographical literature. Consequently, discussions at conferences and in circulated working drafts, if not yet in the more mainstream literature, deal with the definition and redefinition of terms such as patriarchy, sexuality, familial ideologies, male power, and women's oppression.

The second development, which is common to feminist scholarship and politics more generally, is the increased recognition by feminist geographers of the diversity and variety in women's experience of oppression. The crude dichotomy of polar opposites – male and female, nature and culture, hetero- and homo-sexuality, public and private, production and reproduction – are now being challenged in feminist theory and practice. New areas of theory and research are beginning to have an impact upon the discourse of feminist geography. Three areas, in particular, are significant: analyses of cultural representation; work on sexuality, subjectivity, and social relations; and new developments in studies of the interrelationship of race, class, and gender.

Images of nature and cultural representations

A long-standing geographical and contemporary feminist interest in images of nature is one area in which feminist geographers are beginning to develop work that raises questions about the social construction of nature as female or feminine. Despite a considerable scholarly literature in other areas, there is in geography as yet relatively little literature that specifically addresses questions about changing concepts of nature and the oppression of women, or even questions the links between views of nature and more traditional areas of geographic enquiry such as exploration and the exploitation of resources or the form of the built environment. Some indications of the potentiality of this area, however, do exist. In an impressive survey of British poetry and painting, Gold (1984) documented shifts in the image of nature in Western society. Despite radical differences – nature as the all-provider, Mother Earth, to nature as red in tooth and claw – an enduring feature has been the association of the 'natural' with women. In a perceptive but tantalizingly brief passage in his recent book, Smith (1984) commented thus on the significance of the natural/female parallel:

> It is striking that the treatment of women in capitalist society parallels the treatment of nature. As external nature women are objects which mankind attempts to dominate and oppress, ravage and romanticize; they are objects of conquest and penetration as well as idolatry and worship. The language is exact. Women are put on pedestals, but only once their social domination is secure; precisely as with nature, romanticization is then a form of control. But women can never be wholly external since in them resides fertility and the means of biological reproduction. In this sense they are made elements of universal nature, mothers and nurturers, possessors of a mysterious female intuition and so on (Smith 1984, pp. 13–14).

The distinctions and idealized differences between men and women and between nature and culture were made concrete during industrial capitalism's separation of the public world of waged labour and the private world of the home. As we have argued here, feminist geographers have been active in drawing out the significance of these divisions for geographical investigation. A great deal of work remains to be done in exploring and analyzing the complex interconnections between changing material trends and diverse cultural representations in the contemporary world.

Analysis of language and literature is one element that may well repay further study. Kolodny's (1975) work provides an example. She has demonstrated how in US novels and poetry, images of nature to be abused in the exploration and conquest of the land in the push westwards come to dominate the pastoral vocabulary of an earlier period. 'America the Beautiful' was turned into 'America the Raped' and this, she argues, has had important consequences for social interactions between North American men and women:

> Our continuing fascination with the lone male in the wilderness and our literary heritage of essentially adolescent, presexual pastoral heroes, suggests that we have yet to come up with a satisfying model for mature masculinity in this continent; while the images of abuse that have come to dominate the pastoral vocabulary suggest that we have been no more successful in our response to the feminine qualities of nature than we have to the human feminine (Kolodny 1975, p. 147).

Links are also being made between geographers, feminists, and ecologists, both in theory and in practice. The growing political interest in ecology, conservation, anti-nuclear movements, and anti-militarism contains within it a strong strand of radical feminism. 'Ecofeminism' (Caldecott & Leland 1983) has begun to stress the associations between the oppression of women and the exploitation of natural resources, based on the somewhat uncritical acceptance of the nature/culture link and women's purported closeness to nature through their nurturing rôle in reproduction. In contemporary Britain a political expression of this belief is the continuation of the women's encampment at Greenham Common.

Sexuality, subjectivity, and social relations

A second area towards which feminist geographers are drawn is that conjunction of diverse interests in questions of the construction, legitimation, and maintenance of ideologies and beliefs. There is a growing literature on the significance of subjectivity in social relations between women and men that is proving informative in investigations into such areas as the definition and regulation of sexuality, the representation and organization of families and other household structures, and the construction and maintenance of power over 'inferiors' in a variety of situations (see, for example, Beechey & Donald 1985). These are all areas in which spatial variations play a significant rôle and which are becoming part of the refocusing of interest in social geography, as widely defined, on locality-based research. An eclectic range of literature from sociology, anthropology, history, and the approach broadly associated with Foucault is proving a potent source of ideas for feminist geographers (among others). Knopp & Lauria (1985), for example, have drawn on Foucault's (1979) work on sexuality in their contribution to the debate initiated by Foord & Gregson (1986) on the essential characteristics of patriarchal relations. Elsewhere, they have begun to document in empirical detail how sexual predeliction is an important contributory variable to the explanation of inner-city revival in certain US cities (Knopp & Lauria 1985). On the other hand, in a recently published study, Fitzgerald (1987) has written a fascinating account of the initial growth and recent changes in an inner-city gay community in San Francisco. Rises in property values, changes in the ethnic and class composition, and retail changes in the Castro area can only be explained within an understanding of what was happening in the gay community in the city throughout the 1970s. These studies of gay men re-emphasize the importance of the shift in emphasis from women and geography, to gender rôles and finally to gender relations that have been outlined in this chapter. The work, principally carried out by feminists, on gender relations has allowed studies such as those of Knopp and Lauria also to become a part of the subject matter of geography in a way that was not possible a decade ago.

Race, class, and gender

A third area of work that will be of significance in the future – growing out of not only a theoretical awareness of difference and diversity but also often bitter political divisions between feminists – are debates about the interconnections between class, race, and sexual identity. Black women have initiated an important critique of racism within feminist theory and practice (Amos &

Parmor 1981, Bryan *et al.* 1985, Hooks 1982). They have pointed out that stereotypical images of, for example, Afro-Caribbean men as violent or Asian women as passive permeate the women's movement as well as society at large. Black feminists have argued that ethnocentrism permeates the politics of the women's movement and insist that many of its demands – on issues around the family, sexuality, and wage work – are based on the experiences and desires not only of white women but of middle-class white women. The conflicts and contradictions that arise from cross-cutting divisions in society – on class, gender, and racial lines – are now a key area for feminist theory and research. In a recent book, Phillips (1987) has begun to document the implications of these divided loyalties for women's position in contemporary Britain. As yet, this feminist work has paid little attention to the existence of differences between areas or the significance of locality. Although the detailed workplace studies undertaken by feminist economists and sociologists are beginning to show how women are oppressed in the labourmarket in different ways in different areas of the country (Cockburn 1986, Coyle 1982, Game & Pringle 1983, Westwood 1985), there is, as yet, insufficient work which attempts to examine and compare the interconnections of race, class, and gender within different areas. For feminist geographers educated in a tradition that emphasizes the importance of difference and diversity, this is a period of intense intellectual excitement when theoretical and empirical preoccupations appear to be focusing on an agenda that places women's particular oppression at particular times in different places at the forefront of feminist analysis and, increasingly, of geographical analysis. In the areas of economic and social geography there seems to be a rising awareness that gender relations can no longer be discounted or, if they are, that explanation will be impoverished as a result.

The very fact of the inclusion of two chapters in this book dealing broadly with the area of gender relations, as well as of chapters on cultural and social theory, indicates the deep changes that are occurring in contemporary geography. Changes in the practice of geography, however, are slower. To take but one illustration, women's recruitment to all but the most lowly staff positions in British and North American academic departments remains unchanged, despite half of the undergraduates being women. It is now time for practical recognition of gender relations and women's subordination to accompany the theoretical and empirical advances documented here.

References

Amos, V. & P. Parmar 1981. *Resistances and responses: the experience of black girls in Britain*. In *Feminism for girls*, A. McRobbie & T. McCabe (eds). London: Routledge & Kegan Paul.

Asheim, B. T. 1979. Social geography – welfare state ideology or critical social science? *Geoforum* **10**.

Beechey, V. 1983. What's so special about women's employment? Of some recent studies of women's paid work. *Feminist Review* **15**, 23–47.

Beechey, V. & J. Donald 1985. *Subjectivity and social relations*. Milton Keynes: Open University Press.

Birnbaum, B. (n.d.) Women's skills and automation: a study of Women's employment in the clothing industry, 1946–1972. Unpublished paper.

Bowlby, S. R. 1979. Accessibility, shopping provision and mobility. In *Resources and planning*, A. Kirby & B. Goodall (eds). Oxford: Pergamon.
Bowlby, S. 1984. Planning for women to shop. *Environment and Planning D, Society and Space* **2**, 179–99.
Bowlby, S., J. Foord & L. McDowell 1986. The place of gender in locality studies. *Area* **18**, 327–31.
Boys, J. 1984. Making out: the place of women outside the home. In *Making space: women in the man-made environment*, Matrix Book Group (eds). London: Pluto Press.
Brown, L. A., F. B. Williams, C. E. Youngmann, J. Holmes & K. Walby 1972. Day-care centres in Columbus: a locational strategy. Discussion Paper, 26. Columbus, Ohio: Ohio State University, Department of Geography.
Bryan, B., S. Dadzie & S. Scafe 1985. *The heart of the race: black women's lives in Britain*. London: Virago.
Buchanan, K. 1970. *The map of love*. Sydney: Pergamon.
Burnett, P. 1973. Social change, the status of women and models of city form and development. *Antipode* **5**, 57–62.
Burton, L. 1976. The country and the city: the effects of women's changing roles and attitudes on their view of the environment in contemporary fiction. In *Women in society: a new perspective*, P. Burnett (ed.). Mimeograph.
Buttimer, A. 1976. Beyond sexist rhetoric: horizons for human becoming. In *Women in society: a new perspective*, P. Burnett (ed.). Mimeograph.
Caldecott, L. & S. Leland 1983. *Reclaim the Earth*. London: Women's Press.
Castells, M. 1977. *The urban question*. London: Edward Arnold.
Cockburn, C. 1983. *Brothers: male dominance and technical change*. London: Pluto Press.
Cockburn, C. 1986. *Machinery of dominance: women, men and technical know-how*. London: Pluto Press.
Cooke, P. 1981a. Local class structure in Wales. Department of Town Planning. Papers in Planning Research No. 31. Cardiff: University of Wales Institute of Science and Technology.
Cooke, P. 1981b. Inter-regional class relations and the redevelopment process. Department of Town Planning. Papers in Planning Research No. 36. Cardiff: University of Wales Institute of Science and Technology.
Cooke, P. 1986. The changing urban and regional system in the UK. *Regional Studies* **20**, 243–562.
Coveney, L., M. Jackson, S. Jeffreys & P. Mahony 1984. *The sexuality papers: male sexuality and the social control of women*. London: Hutchinson.
Coward, R. 1983. *Patriarchal precedents: women's sexuality today*. London: Routledge & Kegan Paul.
Coyle, A. 1982. Sex and skill in the organisation of the clothing industry. In *Work, women and the labour market*, J. West (ed.). London: Routledge & Kegan Paul.
Duncan, S. S. 1979. Qualitative change in human geography. *Geoforum* **10**, 1–4.
Dunleavy, P. 1980. *Urban political analysis*. London: Macmillan.
Firn, J. R. 1975. External control and regional development: the case of west-central Scotland. *Environment and Planning A* **7**, 393–414.
Fitzgerald, F. 1987. *Cities on the hill*. London: Picador.
Foord, J. & N. Gregson 1986. Patriarchy: towards a reconceptualisation. *Antipode* **18**, 181–211.
Foord, J., L. McDowell & S. Bowlby 1986. For love not money: gender relations in local areas. Discussion Paper 27. Centre for Urban and Regional Studies, University of Newcastle-upon-Tyne.
Foucault, M. 1979. *The history of sexuality*, vol. 1. London: Allen Lane.
Game, A. & R. Pringle 1983. *Gender at work*. London: Pluto Press.
Gold, M. 1984. A history of nature. In *Geography Matters*, D. Massey & J. Allen (eds). Cambridge: Cambridge University Press.

Hanson, S. & P. Hanson 1980. Gender and urban activity patterns in Uppsala, Sweden. *Geographical Review* **70**, 291ff.

Hanson, S. & P. Hanson 1981. The impact of married women's employment on household travel patterns: a Swedish example. *Transportation* **10**, 165–83.

Hayford, A. M. 1974. The geography of women: an historical introduction. *Antipode* **6**, 1–19.

Hillman, M., I. Henderson & A. Whalley 1974. *Mobility and accessibility in the outer metropolitan area.* Political and Economic Planning Report. London: Policy Studies Institute.

Hooks, B. 1982. *Ain't I a woman: black women and feminism.* London: Pluto Press.

Hudson, R. 1980. Regional development policies and female employment. *Area* **12**, 21–33.

Knopp, L. & M. Lauria 1985. Towards an analysis of the role of gay communities in the urban renaissance. *Urban Geography* **6**, 387–410.

Kolodny, A. 1975. *The lay of the land.* Chapel Hill: University of North Carolina Press.

Larimore, A. E. 1978. Humanizing the writing in cultural geography text books. *Journal of Geography* **77**, 183–5.

Lewis, J. 1984. Post-war regional development in Britain: the role of women in the labour market. Unpublished PhD thesis, University of Reading, Department of Geography.

Lewis, J. 1986. Women's employment in the recession. Paper presented at Women and the Man-Made Environment Conference, Amsterdam, 22–5 September.

Libbee, K. S. & M. Libbee 1978. Geographic education and the women's movement. *Journal of Geography* **77**, 176–80.

McDowell, L. 1980. Capitalism, patriarchy and the sexual division of space. Paper presented at a conference on The institutionalisation of sex difference, University of Kent.

McDowell, L. 1983. Towards the understanding of the gender division of urban space. *Environment and Planning D, Society and Space* **1**, 59–72.

McDowell, L. 1986. Beyond patriarchy: a class-based explanation of women's subordination. *Antipode* **18**, 311–21.

McDowell, L. & D. Massey 1984. A woman's place? In *Geography Matters*, D. Massey & J. Allen (eds). Cambridge: Cambridge University Press.

Mackenzie, S. 1980. Women and the reproduction of labour in the industrial city. Working Paper 23. Department of Urban and Regional Studies, University of Sussex.

Mackenzie, S. 1986. Women's responses to economic restructuring: changing gender, changing space. In *The politics of diversity: feminism, Marxism, and Canadian society*, M. Burnett & R. Hamilton (eds). London: Verso.

Mackenzie, S. & D. Rose 1983. Industrial change, the domestic economy and home life. In *Redundant spaces and industrial decline in cities and regions.* J. Anderson, S. Duncan & R. Hudson (eds). London: Academic Press.

Mackenzie, S. & L. Seymour 1976. The role of the family under contemporary urbanism. Paper presented at the Annual Meeting of the Association of American Geographers.

Malos, E. 1980. *The politics of housework.* London: Allison & Busby.

Massey, D. 1979. In what sense a regional problem? *Regional Studies* **13**, 223–43.

Massey, D. 1983. Industrial restructuring as class restructuring: production decentralisation and local uniqueness. *Regional Studies* **17**, 73–91.

Massey, D. 1984. *Spatial divisions of labour: social structures and the geography of production.* London: Macmillan.

Momsen, J. H. & J. Townsend 1987. *Geography of gender in the Third World.* London: Hutchinson.

Monk, J. & S. Hanson 1982. On not excluding half of the human in human geography. *Professional Geographer* **34**, 11–23.

Murgatroyd, L. M. Savage, D. Shapiro, J. Urry, S. Walby, A. Warde & J. Mark Lawson 1985. *Localities, class and gender*. London: Pion.

Newby, H. 1986. Locality and rurality. The restructuring of rural social relations. *Regional Studies* **20**, 209–16.

Pahl, R. 1969. Urban and social theory and research. *Environment and Planning* **1**, 143–54.

Pahl, R. 1975. *Whose city?* Harmondsworth: Penguin.

Pahl, R. 1984. *Divisions of labour*. Oxford: Basil Blackwell.

Palm, R. & A. Pred 1976. A time-geographic approach on problems of inequality for women. Working paper No. 236. Institute of Urban and Regional Development, University of California, Berkeley.

Phillips, A. 1987. *Divided loyalties: dilemmas of sex and class*. London: Virago.

Phillips, A. & B. Taylor 1980. Sex and skill: notes towards a feminist economics. *Feminist Review* **6**, 79–87.

Pickup, L. 1984. Women's gender role and its influence on their travel behaviour. *Built Environment* **10**, 61–8.

Redclift, N. & E. Mingione 1984. *Beyond employment*. Oxford: Basil Blackwell.

Rose, D. 1981. Home ownership and industrial change. University of Sussex Urban and Regional Studies Working Paper 8.

Rowbotham, S. 1973. *Woman's consciousness, man's world*. Harmondsworth: Penguin.

Rubery, J. 1980. Structured labour markets, worker organisations and low pay. In *The economics of women and work*, A. H. Amsden (ed.). Harmondsworth: Penguin.

Sayer, A. 1985. Industry and space: a sympathetic critique of radical research. *Environment and Planning D, Society and Space* **3**, 3–29.

Seager, J. & A. Olson 1986. *Women in the world: an international atlas*. London: Pluto Press.

Smith, D. 1973. *The geography of well-being in the United States*. New York: McGraw-Hill.

Smith, D. 1977. *Human geography: a welfare approach*. London: Edward Arnold.

Smith, N. 1984. *Uneven development*. Oxford: Basil Blackwell.

Synge, J. 1973. Scottish regional and sex differences in school achievement and entry to further education. *Sociology* **7**, 107–116.

Synge, J. 1974. Post secondary education and rural–urban migration. *Scottish Educational Studies* **6**, 13–18.

Tivers, J. 1977. Constraints on urban activity patterns: women with young children. Occasional Paper 6. Department of Geography, King's College, University of London.

Tivers, J. 1986. *Women attached. The daily lives of women with young children*. Beckenham: Croom Helm.

Urry, J. & A. Warde 1985. Introduction. In *Localities, class and gender*, L. Murgatroyd, M. Savage, D. Shapiro, J. Urry *et al.*, 1–35. London: Pion.

Walby, S. 1986. *Patriarchy at work*. Cambridge: Polity Press.

Weeks, J. 1985. *Sexuality and its discontents*. London: Routledge & Kegan Paul.

Weeks, J. 1986. *Sexuality*. London: Tavistock Publications.

Women and Geography Study Group (WGSG) of the Institute of British Geographers 1984. *Geography and Gender*. London: Hutchinson/Explorations in Feminism Collective.

Zelinsky, W., J. Monk & S. Hanson 1982. Women and geography: a review and prospectus. *Progress in Human Geography* **6**, 317–66.

8 *Geography, race, and racism*

Peter Jackson

Until recently geographers have used the terms race and race relations uncritically, following the commonsense language of political discourse and academic social science. Two essays by Lawrence (1982a, 1982b) in *The empire strikes back* provide a critical reading of this literature and reveal that, despite their liberal intentions, social scientists have often simply reflected the racist categories and prejudices of the society in which their work is embedded.

Geographers have been no exception to this rather dismal picture, although they have tended to indulge their own particular preoccupation with spatial aspects of race relations such as immigration, residential segregation, and ethnic assimilation (e.g. Peach 1968, 1975, 1985, Robinson 1982, 1986). Geographical work has been concerned mainly with measuring patterns of minority-group concentration and dispersal generally, failing to ask questions about the *social* significance of *spatial* segregation (cf. Dennis 1980, Harris 1984). Recent attempts to recast this work in terms of broader theories of ethnic pluralism (Clarke *et al.* 1984) have not resolved the issue satisfactorily, failing to recognize that questions of ethnicity and ethnic identity involve social relations in which differences of power are fundamentally at stake. Instead, geographers have persisted with untenable theories of ethnic assimilation and cultural pluralism despite criticisms of these concepts on both theoretical and political grounds (e.g. Blaut 1983, Yinger 1981). As Doherty argued some 15 years ago, such studies concentrate on black people as a 'problem' rather than on the values and structure of the wider society, failing to identify the historical roots of contemporary racism or to examine the social and economic basis of racial prejudice and discrimination (Doherty 1973). Even among radical geographers, Blaut is the exception in having given sustained attention to the closely related issues of racism and nationalism and to their geographical manifestations in ghettoization and imperialism (e.g. Blaut 1970, 1974, 1983, 1987).

The development of a radical social geography of race and racism depends on the evolution of a more adequate theoretical foundation. Despite a long-established critique of the Chicago School by Castells (1977) and others, ecological analyses remain popular. It is still possible to read studies of neighbourhood change (e.g. Taub *et al.* 1984) couched in the language of 'invasion/succession' popularized long ago by Park *et al.* (1925). Geographers have compounded the error by an incomplete reading of Park's work that emphasizes the quantitative and ecological to the relative detriment of the qualitative and ethnographic (Bulmer 1984, Jackson & Smith 1981, 1984). In moving away from ecological forms of analysis, geographers drew extensively on Weberian theories of consumption classes as developed by Rex & Moore

(1967). A variety of managerialist approaches developed from this early work, reflected in contemporary studies of race and housing that show how institutional racism severely restricts black people's access to good quality housing (e.g. Phillips 1986). Few geographers followed Harvey's lead, exploring Marxist theories of the intersection of race and class, first advocated in his analysis of ghetto formation in *Social justice and the city* (1973) and subsequently elaborated in his *The urbanization of capital* (1985). Before we can discuss the specifically geographical contours of race and racism, however, we need to develop a greater sensitivity to the theoretical problem involved in defining these complex terms.

This chapter begins by defining race as a social relation between groups of people across deeply entrenched lines of inequality. It then retheorizes the significance of space in the development of particular social relations of race by exploring some specifically territorial forms of racial oppression. Suggesting that *racism* is the problem rather than race, the chapter concludes by examining the variety of ways in which racist oppression can be challenged, employing the concept of *resistance* first proposed by Gramsci and later developed by Hall and others (Hall & Jefferson 1976).

Race as social relation

Different races have been distinguished for generations, with social distinctions drawn on the basis of alleged physical attributes. The development of a critical approach to race demands that we challenge the inference that such distinctions are rooted in an unalterable human nature. Significant continuities can be shown to exist between the forms of racist imagery employed in different historical periods. For example, the Biblical representation of black people as the descendants of Ham, whose blackness symbolized God's punishment for over-indulgence, can be related to subsequent publications like *The geographical history of Africa*, published in Arabic in the early 16th century, which associate black people with courage, pride, guilelessness, cruelty, and easily aroused passions. Such continuities, however, must be accounted for historically. Their existence does not constitute evidence for theories of biological immutability.

The historical record is complete with contradictory images of black people, some sources depicting blacks as noble savages while others represent them as comic buffoons (Street 1975, Mair 1986). Such stereotypes often reveal more about the dominant society and its current anxieties than they do about the minority group itself. The genesis of particular stereotypes and their persistence can only be explained in terms of their material context. Shakespeare's *Othello* provides a classic example. It was first produced in London in 1604, a time of growing unemployment ('want of service') when royal proclamations lamented the existence of 'divers blackmoors brought into this realme of which kinde of people there are already here to manie', leading to calls for their repatriation ('to bring them home to their native country') (Cowhig 1985, Fryer 1984, Greater London Council 1986). Racist images are highly refractive but, like other ideological representations, an understanding of them depends on recognizing the material interests that they serve (Urry 1981).

Gobineau's turn-of-the-century classification of human beings into three races (black, white, and yellow), each supposedly endowed with particular social characteristics (such as a gift for leadership in the case of whites and an inherent sensuality in the case of blacks), may now strike us as simply ludicrous. Travellers' reports from the British Empire in the 18th and 19th centuries, such as the following account from the *Gentleman's Magazine* (1788), now also seem thoroughly disingenuous:

> The Negro is possessed of passions not only strong but ungovernable; a mind dauntless, warlike and unmerciful; a temper extremely irrascible; a disposition indolent, selfish and deceitful; fond of joyous sociality, riotous mirth and extravagant shew . . . a terrible husband, a harsh father and a precarious friend (Walvin 1982, p. 60).

The influence of social Darwinism was rampant in Britain and North America little more than a hundred years ago. Spencer's *The study of sociology* (1876), was still widely read in the United States in the 1890s (Hofstadter 1955, Peel 1972). The emerging discipline of geography could not resist the pernicious influence of scientific racism. Livingstone's (1984) portrait of one of the forefathers of American geography, Nathaniel S. Shaler, shows how easily 'scientific' analyses of blacks, Jews, American Indians, and European peasants shaded into areas of current political debate about eugenics, immigration restrictions, and religious intolerance. Semple's *Influences of geographic environment* (1911) betrayed a similar belief in the different 'mental energies' and 'temperaments' of various racial and ethnic groups with the northern peoples of Europe characterized as 'energetic, provident, serious, thoughtful rather than emotional, cautious rather than impulsive', while the 'southerners of the subtropical Mediterranean' were thought to be 'easy-going, improvident except under pressing necessity, gay, emotional, [and] imaginative', qualities which degenerated into 'grave racial faults' among 'the negroes of the equatorial belt' (quoted in Peet 1985, p. 322).

We should not be trapped into regarding racial categorization as an intellectual puzzle within a narrowly conceived history of ideas (cf. Banton 1977). Racist ideologies are of immediate political significance because of the *practices* they inform. Contemporary British politics provides a telling example. The birth of a new racism has been reported (Barker 1981) combining aspects of ethnology and sociobiology with virulent strains of nationalism and xenophobia (cf. van den Berghe 1978). Characteristic of the new racism is a renewed tendency to *naturalize* racial distinctions. Social differences are said to originate in biological or inherent cultural factors belonging to the realm of nature and not, therefore, to be changeable through political action or social reform. The inevitability of racial conflict and the natural antipathy and unassimilability of different races are no longer ideas confined to right-wing extremists like Enoch Powell. Mainstream politicians including the British Prime Minister, Margaret Thatcher, have voiced fears of the 'swamping' of British culture by immigrants, legitimizing those who have reacted with hostility and outright violence (Miles & Phizacklea 1984).

Hall (1978) reminds us that race should not be regarded as a permanent human or social deposit whose origins lie in natural, biological, or genetic differences. It should instead be recognized as embodying a social relation between individuals and groups characterized by fundamental inequalities in power. Racial distinctions are socially constructed under historically specific material circumstances. Racism takes different forms at different times, each 'historically specific and articulated in a different way with the societies in which they appear' (Hall 1978, p. 26). It is possible to identify many different forms of racism, historically and at the present time, distinguishing between individual, popular (or commonsense) and institutional forms of racism:

> Racism does not stay still; it changes shape, size, contours, purpose, function – with changes in the economy, the social structure, the system and, above all, the challenges, the resistances to that system (Sivanandan 1983, p. 2)

While it is vital to understand racism historically, it is important to reject the reduction of racist attitudes to a set of stereotypes passively inherited from the days of slavery and Empire. Britain's imperial past and the legacy of slavery in Britain and the United States provide a reservoir of racist imagery drawn on selectively according to present circumstances. But racism is best understood as an *active creation* aimed at making sense of everyday reality. Its various forms must be defined in relation to specific economic circumstances prevailing at the time (Miles & Phizacklea 1981). We may therefore define racism as an ideology ascribing negatively evaluated characteristics to a group of people identified as biologically distinct. Racism further refers to negative beliefs, held by one group about another, identifying and setting them apart by attributing significance to some biological or other 'inherent' characteristic (Miles 1982, p. 78).

The territorial basis of racial oppression

Social geographers are now moving away from an exclusive interest in patterns of immigration, segregation, and assimilation to focus on racism and its geographical ramifications (Jackson 1985, 1987). In so doing, the question arises as to whether there is, in fact, a specific geography of race and whether race relations should be constituted as a separate field for geographical study, a tendency that has been roundly criticized in sociology (e.g. Bourne & Sivanandan 1980, Lawrence 1982b, Miles 1984). It is clearly possible to identify certain distinctively territoral forms of racism, whether or not this is regarded as a separate subfield of race relations (cf. Leach 1973). It may also be possible to define a more general sense in which the social construction of race and racism are *spatially constituted*, to employ the language of contemporary social theory (Gregory & Urry 1985). The spatial structuring of contemporary racism will be explored through two examples: the geography of apartheid in South Africa and the geography of racist attacks in Britain.

The geography of apartheid

Geographers have undertaken timely and significant research on the political geography of South Africa examining the origins of racist oppression (e.g. Mabin 1986a, 1986b), the place of South Africa within the international division of labour (e.g. O'Keefe 1983), and the contemporary expression of segregation-

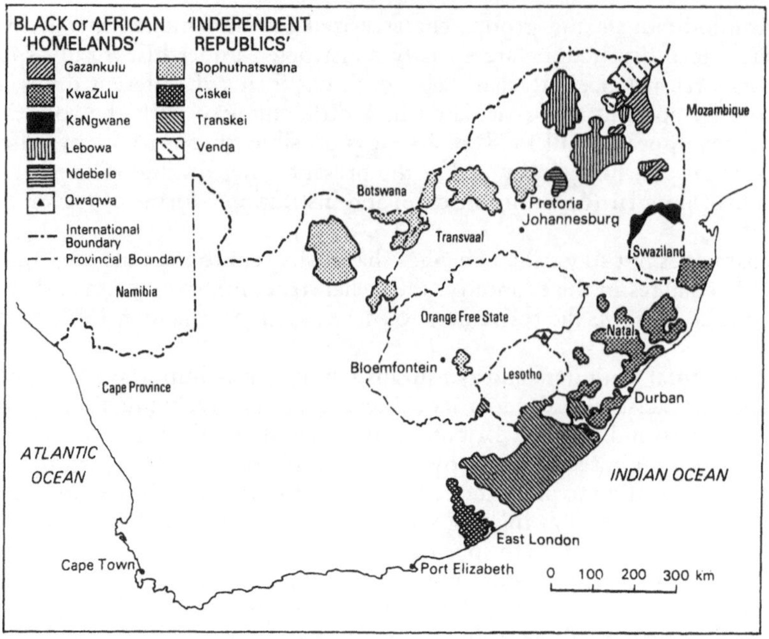

Figure 8.1 Apartheid at the national scale according to the South African government's 1975 consolidation proposals (source: Smith 1983).

ist legislation for particular groups in particular cities (e.g. Western 1981). Despite considerable ideological differences concerning the relative importance of race and class in analyzing the emergence, persistence, and ultimate transformation of apartheid, a strong and coherent tradition of geographical research has emerged (Lemon 1976, Smith 1982, 1983) which has influenced the way the geography of South Africa is taught (see, for example, Gill (1983) and a special issue of *Contemporary Issues in Geography and Education* (1985)).

The idea of race provides the organizing principle of social and political life in South Africa. The four race groups (Black, White, Coloured, and Asian), with various tribal subdivisions, are political constructions with the force of law. A wide range of policies and practices are informed by the philosophy of apartheid (or separate development) having as their rationale the efficient administration of the Republic in the interests of the Whites who currently form little more than 10 per cent of the total population. Apartheid has a strongly spatial element such that, in the twisted logic of former Prime Minister Vorster, if a White man were to wake up one morning and find himself a Black man, 'the only major difference would be geographical' (quoted in the *Johannesburg Star*, 3 April 1973). Such a belief is impelled by the 'separate but equal' ideology of apartheid. The reality is, of course, very different.

Apartheid operates at a variety of spatial scales from the segregation of lavatories and other public facilities (so-called 'petty apartheid', now largely dismantled) to the subdivision of the national territory into Bantustans (Black or African homelands) and independent republics (Fig. 8.1). The official

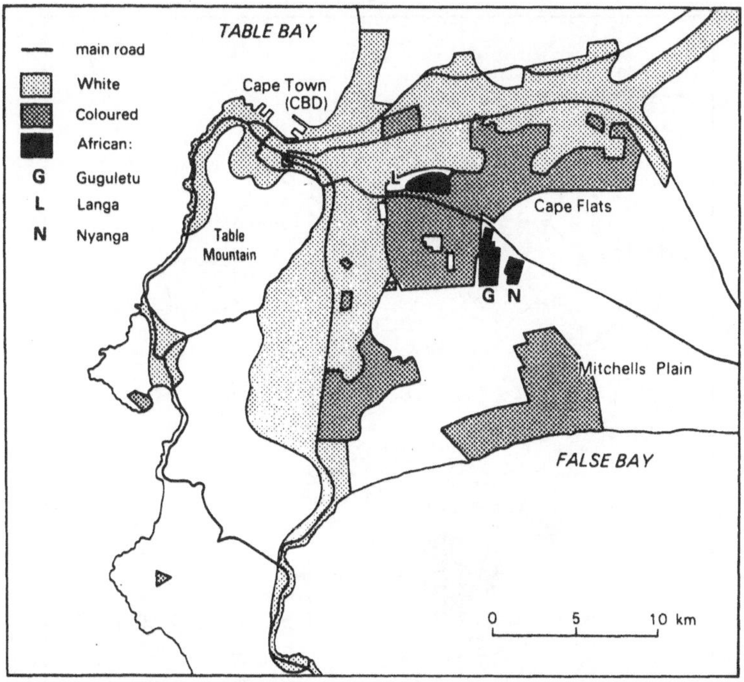

Figure 8.2 Residential areas of different race groups in the Cape Town metropolitan area (source: Smith 1983).

intention of this territorial subdivision is that each of the homelands, set aside for each major tribal group, should eventually become independent in a formal political sense, leaving the Whites in control of the rest of South Africa. So far only Bophuthatswana, Ciskei, Transkei, and Venda have achieved this status.

The Group Areas Act of 1950 gave legal status to residential segregation. Since then a variety of reforms have been enacted, changing but not dismantling the geography of apartheid at the national scale. The same principle is applied at the urban scale where residential areas are designated for each of the different race groups (Fig. 8.2). In Cape Town, for example, large areas are set aside for the Coloured population in the Cape Flats and in the new town of Mitchells Plain. The Blacks are concentrated in three townships, reserving the central parts of the city and the attractive residential areas on the eastern flank of Table Mountain for the Whites. In reality, throughout South Africa about half the Black population lives in the areas designated for Whites, but the principle of segregation is upheld in law and could until recently be implemented with inhuman severity through the so-called 'Pass Laws'. The operation of other measures, including the enforced relocation or wholesale destruction of entire neighbourhoods, has been documented by Western in his study of *Outcast Cape Town* (1981).

At every scale, the practical effect of apartheid is to maintain the economic and political supremacy of the Whites. This may be accomplished with extreme crudity, through the relocation strategies and pass laws discussed above,

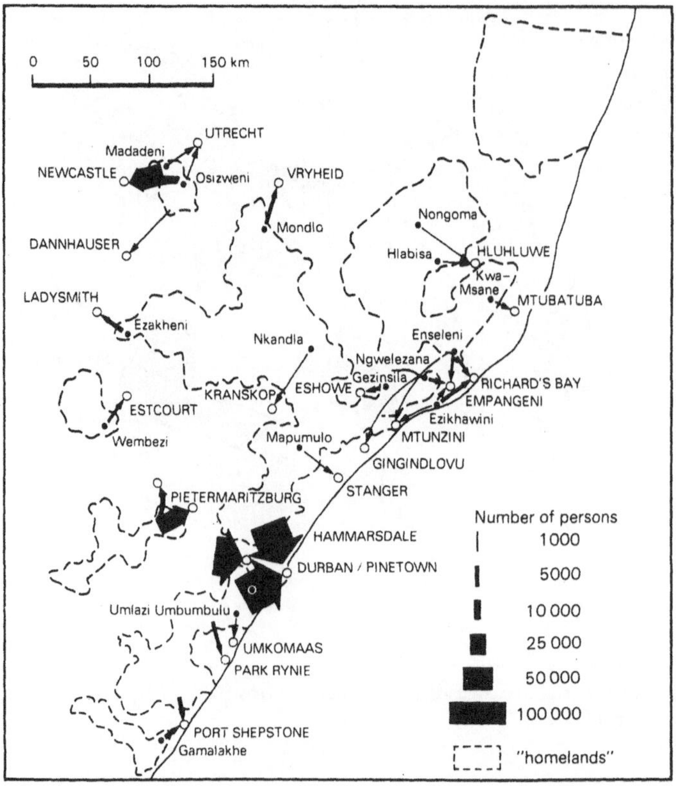

Figure 8.3 Frontier commuting from Kwa Zulu into 'White' South Africa (source: Smith 1983).

through restrictions on intermarriage, or through modifications to the group areas. In 1975, for example, the boundaries of the homelands were redrawn to exclude important industrial areas, most mineral resources, all but the smallest White settlements, and practically all transportation lines (Smith 1983). In other cases the political effect of apartheid can be quite subtle. An example is provided by the process of 'frontier commuting' in Durban, where Black workers reside in townships such as Umlazi or KwaMashu in the KwaZulu homeland but commute into the city for employment (Fig. 8.3). This enforced separation of home and work has a number of effects: it ensures an abundant supply of cheap Black labour to White employers; it reduces the potential for labour militancy by obliging the workforce to commute from distant areas; and it transfers the financial responsibility for the social reproduction of the labourforce to the homelands where they are required to reside (Smith 1983).

South Africa provides a clear example of the relationship between racist ideology and discriminatory practice. Indeed, it is hard to see how apartheid could be maintained without recourse to a spatial framework such as that enacted by the Group Areas Act. Segregation may, however, provide a basis for racist ideologies in other societies, like Britain, where the legislative framework is less overtly racist (Smith 1988). The geography of racist attacks in Britain provides a further example of the need for research on the spatial constitution of racism in all its forms.

The geography of racist attacks
Despite a reputation for fair play and racial tolerance, Britain has a long history of hostility towards foreigners, including Huguenot, Jewish, Irish, Italian, Asian, and Afro-Caribbean immigrants (Fryer 1984). The extent and severity of this hostility, which may lead to harassment or physical violence, has only recently been acknowledged (e.g. Home Office 1981, Greater London Council 1984). There is still a considerable reluctance in official circles to examine the rôle of extreme right-wing groups such as the National Front and the British Movement in fomenting the kind of antagonism towards black people that encourages verbal and physical violence, despite growing evidence of such a connection (e.g. Bethnal Green and Stepney Trades Council 1978, Doherty 1983).

The spatial distribution of racial violence has rarely been examined. It is usually assumed that such attacks are confined to the most depressed areas of Britain's ailing inner cities, particularly in London's East End. A report by the Greater London Council Police Committee included a map of the London Borough of Tower Hamlets depicting the incidence of racial harassment in relation to the Council's housing estates (Greater London Council 1984), while the Bethnal Green and Stepney Trades Council argued in *Blood on the streets*, that

> Beneath the headlines is an almost continuous and unrelenting battery of Asian people and their property in the East End of London. The barrage of harassment, insult and intimidation, week in week out, fundamentally determines how the immigrant community here lives and works, how the host community and the authorities are viewed and how Bengali people in particular think and act (Bethnal Green and Stepney Trades Council 1978, p. 3).

A number of specific reasons exist for the extremely high rate of racist violence in London's East End. Husbands (1983) uses Suttles's (1968) concept of the 'defended neighbourhood' to explain the indigenous white population's reaction to declining housing and employment opportunities which coincided with large-scale immigration from the New Commonwealth and Pakistan. The scapegoating of coloured immigrants was almost inevitable. But it is dangerous to suppose that such geographically specific circumstances provide a complete explanation for the pattern of racist attacks. Recent evidence suggests that the incidence of racist attacks is more widespread than may have been previously supposed.

In London, for example, detailed Metropolitan Police evidence, reported in *The London Standard* (20 May 1986), showed that the East End Borough of Tower Hamlets had the highest number of 'racial incidents' (255) in 1985, but that other police districts throughout the metropolis, such as Havering, Barking, and Newham (254), Brent and Harrow (177), Ealing and Hillingdon (130), Greenwich and Bexley (123) and Croydon and Sutton (107), were not far behind (Fig. 8.4). These data include incidents reported to the police, ranging in

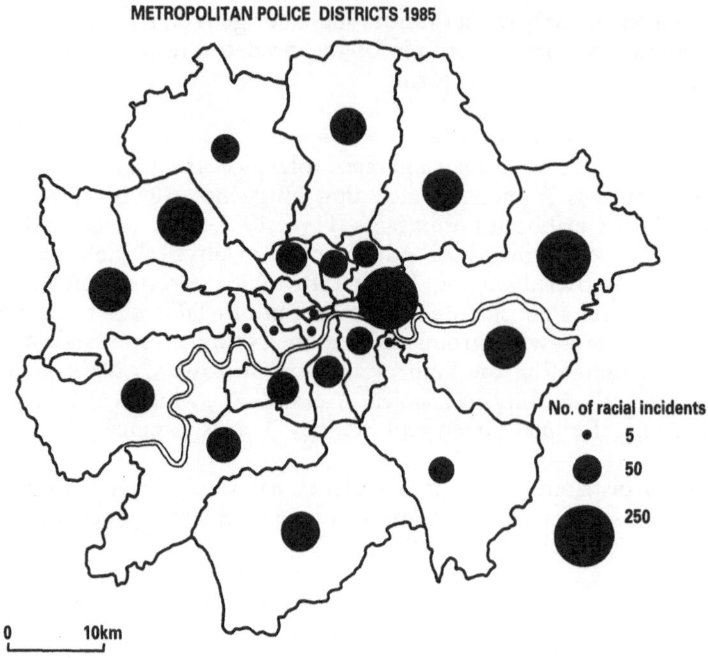

Figure 8.4 Distribution of racial incidents in London by Metropolitan police districts, 1985 (source: *London Standard*, 20 May 1986).

severity from verbal abuse and racist grafitti to criminal damage and physical assault. They do not support an explanation couched purely in terms of specific local circumstances. The precise geography of racist attacks in London requires further research but seems to be underpinned by a more pervasive racism affecting the entire metropolis.

A recent report on racist attacks in Britain supports this conclusion, arguing that: 'Few areas in Britain can now be regarded as completely safe for black residents' (Gordon 1986, p. 15). Racist attacks have been reported from the Prime Minister's parliamentary constituency of Finchley to Shrewsbury in rural Shropshire. No immediate correlation appears with size of black population or severity of local economic and social problems. Racist violence in Britain is not restricted to a few isolated incidents of irrational violence, attributable to the personal prejudices of bigoted individuals. It is not occasional or even particularly localized. It is persistent and widespread, reflecting the inequality and racism of British society in general.

Racism and resistance

Racism is part of the dominant ideology of many advanced capitalist societies, yet its hegemony is neither complete nor uncontested. Challenges to the forces of racist oppression can be theorized in terms of the Gramscian notion of resistance. According to Gramsci, the culture of the ruling class is never homogeneous, but is layered, reflecting different fractional interests within that class. Working-class racism is an equally inconsistent ideology. As a survey of working-class explanations of social and economic decline in the London Borough of Brent discovered: 'Racist ideas abound along with many others often as equally vague and inconsistent as the racist ones, or indeed, with self-professed ignorance' (Miles & Phizacklea 1981, pp. 95–6). A contradictory and inconsistent racism is faced by challenges which are various and discontinuous. According to the Centre for Contemporary Cultural Studies in Birmingham (Hall & Jefferson 1976, Centre for Contemporary Cultural Studies 1982), subordinate groups develop a repertoire of oppositional strategies over time that challenges the authority and power of dominant groups to impose their preferred reading of the world. Sometimes these strategies take a directly *instrumental* form (in the case of strikes, lock-outs, marches, and other forms of political activity). More often they are *symbolic* or ritualistic (as in the case of youth subcultures and the symbolic forms of reggae and Rasta).

Such cultural and political strategies may be attempts by excluded groups to win physical and social space for themselves and their own forms of life. For example, ghetto space represents a physical area over which an excluded group exercises a degree of control partly through that area's reputation for crime and danger. Social space includes those situations in which a network of kinship and friendship helps define an alternative moral order to the oppression imposed by a racist society. While these two domains are conceptually distinct, in practice they frequently interact as groups contest social space 'on the ground'. Only rarely do rituals of resistance remain purely symbolic, a totally 'imaginary' resolution of material contradictions (Hall & Jefferson 1976). Since most oppositional strategies are defined as illegitimate by the state, they often take place in public (as marches and protests, sit-ins and demonstrations, for example). It is therefore possible to trace a *geography of resistance* in terms of the simple spatial distribution of the practices we define as resistance and, more significantly, in terms of the spatial constitution of resistance itself. Another dimension to the geography of resistance is suggested by Miles (1978) who argues that the cultural form of Rastafarianism in Britain is significantly different from its expression in the Caribbean. It is not just a passive cultural import from Jamaica but a creative construction by a section of mainly male, black youth, actively modifying the cultural symbols and practices of Jamaica within a specifically English urban context, the new Babylon of their own experience (Miles 1978).

As James has shown in his succinct account of the *History of Negro revolt* (1985), there is a long tradition of black people's struggles against racist oppression from the San Domingo Revolution of 1791 to the American civil rights movement of the 1960s. The line he traces can be continued to the present day, as in Marable's history of *Black American politics* (1985), or the Bhavnanis' account of racism and resistance in Britain (Bhavnani & Bhavnani 1985). In each case, resistance has a particular geographical configuration. This can be illustrated with reference to the geography of race-related urban unrest in Britain and the United States.

Riots and rebellion

Between 1964 and 1968 American cities were torn apart by a series of riots in which at least 220 people were killed. In 1969 alone some 500 civil disorders

Figure 8.5 Distribution of major riots in US cities, 1967 (source: Rose 1971).

were recorded representing the most violent phase of the civil rights movement following the assassination of the Reverend Martin Luther King in 1968. The wave of urban riots began in New York (in the black ghettos of Harlem and Bedford–Stuyvesant) and spread to over 200 cities across the US during the next five years.

Although the riots reflected a deep-seated malaise among poor urban blacks concerning the persistence of discrimination, segregation, and inequality, the triggering incident (as in British cities during the 1980s) was in many cases confrontation with the police. The Kerner Report, charged with investigating the causes of the US riots, concluded that the major grievances of residents in the worst affected areas were police brutality, unemployment, and poor housing (National Advisory Commission on Civil Disorders 1968). Riots reflected the frustrations of ordinary black people at the failure of American society to incorporate them in the general prosperity of the postwar years. The Report was unequivocal in blaming white Americans who, it argued, had never fully understood that 'white society is deeply implicated in the ghetto. White institutions created it, white institutions maintain it, and white society condones it' (National Advisory Commission on Civil Disorders 1968, p. 1). Paradoxically, it was the existence of the ghetto, as a territorial entity, rather than the level of poverty and deprivation itself, that provided the physical basis for such dramatic forms of black urban protest. For the 1960s had actually seen some improvement in social conditions in the ghetto, associated with the War on Poverty and reflected in the passage of the Civil Rights Act in 1964.

At the time, Rose lamented the minimal contribution of geographers to an understanding of ghetto violence (Rose 1971, p. 84). His own analysis of the riots demonstrated the extent to which race-related civil unrest had diffused throughout the hierarchy of cities with sizeable black ghettos. As late as 1967 some cities with large black ghettos, such as Washington, Baltimore, St Louis, Pittsburgh, and Gary, had not experienced major riots (Fig. 8.5). But in 1968,

Figure 8.6 Distribution of major riots in US cities, 1968 (source: Rose 1971).

as the total number of violent incidents peaked, riots occurred throughout the nation, in Southern as well as Northern cities, in cities without large black ghettos, and in smaller centres as well as big cities (Fig. 8.6). Despite this widespread spatial pattern, Rose argues that the social and political significance of ghetto violence was geographically differentiated. The riots represented a form of social protest originating in the Northeast and taking several years to diffuse to the South where different techniques, such as sit-ins, had been developed in response to regionally-specific conditions (like overt segregation). Collective violence seems to have originated as a technique to deal with covert discrimination such as prevailed in the North, only subsequently transcending regional boundaries (Rose 1971, p. 93).

During the 1960s the violence was mainly confined to the ghetto areas themselves, often being directed against white-owned property. There is also ample evidence for white hostility to the spread of the black population beyond the confines of the ghetto (Drake & Cayton 1962, Morrill 1965, Spear 1967, Osofsky 1971). The history of violence associated with the spread of Chicago's black ghetto is particularly well documented. From the 1919 riot to the recent disturbances in the suburb of Cicero, there is a clear pattern of racially motivated violence against the incursion of black people into formerly white areas, whether the 'invasion' was led by the blockbusting activities of private real estate agents or by construction of public housing projects. The vast majority of violent incidents occurred on the expanding edges of the black ghetto or in more distant areas associated with conflicts over the siting of public housing (Fig. 8.7).

The level of violence appeared to peak in 1919 and in the 1960s, but recent research has shown that the intervening years constituted an 'era of hidden violence' associated with the highly-charged atmosphere of rapid neighbourhood change (Hirsch 1983). The same research reveals that the development of Chicago's 'second ghetto', on the city's West Side, was not a casual and

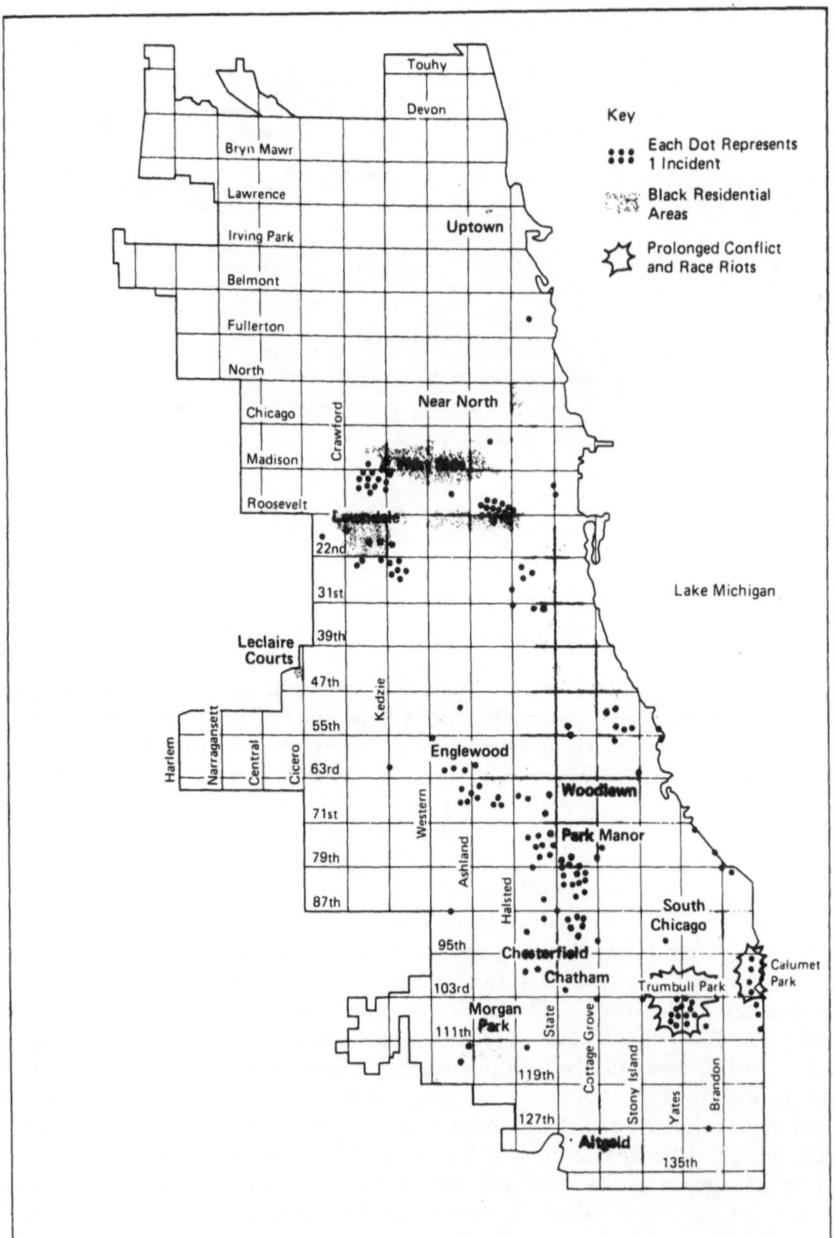

Figure 8.7 Location of racial violence in Chicago, 1956–7 (source: Drake & Cayton 1962).

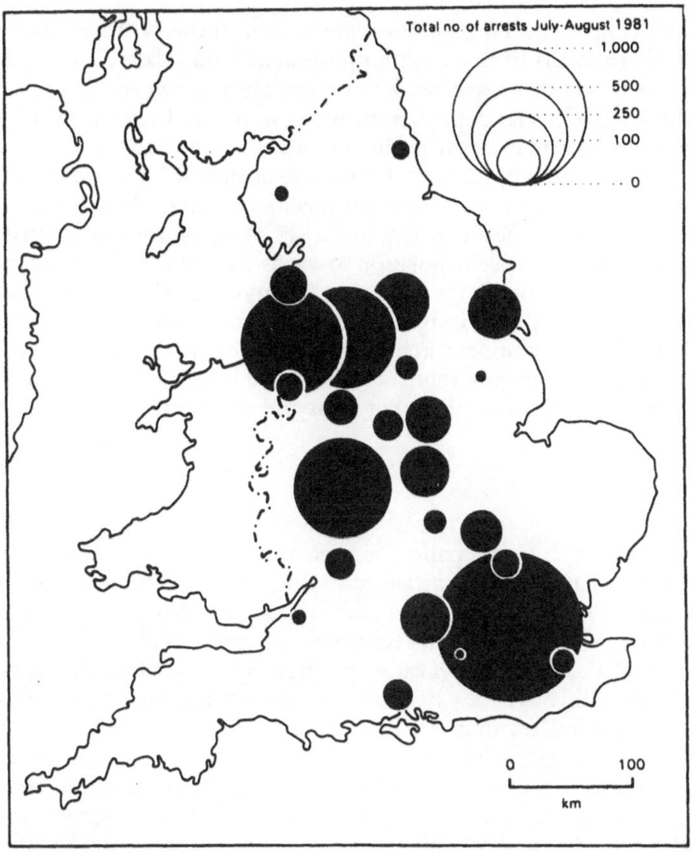

Figure 8.8 Numbers arrested in serious incidents of public disorder in England and Wales, July–August 1981 (source: Peach 1985).

unplanned process but involved deliberate policy decisions on the part of the Chicago Housing Authority with the active connivance of the Democratic Party machine dominated by Mayor Daley and his largely white-ethnic followers. The violence that greeted the expansion of the ghetto into formerly white neighbourhoods was given the tacit support and political sanction of city government. The ghetto has not merely persisted, therefore. It has been periodically renewed and strengthened, reinforced by public money, and fortified by the power of the state. American research also reveals that the cities most severely affected by riots were characterized by above-average rates of unemployment, large and rapidly increasing black populations, poor educational achievement and low incomes, and poor housing conditions with low owner-occupancy rates (Downes 1969).

In Britain, the geographical incidence of civil unrest in the early 1980s was somewhat more complicated (Peach 1985). The 'disorders', as they came to be officially described, began in 1980 in the St Paul's district of Bristol. The following year there were major riots in Brixton (south London), Toxteth (Liverpool) and Moss Side (Manchester), with smaller riots in several other

British cities (Fig. 8.8). These cities were among those with the worst inner-city problems in terms of housing, employment, and related social conditions (Hamnett 1983), yet there was no direct correlation between the propensity to riot and the level of social deprivation. In his influential report on the Brixton disorders, Lord Scarman argued that 'there are, undoubtedly, parts of the country which are equally deprived where disorders did not occur' (Scarman 1981, para. 2.38). He concluded that no specific aspect of deprivation could be described as a *cause* of the disorders but a set of social *conditions* existed that, taken together, created a predisposition towards violent protest. In all the recent research into the origins of the riots (e.g. Benyon 1984, Cowell *et al.* 1982, Joshua *et al.* 1983, Kettle & Hodges 1982), too little work of a specifically geographical kind has been undertaken (but see Burgess 1985). This is particularly regrettable as various events that the press have labelled 'riots' have included quite different occurrences in different areas and at different times (Keith 1987).

Conclusion

This chapter began by criticizing the theorization of race, suggesting that it should be seen as a *social construction* reflecting material conditions structuring the social relations between groups and individuals at particular times and in particular places. Like other social scientists, geographers have perpetuated the 'commonsense' racism that pervades political and academic discourse about race. Only belatedly have they begun to realize that racism is a proper object of enquiry for a profession that is still predominantly white, middle-class, and male. This point is more forcefully made by Brown who argues that 'white academics with an interest in race must relinquish their self-appointed role as the "translators" of black cultures, in favour of analyses of white society, i.e. of racism' (Brown 1981, p. 198). Whether in teaching or research, as Hall reminds us, challenging racism is not a moral duty that our privileged intellectual position obliges us to undertake; it is one of the most important ways of understanding how society actually works and how it has arrived where it is today (Hall 1981, p. 69).

Racism has been defined as an ideology that serves to legitimize the adverse treatment of certain racialized sections of migrant labour (Miles 1982). The argument advanced in this chapter is broadly sympathetic to this approach, although it should not be implied that the political, cultural, and ideological dimensions of racism can simply be read off from the economic. Given our earlier remarks about the historically and geographically differentiated social construction of race, it is possible to argue that while racism may be determined economically, it is contested politically and experienced culturally.

For those committed to the development of an explicitly anti-racist social science, one of the key issues concerns the rôle of race within the broader structure of social relations. It is no longer acceptable simply to assert the theoretical primacy of class and to relegate race to a subordinate position, performing a purely ideological rôle in *interfering with* the development of a common class consciousness among particular groups of people. Castells has begun to address this issue in his cross-cultural theory of urban social movements which argues that

> although class relationships and class struggle are fundamental in understanding the process of urban conflict, they are by no means the only or even the primary source of urban social change. Our theory must recognize other sources of urban social change: the autonomous role of the state, gender relationships, ethnic and national movements, and movements that specifically define themselves as citizens' movements (Castells 1983, p. xviii).

As Castells's work itself now implies, however, these questions are more likely to be resolved through a programme of theoretically-informed empirical research than by abstract theoretical speculation. Several studies confirm this view and point to the essentially contingent relationship between race and class. Katznelson's (1981) account of the making and remaking of northern Manhattan, for example, reaches the general conclusion that American urban politics have been governed by rules stressing ethnicity, race, and territoriality, and that class is virtually absent from people's mutual discourse outside of the workplace. He accounts for this in terms of urban America's specific historical geography which led to a marked separation of home and work and to the development of a system of 'city trenches' that confined conflicts within particular domains rather than allowing more general crisis tendencies to emerge.

In Britain, the attempt to identify the rôle of race within the changing spatial division of labour has been rather less well developed (though see Cross 1985). Instead, attention has been focused on the politicization of race and on what one commentator has described as the shift from immigration controls to 'induced repatriation' (Sivanandan 1982). Few politicians in the major parties openly advocate repatriation as a remedy for what they perceive as Britain's race relations problems. It can, however, be argued that recent changes in immigration and nationality laws (including the introduction of compulsory visa requirements for citizens of a number of New Commonwealth countries), the selective operation of the so-called 'sus' law (used to detain people on the suspicion that they have committed an offence), hard-line policing in black residential areas, 'fishing raids' by immigration officials, and the failure of the state to treat the issue of racist attacks with sufficient urgency, differ little in their effects from a policy of induced repatriation.

Future research in geography should be addressed towards the identification of the times and places that were critical to the formation of race relations as a political issue. The politicization of race and the racialization of politics are two fields for further investigation that offer specific opportunities for geographical research. Despite the recent spate of historical work on the history of black people in Britain (Fryer 1984, Ramdin 1987), too little attention has been paid to the question of resistance in general and to the *sites of struggle* in particular, in both the social and spatial sense of that evocative phrase. There is scope for original geographical work, both in surveying the social geography of black people's struggles in general and in detailing their local contours at the ethnographic level. Much of this work can be accommodated within a political-economy framework, starting with an examination of the rôle of migrant labour in the changing international division of labour but, as this chapter has sought to demonstrate, the social construction of race and the geography of racism also affords several opportunities for future research within the field of cultural studies and only a radically reconstituted cultural geography (Cosgrove & Jackson 1987) can adequately address these issues.

Acknowledgement

I would like to thank Richard Peet and Nigel Thrift for their comments on an earlier draft of this chapter and Lauren McClue who drew or redrew the illustrations.

References

Banton, M. 1977 *The idea of race*. London: Tavistock Publications.
Barker, M. 1981. *The new racism: conservatives and the ideology of the tribe*. London: Junction Books.
Benyon, J. (ed.) 1984. *Scarman and after: essays reflecting on Lord Scarman's report, the riots and their aftermath*. Oxford: Pergamon.
Berghe, van den, P. L. 1978. Race and ethnicity: a socio-biological perspective. *Ethnic and Racial Studies* **1**, 401–11.
Bethnal Green and Stepney Trades Council 1978. *Blood on the streets: a report on racist attacks in East London*. London: Bethnal Green & Stepney Trades Council.
Bhavnani, K. K. & R. Bhavnani 1985. Racism and resistance in Britain. In *A socialist anatomy of Britain*, D. Coates, G. Johnston & R. Bush (eds), 146–59. Cambridge: Polity Press.
Blaut, J. M. 1970. Geographic models of imperialism. *Antipode* **2**, 65–85.
Blaut, J. M. 1974. The ghetto as an internal neo-colony. *Antipode* **6**, 37–41.
Blaut, J. M. 1983. Assimilation versus ghettoization. *Antipode* **15**, 35–41.
Blaut, J. M. 1987. *The national question and colonialism*. London: Zed Press.
Bourne, J. & A. Sivanandan 1980. Cheerleaders and ombudsmen: the sociology of race relations in Britain. *Race and Class* **21**, 331–52.
Brown, K. R. 1981. Race, class and culture: towards a theorisation of the 'choice/constraint' concept. In *Social interaction and ethnic segregation*, P. Jackson & S. J. Smith (eds), 185–203. London: Academic Press.
Bulmer, M. 1984. *The Chicago School of sociology: institutionalization, diversity and the rise of sociological research*. Chicago: University of Chicago Press.
Burgess, J. A. 1985. News from nowhere: the press, the riots and the myth of the inner city. In *Geography, the media and popular culture*, J. A. Burgess & J. Gold (eds), 192–228. London: Croom Helm.
Castells, M. 1977. *The urban question*. London: Edward Arnold.
Castells, M. 1983. *The city and the grassroots*. London: Edward Arnold.
Centre for Contemporary Cultural Studies 1982. *The empire strikes back: race and racism in 70s Britain*. London: Hutchinson/Centre for Contemporary Cultural Studies.
Clarke, C. G., D. Ley & C. Peach (eds) 1984. *Geography and ethnic pluralism*. London: Allen & Unwin.
Contemporary Issues in Geography and Education 1985. South Africa: apartheid capitalism (special issue). Vol. 2.
Cosgrove, D. E. & P. Jackson 1987. New directions in cultural geography. *Area* **19**, 95–101.
Cowell, D., F. Jones & J. Young (eds) 1982. *Policing the riots*. London: Junction Books.
Cowhig, R. 1985. Blacks in English Renaissance drama and the role of Shakespeare's Othello. In *The black presence in English literature*, D. Dabydeen (ed.), 1–25. Manchester: Manchester University Press.
Cross, M. 1985. Black workers, recession and economic restructuring in the West

Midlands. Paper presented at a conference on Racial Minorities, Economic Restructuring and Urban Decline, Centre for Research in Ethnic Relations, University of Warwick.

Dennis, R. J. 1980. Why study segregation? More thoughts on Victorian cities. *Area* **12**, 313–17.

Doherty, J. 1973. Race, class, and residential segregation in Britain. *Antipode* **5**, 45–51.

Doherty, J. Racial conflict, industrial change and social control in post-war Britain. In *Redundant spaces in cities and regions? Studies in industrial decline and social change*, J. Anderson, S. Duncan & R. Hudson (eds), 201–39. London: Academic Press.

Downes, B. T. 1969. Social characteristics of riot cities. *Social Science Quarterly* **50**, 514–16.

Drake, St C. & H. R. Cayton 1962. *Black metropolis: a study of Negro life in a Northern city* (revised edn). New York: Harper & Row.

Fryer, P. 1984. *Staying power: the history of black people in Britain*. London: Pluto Press.

Gill, D. (ed.) 1983. *Racist society: geography curriculum*. London: University of London, Institute of Education.

Gordon, P. 1986. *Racial violence and harassment*. London: Runnymede Trust.

Greater London Council 1984. *Racial harassment in London*. London: Greater London Council Police Committee.

Greater London Council 1986. *A history of the black presence in London*. London: Greater London Council.

Gregory, D. & J. Urry (eds) 1985. *Social relations and spatial structures*. London: Macmillan.

Hall, S. 1978. Racism and reaction. In *Five views of multi-racial Britain*, Commission for Racial Equality (eds), 23–35. London: Commission for Racial Equality.

Hall, S. 1981. Teaching race. In *The school in the multicultural society*, A. James & R. Jeffcoate (eds), 58–69. London: Harper & Row.

Hall, S. & T. Jefferson (eds) 1976. *Resistance through rituals: youth subcultures in post-war Britain*. London: Hutchinson.

Hamnett, C. 1983. The conditions in England's inner cities on the eve of the 1981 riots. *Area* **15**, 7–13.

Harris, R. 1984. Residential segregation and class formation in the capitalist city: a review and directions for research. *Progress in Human Geography* **8**, 26–49.

Harvey, D. 1973. *Social justice and the city*. London: Edward Arnold.

Harvey, D. 1985. *The urbanization of capital*. Oxford: Basil Blackwell.

Hirsch, A. R. 1983. *Making the second ghetto: race and housing in Chicago, 1940–1960*. Cambridge: Cambridge University Press.

Hofstadter, R. 1955. *Social Darwinism in American thought* (revised edn). Boston: Beacon Press.

Home Office 1981. *Racial attacks*. London: Home Office.

Husbands, C. T. 1982. East End racism, 1900–1980: geographical continuities in vigilantist and extreme right-wing political behaviour. *London Journal* **8**, 3–26.

Husbands, C. T. 1983. *Racial exclusionism and the city: the urban support of the National Front*. London: Allen & Unwin.

Jackson, P. 1985. Social geography: race and racism. *Progress in Human Geography* **9**, 99–108.

Jackson, P. (ed.) 1987. *Race and racism: essays in social geography*. London: Allen & Unwin.

Jackson, P. & S. J. Smith 1981. Introduction. In *Social interaction and ethnic segregation*, Jackson, P. & S. J. Smith (eds), 1–17. London: Academic Press.

Jackson, P. & S. J. Smith 1984. *Exploring social geography*. London: Allen & Unwin.

James, C. L. R. 1985. *A history of Negro revolt* (revised edn). London: Race Today Publications.

Joshua, H., T. Wallace & H. Booth 1983. *To ride the storm: the 1980 Bristol 'riot' and the state*. London: Heinemann.

Katznelson, I. 1981. *City trenches: urban politics and the patterning of class in the United States*. Chicago: University of Chicago Press.
Keith, M. 1987. 'Something happened': the problem of 'explaining' the 1980 and '81 riots in British cities. In *Race and racism: essays in social geography*, P. Jackson (ed.). London: Allen & Unwin.
Kettle, M. & L. Hodges 1982. *Uprising! The police, the people, and the riots in Britain's cities*. London: Pan Books.
Lawrence, E. 1982a. Just plain common sense: the 'roots' of racism. In *The Empire strikes back: race and racism in 70s Britain*, Centre for Contemporary Cultural Studies (eds), 47–94. London: Hutchinson.
Lawrence, E. 1982b. In the abundance of water the fool is thirsty: sociology and black 'pathology'. In *The Empire strikes back: race and racism in 70s Britain*, Centre for Contemporary Cultural Studies (eds), 95–142. London: Hutchinson.
Leach, B. 1973. The social geographer and black people: can geography contribute to race relations? *Race* **15**, 230–41.
Lemon, A. 1976. *Apartheid: a geography of separation*. Farnborough: Saxon House.
Livingstone, D. N. 1984. Science and society: Nathaniel S. Shaler and racial ideology. *Transactions of the Institute of British Geographers* **NS 9**, 181–210.
Mabin, A. 1986a. Labour, capital, class struggle and the origins of residential segregation in Kimberley, 1880–1920. *Journal of Historical Geography* **12**, 4–26.
Mabin, A. 1986b. At the cutting edge: the new African history and its implications for African historical geography. *Journal of Historical Geography* **12**, 74–80.
Mair, M. 1986. *Black rhythm and British reserve: interpretations of black musicality in racist ideology since 1750*. Unpublished PhD dissertation, University of London.
Marable, M. 1985. *Black American politics: from the Washington marches to Jesse Jackson*. London: Verso.
Miles, R. 1978. *Between two cultures? The case of Rastafarianism*. Working Papers on Ethnic Relations 10. Bristol: Research Unit on Ethnic Relations/Social Science Research Council.
Miles, R. 1982. *Racism and migrant labour*. London: Routledge & Kegan Paul.
Miles, R. 1984. Marxism versus the sociology of 'race relations'. *Ethnic and Racial Studies* **7**, 217–37.
Miles, R. & A. Phizacklea 1981. Racism and capitalist decline. In *New perspectives in urban change and conflict*, M. Harloe (ed.), 80–100. London: Heinemann Educational Books.
Miles, R. & A. Phizacklea 1984. *White man's country: racism in British politics*. London: Pluto Press.
Morrill, R. L. 1965. The Negro ghetto: problems and alternatives. *Geographical Review* **55**, 339–69.
National Advisory Commission on Civil Disorders 1968. *Report*. Washington, DC: Government Printing Office (the Kerner Report).
O'Keefe, P. 1983. The changing international division of labour and southern Africa: an overview. *Antipode* **15**, 5–7.
Osofsky, G. 1971. *Harlem: the making of a ghetto, 1890–1930*, 2nd edn. New York: Harper & Row.
Park, R. E., E. W. Burgess & R. D. McKenzie 1925. *The city*. Chicago: University of Chicago Press.
Peach, C. 1968. *West Indian migration to Britain: a social geography*. London: Oxford University Press for the Institute of Race Relations.
Peach, C. (ed.) 1975. *Urban social segregation*. London: Longman.
Peach, C. 1985. Immigrants and the 1981 urban riots in Britain. In *Contemporary studies of migration*, P. E. White & G. A. van der Knaap (eds), 143–54. Norwich: GeoBooks.
Peach, C., V. Robinson & S. Smith (eds) 1981. *Ethnic segregation in cities*. London: Croom Helm.

Peel, J. D. Y. (ed.) 1972. *Herbert Spencer on social evolution*. Chicago: University of Chicago Press.

Peet, R. 1985. The social origins of environmental determinism. *Annals of the Association of American Geographers* **75**, 309–33.

Phillips, D. 1986. *What price equality? A report on the allocation of G.L.C. housing in Tower Hamlets*. GLC Housing Research and Policy Document no. 9. London: Greater London Council.

Ramdin, R. 1987. *The making of the black working class in Britain*. London: Gower.

Rex, J. & R. Moore 1967. *Race, community and conflict: a study of Sparkbrook*. London: Oxford University Press for the Institute of Race Relations.

Robinson, V. 1982. The assimilation of South and East African Asian immigrants in Britain. In *Demography of immigrants and minority groups in the United Kingdom*, D. A. Coleman (ed.), 143–68. London: Academic Press.

Robinson, V. 1986. *Transients, settlers, and refugees: Asians in Britain*. Oxford: Clarendon Press.

Rose, H. M. 1971. *The black ghetto: a spatial-behavioral perspective*. New York: McGraw-Hill.

Scarman, Lord 1981. *The Brixton disorders*, 10–12 April 1981. London: HMSO. Cmnd. 8427.

Sivanandan, A. 1982. *A different hunger: writings on black resistance*. London: Pluto Press.

Sivanandan, A. 1983. Challenging racism: strategies for the '80s. *Race and Class* **25**, 1–11.

Smith, D. M. (ed.) 1982. *Living under apartheid: aspects of urbanization and social change in South Africa*. London: Allen & Unwin.

Smith, D. M. 1983. *Update: apartheid in South Africa*. London: Department of Geography and Earth Science, Queen Mary College, Special Publication no. 6.

Smith, S. J. 1988. *White supremacy in Britain? Critical interpretations of racial segregation*. Cambridge: Polity Press.

Spear, A. H. 1967. *Black Chicago: the making of a Negro ghetto, 1890–1920*. Chicago: University of Chicago Press.

Street, B. V. 1975. *The savage in literature*. London: Routledge & Kegan Paul.

Suttles, G. D. 1968. *The social order of the slum*. Chicago: University of Chicago Press.

Tabu, R. P., D. G. Taylor & J. D. Dunham 1984. *Paths of neighborhood change: race and crime in urban America*. Chicago: University of Chicago Press.

Urry, J. 1981. *The anatomy of capitalist societies*. London: Macmillan.

Walvin, J. 1982. Black caricature: the roots of racialism. In *'Race' in Britain: continuity and change*, C. Husbands (ed.), 50–72. London: Hutchinson.

Western, J. 1981. *Outcast Cape Town*. London: Allen & Unwin.

Yinger, J. M. 1981. Towards a theory of assimilation and dissimilation. *Ethnic and Racial Studies* **4**, 249–64.

9 *Marxism, culture, and the duplicity of landscape*

Stephen Daniels

In an area of overlap between radical and humanistic geography a new cultural geography is emerging (Ley 1985, Johnston *et al.* 1986, Cosgrove & Jackson 1987). A number of features distinguish this new cultural geography from the old, from the tradition deriving from the work of Sauer. There is a humanistic emphasis on the symbolic as well as on the material dimension of culture – on painting, literature, and the mass media as sources as well as on more palpable artefacts like fences and farm buildings (Cosgrove & Daniels 1987). Moreover, the very distinction between the material and the symbolic is brought into question with the development of the analogy of all artefacts – from poems to maps to fields of crops – as cultural texts or representations (Daniels & Cosgrove 1987). There is a radical emphasis on culture as a medium of social power, in the making of élite or official authority, for example in landscape parks (Daniels 1982a) and disciplinary institutions (Driver 1985), and in the exercise of power apart from, or against that authority, for example, in urban graffiti (Ley & Cybriwsky 1974) and rock music lyrics (Jarvis 1985). As these examples suggest, there has been a broadening of the purview of cultural geography from predominantly agricultural landscapes (often by implication stable and premodern ones) to include more explicitly dynamic landscapes (modern or postmodern cities and suburbs) (Ley 1987) and an endeavour to explicate the conflict and tension in rural landscapes that seem on the face of it to have none (Daniels 1981, 1982a, 1982b). Landscape, the central concept of traditional cultural geography, does not easily accommodate political notions of power and conflict, indeed it tends to dissolve or conceal them; as a consequence the very idea of landscape has been brought into question (Punter 1982a, 1982b, Cosgrove 1985).

Theoretically less shy than their forebears, new cultural geographers have adapted much of their theorizing from Western Marxist thought on culture (Cosgrove 1983, Thrift 1983). The first part of this chapter will look closely at the changing status and meaning of culture in Western Marxism. This will involve clarifying, indeed celebrating, the fluctuations and tensions in Marxist writings and by extension in the very concept of culture itself; in Williams's words, 'one of the two or three most complicated words in the English language' (Williams 1983, p. 87). Williams's name will recur throughout this chapter along with that of another Marxist cultural critic, Berger, whose work is, in many ways, complementary to Williams's. Their writings have already coloured the new cultural geography and in the second part of this chapter I will show how they could give it more shape – not by colonizing cultural geography more completely, but by opening up the broad domain of geographical experience and imagination which I see as central to their work. This will involve making more of a *rapprochement* with traditional Sauerian cultural geography – in emphasizing observation, in emphasizing the importance of education, in reinstating the biophysical world, and in reinstating the idea of landscape, not despite its difficulty as a comprehensive or reliable concept, but because of it.

Marxism and culture

Marx made little of what we now call culture and progressively less as economics became the main focus of his thought. And yet this century many Western Marxists have reversed Marx's trajectory, turning away from (narrowly conceived) economic structures and (economically structured) class relations to a range of topics – including education, the family, ethnicity, art, the mass media – which are now collectively labelled as 'cultural'. The intention and outcome of much empirical and theoretical work by Western Marxists has been to grant culture more substance and more power in the constitution of society than did Marx in his lifetime or have classical Marxists since (Anderson 1976).

This shift in priority may be seen to express a shift in the constitution of Western society. In Marx's lifetime, especially when he was writing *Capital*, conspicuous increases in large-scale commodity production convinced observers of varying ideological persuasions that the economy formed a primary and autonomous sphere, driving, if not generating, the rest of society. In the classical Marxist formulation the mode of production with its constituent forces and (class) relations formed the base or infrastructure of society which determined the existence and shape of the superstructure. While formal institutions like the state and the church were seen to be less distant from the base and so less ethereal than activities like art and literature, the superstructure overall was regarded as ideological, a fabric of appearances projected by and, in turn, obscuring the brute force of economic reality (Williams 1977, p. 92). This century it has proved increasingly implausible to conceptualize Western society in this way. The resilience and power of bourgeois institutions (if only in thwarting the theoretical destiny of the working class) could not be reduced to relations arising from the prevailing mode of production. Some of these institutions and their associated ideas and practices – like state education, advertising, nationalist feeling, tradition – only became empowered and conceived of as culture after Marx's death. To explain their power Western Marxism has absorbed a variety of non-Marxist thought – sociological, psychoanalytical, existentialist – along with some of its cultural pessimism. Concepts like Gramsci's 'hegemony', Althusser's 'ideological state apparatuses', Lefebvre's 'everyday life', and Williams's 'cultural materialism' all express a recognition of how culture, no longer merely as part of a superstructure, could buttress, indeed could effectively produce, the status quo (Jay 1984, pp. 82–110). Williams declares:

> From castles and palaces and churches to prisons and workhouses and schools; from weapons of war to a controlled press: any ruling class, in variable ways though always materially, produces a social and political order. These are never superstructural activities. They are the necessary material production within which an apparently self-subsistent mode of production can alone be carried on. The complexity of this process is especially remarkable in advanced capitalist societies where it is wholly beside the point to isolate 'production' and 'industry' from the comparably material production of 'defence', 'law and order', 'welfare', 'entertainment', and 'public opinion' (Williams 1977, p. 93).

Williams identifies this process of cultural materialism in precisely the period in which commodity production was singled out for structural priority. 'The task is not to see how the industrial revolution affected other sectors [of society], but to see that it was an industrial revolution in the production of culture as much as an industrial revolution in the production of clothing . . . or in the production of light, of power, of building materials' (Williams 1979a, p. 144). The model of base and superstructure is, he maintains, 'essentially a bourgeois formula; more specifically a central position of utilitarian thought'. Without sacrificing his 'sense of the commanding importance of economic activity and history' Williams has attempted 'to develop a different kind of theory of social totality; to see the study of culture as the study of relations between elements in a whole way of life' (Williams 1980, p. 20).

This enlarging and empowering of the concept of culture as a way of life owes much to the study this century of societies in which capitalist forms of production and exchange are only weakly developed, if at all, and in which the material and symbolic dimensions of production and exchange seem to be inextricably intertwined. In the late 1950s and early 1960s Williams and other New Left scholars, notably Thompson, brought this anthropological notion of culture critically to bear on the English patrician notion of culture as a set of timeless and elevated standards opposed to the vulgar ways of the majority (Johnson 1979a, 1979b). They saw this patrician notion as a confiscation of culture and their work as a form of recovery (Williams 1979a, p. 155). Thompson emphasized culture as the medium in which the English working class 'made itself' and the ingredients of this medium were characteristically patrician notions like tradition and experience as well as characteristically plebeian ones like strikes and demonstrations. Thompson was also intent to mobilize this notion of culture against the utilitarian debasement of culture; moreover, he included Marx and Engels in the utilitarian tradition which supposed that the English working class was spontaneously generated by the factory system, stripped to bare necessity, with nothing to lose but its chains (Thompson 1963). Like many New Left Marxists, Thompson has distanced himself from the Marx of *Capital*, whom he reproaches for producing a socialist version of Economic Man, to embrace a younger, more holistic Marx who craved a 'unitary knowledge of society' and also a more elderly Marx who 'in his increasing preoccupation in his last years with anthropology was resuming the projects of his Paris youth, if he was always trying to shove them back into an economic frame of reference' (Thompson 1978a, pp. 355, 364). In an essay on his own engagement with anthropology, Thompson comes to similar conclusions as Williams about the shortcomings of the structural formula of classical Marxism, even for the utilitarian world of capitalism: 'There is no way in which I find it possible to describe Puritan or Methodist work discipline as an element in the "superstructure" and then work itself in a "basis" somewhere else' (Thompson 1978b, p. 262). Culture has, as it were, dissolved the categories of classical Marxism.

Writing in 1961, Williams found society, in both bourgeois and classical Marxist thought, reduced to two spheres of interest, 'the system of decision (politics) and the system of maintenance (economics)'. This excluded or marginalized another two spheres which Williams regarded as no less important, 'the system of learning and communication' and 'the complex of relationships based on the generation and nurture of life' (Williams 1965, p. 133), spheres which were predicated on the notion of culture as a reproductive process, as cultivation, in both a mental and a physical sense (Williams 1979a, pp. 154–5). The process of learning has always been a central theme in Williams's work and in the English romantic tradition – featuring such figures as William Morris, D. H. Lawrence, and John Ruskin – which New Left scholars have grafted on to Marxism. This has given New Left Marxism a more gradualist inflection than classical or structural varieties. The most educational of Williams's books is entitled *The long revolution* (Williams 1965) and implicit in Williams's notion of culture is one of the strands he identifies in the concept of 'evolution', 'the unrolling of something already implicitly formed (like a national *way of life*)' (Williams 1983, p. 122; emphasis in original). This emphasis on culture as process implies a more organic, less mechanical, modelling of the constitution of society and social change. Thompson, for instance, has toyed with the metaphor of kernel and husk to replace that of base and superstructure (Thompson 1978b); also he has likened what he sees as the reductionism of economistic, structural Marxism to the defoliation of a quintessentially English landscape: 'the boughs are bared of all culture, the fields have been stripped of every green blade of human aspiration' (Thompson 1978a, p. 359). Williams's second, and more explicitly organic cultural sphere, 'the complex of relationships based on the generation and nurture of life', remained undeveloped in his own writings until the mid-1970s. It was then realized less in response to feminism than to 'green socialism', 'the most hopeful social and political movement of our time' (Williams 1984a, p. 219). It was also provoked by what Williams saw as idealist tendencies in Marxist thinking on learning and communication.

The conception of culture as a '*signifying system* through which . . . a social order is communicated, reproduced, experienced and explored' is, Williams finds, symptomatic of the 'extended communicative character' of 'advanced capitalist' societies (Williams 1981, p. 13; emphasis in original). Writing from within such societies in which image making is replacing the production of hard commodities, and in which potentially anything – nature, sex, other people – can come to us as an image, the notion of a realm of cultural appearances now seems to many Western Marxists less superficial and insubstantial than once it did. Williams observes how advertising this century has moved from the frontier of selling goods and services to finance a whole range of communication and to become fully implicated in the organization of people's lives; advertising is not just the 'official art' of monopoly capitalism but its 'magic system' (Williams 1980, pp. 170–95). For Berger advertising is central to the hyper-visual nature of publicity, which 'is the life of this culture – in so far as without publicity capital would not survive'. 'In no other form of society in history has there been such a concentration of images, such a density of visual messages', declares Berger, and 'we accept the total system of publicity images as we accept an element of climate' (Berger 1972a, pp. 129–30).

Other societies, at other times, have not been image poor – far from it. And arguably a modern consciousness of imagery has brought into focus that of the past. In his introduction to *The language of images* (1980) Mitchell observes that

> we inhabit a world so inundated with composite pictorial-verbal forms (film, television, illustrated books) and with the technology for the rapid, cheap production of words and images (cameras, xerox machines, tape recorders) that nature threatens to become what it was for the Middle Ages: an encyclopaedic illuminated book overlaid with ornamentation and marginal glosses, every object converted into an image with its proper label or signature (Mitchell 1980, p. 359).

Within English Marxist historiography there has since the 1970s been an emphasis on the rôle of symbolic imagery in the constitution of class relations and the exercise of power, for example in the iconography of class conflict during the Civil War (Hill 1977, pp. 171–81), in the 'theatre' and 'counter-theatre' of Georgian land and law (Thompson 1975), in the literary images of country and city in the social history of rural and urban development (Williams 1973), and in the 'invention of tradition' (Hobsbawm 1983). Moreover, a recognition of images *in* history has, perhaps inevitably, accompanied a recognition *of* images of history, the metaphors and analogies which configure historical discourse, including that of Marx (Mitchell 1986, pp. 160–208, White 1973, pp. 250–330, La Capra 1983). From this perspective the writing of Marx is seen still less as a body of scientific doctrine to be consolidated, proved, or disproved, and even more as a discourse which is part of an imaginative (and not necessarily Marxist) tradition. Thus, for Berman the epistemological and political importance of *The Communist Manifesto* is not its causal logic but its imaginative rhetoric, enshrined in its 'central image', 'All that is solid melts into air' (Berman 1983). Placed in the context of other modernist texts and landscapes, Berman finds in *The Communist Manifesto* a language for helping to make sense of, and positively to exploit, the 'maelstrom of perpetual disintegration and renewal' that (conspicuously in modern cities) defines modernity, not (as his more classical Marxist critics have complained (Anderson 1984)) the material for forging a weapon to abolish it. Moreover, for Berman, the 'modernist melting vision' of *The Communist Manifesto* 'everywhere pulls like an undertow against the more "solid" Marxian visions we know so well'; 'Reading *Capital* won't help us if we don't also know how to read the signs in the street' (Berman 1984, p. 123).

In acknowledging the iconographic power and even constitution of society and social description, Western Marxists have been both iconoclastic – in attempting to dismantle the illusory power of images to reach an actual or at least authentic world, and idolatrous – in accepting that such a world is no longer tenable and that the appropriate strategy is to harness the power of imagery against capital or at least in a way that will make sense of its workings. So, striking an iconoclastic posture, Williams declares that the 'magic [of advertising] obscures the real sources of general satisfaction' (Williams 1980, p. 189) and similarly Berger concludes that the contrast between advertising's 'interpretation of the world and the world's actual condition is a very stark one' (Berger 1972a, p. 151). And yet Berger also, in the same essay, strikes an idolatrous posture in emphasizing the emancipatory potential of modern modes of reproduction, notably photography and film. Berger is here echoing Benjamin. In an essay of 1936 Benjamin recalled how 'our taverns and our metropolitan streets, our offices and furnished rooms, our railroad stations and our factories appeared to have locked us up hopelessly. Then came the film and burst this prison-world asunder' and through such techniques as close-up and slow motion 'manages to assure us of an immense and unexpected field of action' and, like that other modern invention psychoanalysis, 'introduces us to an unconscious optics' (Benjamin 1973, pp. 238–9). For Berger modern means of reproduction have liberated art from its preserve (often physically in the form of palaces and country houses) in the culture of the ruling classes. And while most reproductive contexts – art books, advertisements, gilt frames in living rooms – are used to 'bolster the illusion that nothing has changed', if the 'new language of images were used differently [in such provocative techniques as photomontage] it would, through its use, confer a new kind of power . . . we could begin to define our experiences more precisely . . . not only personal experience, but also the essential historical experience of our relation to the past' (Berger 1972a, p. 33).

Marxism and postmodernism

In theorizing centrality of the mass media to advanced capitalism some postmodernist Marxists maintain that it is now no longer possible to sustain a categorical distinction between the real and the imaginary. Debord observes that in advanced capitalist societies

> all of life presents itself as an immense accumulation of *spectacles*. Everything that was directly lived has moved away into a representation . . . . One cannot abstractly contrast the spectacle to actual social activity: such a division is itself divided. The spectacle which inverts the real is in fact produced. Lived reality is materially invaded by the contemplation of the spectacle while simultaneously absorbing the spectacular order, giving it positive cohesiveness . . . reality rises up within the spectacle and the spectacle is real (Debord 1983, props 1, 8).

And for Baudrillard, imagery in advanced capitalism neither masks nor reflects a 'basic reality', 'it bears no relation to reality at all: it is its own pure simulacrum'. 'The territory no longer precedes the map or survives it', he maintains, 'henceforth it is the map which precedes the territory.' Disneyland is for Baudrillard 'a perfect model of all the entangled orders of simulation', a 'play of illusions and phantasms' that has patrons 'revelling in the real America' (Baudrillard 1983, pp. 11, 25).

In the fullest Marxist version of postmodernism, Jameson (1984) defines consumer capitalism as a 'culture of the image' or, more exactly, of the 'simulacrum': 'the culture of the simulacrum comes to life in a society where exchange value has been generalized to the point where the very memory of use-value is effaced.' In such a society there is 'a prodigious expansion of culture throughout the whole social realm, to the point where everything in our social life – from economic value and state power to practices and the very structure of the psyche itself – can be said to have become "cultural" in some original and as yet untheorized sense.' We are 'so deeply suffused and infected' by this postmodern condition that 'old-fashioned ideological critique' is no longer tenable; Jameson casts doubt even on the notions of history, morality, and experience so central to English New Left Marxism and especially to what he calls 'the redemptive historiography of E. P. Thompson'. The past itself has 'become a vast collection of images, a multitudinous photographic simulacrum'. In the postmodern condition the past as referent or signified is bracketed or effaced leaving only texts of free-floating signifiers and the historian, relieved of the burden of recovering the real world, free to indulge in 'textual play'.

In dissolving culture into representation, and in questioning the truth-content of any representation, this post-structuralist postmodernism is opposed to the sturdy notion of culture within New Left Marxism with its ingredients of morality, experience, and tradition; and it has been vigorously resisted (Latimer 1984). For Williams much Marxist culture theory is now 'rabid idealism'; it is in danger of reaching the point at which 'the epistemological wholly absorbs the ontological' with its assumption 'that it is only in articulation that we live at all'. The central concept of Williams's thought, 'structure of feeling', is an 'area of tension' between articulation and 'experience', especially experience of the natural world. Williams finds it necessary to reaffirm 'an absolute founding presumption of materialism: namely that the natural world exists whether anyone signifies it or not' (Williams 1979a, pp. 167–8). So while in his engagement with classical Marxists Williams has dissolved the 'economy' as an independent and prior material realm, he has more recently, in response to post-structural Marxism, reconstituted nature as that realm (if that is not too post-structuralist a way of putting it). In terms of the history his own thought has developed the biophysical dimension of the four-fold conception of society sketched in *The long revolution.* This has for Williams implied a strengthening of the notion of culture.

Drawing on the writings of the Italian Marxist Timpanaro, Williams declares that 'the deepest cultural significance of a relatively unchanging biological human condition is probably to be found in some of the basic material processes of the making of art: in the significance of rhythms in music and dance and language, or of shapes and colours in sculpture and painting'; these 'are at times the most powerful elements of the work' (Williams 1980, p. 113). Berger also now emphasizes a biophysical materialism. He finds that the making of images in art as opposed to advertising involves the transformation of energy through materials – subject matter and style but pre-eminently physical materials like pigment. It is an expressive notion of imagery and the imagination that invests appearances with shape and texture and that is informed by a robust sense of desire; 'Ideology partly determines the finished result [of an artistic image]', acknowledges Berger, 'but it does not determine the energy flowing through the current. And it is with this energy that the spectator identifies.' It involves 'cheating the visible' of the status quo: 'every image used by a spectator is a *going further* than he could have achieved alone, towards a prey, a Madonna, a sexual pleasure, a landscape, a face, a different world' (Berger 1985, p. 203; emphasis in original). In maintaining that it is still possible for art and the aesthetic dimension to sustain creative independence from, and critical purchase upon, other sorts of commercially or politically propagandist imagery, Berger and Williams reinstate an exclusive but not necessarily élitist sense of culture.

This recent ecological strain in the writings of Williams and Berger has been reformulated as an alternative postmodernist Marxism by Fuller. Fuller opposes it both to the 'mega-visual tradition of monopoly capitalism' and, as he sees it, to its reproduction in Marxist cultural theory. In contrast to the playful postmodernism of Jameson, predicated on a consumer 'culture of euphoric surfaces', Fuller's is a postmodernism of moral seriousness, predicated on a notion of cultural depth (Jencks 1986, p. 48). Fuller celebrates such qualities as tradition, locality, ecological harmony, and a sense of community; he enlists William Morris and John Ruskin as the twin prophets of a redemptive postmodernism that is at once patriotic and nostalgic: 'despite the advance of American culture and the microchip', Fuller is glad to say, 'this yearning for a "fair broad land of lovely villages" persists. Antiquaries, poets and painters continue to turn to a vision of "true England" for inspiration and consolation' (Fuller 1980; 1983, pp. 2–19; 1985, pp. 77–91, 277–83). As Baudrillard says, 'when the real is no longer what it used to be, nostalgia assumes its full meaning' (Baudrillard 1983, p. 12).

Culture in radical geography

Culture has relatively little weight in the corpus of radical geography – or at least in how that corpus has hitherto been constructed. Recurring topics like industrialization, housing, and underdevelopment have been approached mainly from a classical or structural Marxist perspective of political economy. Such topics have been the staple of human geography generally and the polemical thrust of radical geography has been mainly directed against prevailing liberal political-economic perspectives. This engagement has been largely conducted on a conceptual terrain of space and location and in a theoretical climate of modernization which is seen precisely to erode the relevance of more morphological notions like landscape or region and so to reduce culture to a residual category. It is, therefore, not surprising that the (consciously anti-modernist) tradition of Sauerian cultural geography has scarcely been addressed by radical geography. Furthermore, the pre-occupation with culture of humanistic geographers, and their tendency to downplay its historical or material dimensions (Daniels 1985), has perhaps weakened the status of culture within radical geography still further.

But there are now the indications of a radical geographical engagement with culture, in a set of overlapping writings which addresses the politics of reproduction. These emphasize the contextual as well as the ideological dimension of notions like nature (Sayer 1979, Olwig 1984), landscape (Cosgrove 1984), region (Gregory 1982) and locality (Massey 1984), and some draw on the English Marxist tradition I have discussed (Cosgrove 1983, Thrift 1983). Furthermore, there is a resonance between the tradition which Williams and Thompson invoke and a tradition of radical, but not necessarily Marxist, geography which is now being more fully recovered. In the writings of Reclus and Kropotkin, as in the writings of Morris and Ruskin, there is an emphasis on the aesthetics and morality of landscape, on the organicism and the historicity of nature, and on the critical importance of education (Galois 1977, Reclus 1977, Stoddart 1986, pp. 128–41). Culture is central to two recent books written from a Marxist geographical perspective, *Consciousness and the urban experience* (Harvey 1985) and *Social formation and symbolic landscape* (Cosgrove 1985). I will discuss these books in counterpoint to two Marxist texts (by an art critic and a sociologist) which contest their assumptions. This will be to develop geographically the issues I raised in the first section of this chapter and to prepare the ground for the third section on geographical experience and imagination in which I return to the work of Williams and Berger.

First published in 1979, Harvey's essay 'Monument and myth', an account of the political symbolism of the Basilica of Sacré-Coeur, seemed at the time a departure from the political-economic perspective in Marxist geography which Harvey himself had helped to develop. Essentially, the essay is about the politics of historical representation, in the built environment and in writing, and Harvey's own narrative is intended to recover a radical history the building of the Basilica has obscured (Harvey 1979). In terms of Marx's own writings 'Monument and myth' is written more in the historiographic tradition of *The Eighteenth Brumaire* and the *Communist Manifesto* than in the political-economic tradition of *Capital*. Its republication in 1985 along with an extended essay on 19th-century Paris and two more theoretical essays on 'Consciousness and the urban experience' (Harvey 1985), adds up to a substantial engagement with the culture and cultural imagery of capitalism. In order to specify the nature of this engagement, it is instructive to consider another book which also addresses cultural imagery in 19th-century Paris from a Marxist perspective, Clark's *The painting of modern life* (1984), and Harvey's response to it.

A central issue of Clark's book is the adequacy of landscape to represent the modernization of land and life in the areas of 19th-century Paris and its environs which experienced the extensive environmental and economic redevelopments associated with the name of Baron Haussmann. This is not for Clark a question of a representational response to a more concrete reality; 'Haussmannization' is itself a representation, 'a representation [of capital] laid out upon the ground in bricks and mortar'. It made Paris belong to (bourgeois) Parisians 'now simply *as an image*, something occasionally and casually consumed in spaces expressly designed for the purpose – promenades, panoramas, outings on Sundays, great exhibitions, and official parades' (Clark 1984, p. 36; emphasis in original). Haussmann's rebuilding was a 'spectacle' in the postmodernist sense of the term, not, as it seemed, with its legible system of boulevards, water supply, sewers, parks, and administrative buildings, 'an image mounted securely and finally in place'; but 'an account of the world competing with others'. For Clark, much Impressionist landscape painting was illusory or ideological in this same sense of representing relations between features of the modern world as stable, secure, and clearly legible. Only Manet articulated modernism's other form of life (the form which is now conventionally labelled postmodern), the provisionality, contingency, and seepage in a dynamic field of images and signs; and Manet resolved them into a pictorial image by the 'flatness' of his style, putting 'landscape in doubt' (Clark 1984, pp. 174–84). In using ideology in its illusory sense Clark invites the conclusion that some representations are disguising more material fundamentals; but he resists this inference. Society itself is for Clark, 'a hierarchy of representations', a class-bound society like modernizing Paris 'a battlefield of representations'. The 'economy . . . is in itself a realm of representations'; 'how else', he asks, 'are we to characterize money, for instance, or the commodity form or the wage contract.' This is not to say that all representations have the same gravity.

> It is one thing (and still necessary) to insist on the determinate weight in society of those arrangements we call economic; it is another to believe that in doing so we have poked through the texture of signs and conventions to the bedrock of matter and action upon it (Clark 1984, p. 6).

Harvey too emphasizes the importance of representation in 19th-century Paris (the 'swirl and confusion of images, representations and political rhetoric') (Harvey 1985, pp. 182–3), and its importance in modern capitalism generally. He probes the representational power of money, in particular its power to dissolve more stable, hierarchical modes of consciousness and experience like landscape. He cites Henry James's lament for landscape, its familiarity, security, and solidity, in the face of its erosion in a fiercely financial climate (Harvey 1985, p. 28), a lament we can trace back at least through the 18th and 17th centuries in England and which is probably built into the very idea of landscape (Daniels 1982b, 1986). But, in criticism of Clark, Harvey insists on a structural basis to capitalism and to the images which form within it: 'There is something wrong, I believe, with regarding money or the labour process as just one kind of representation among many'; such a post-structuralist view as Clark's can get little purchase on 'the true dynamic of the transformation of class relations' (Harvey 1987, p. 320). For Harvey the structural basis of modern capitalism lies in the process of capital accumulation, and the point of taking the spectacle of capitalism seriously is to see through it to this basis (Harvey 1985, p. 200) to the realm which Clark caricatures as 'the bedrock of matter and action upon it'. If Clark is (ironically) aware of the post-structuralist aspects of his text, the piling of representation upon representation (Clark 1984, p. 10), Harvey's long chapter on 'Paris, 1850–1870' is cast in a more classical mould: cultural issues are addressed consequent to a description of 'a first-order material fact . . . the actual order of space', the articulation in the built environment of the process of capital accumulation and circulation (Harvey 1985, p. 76) – much as in traditional regional geography texts cultural issues are addressed consequent to the first-order material facts of physiographic relief and drainage.

The most sustained examination of the adequacy of landscape by a Marxist geographer is Cosgrove's *Social formation and symbolic landscape* (1984). Cosgrove locates the origins of the idea of landscape in the early capitalist areas of Italian Renaissance city–states: 'the city is the birthplace of both capitalism and landscape.' Landscape is for Cosgrove a 'way of seeing' that, with its basis in the mathematical technique of linear perspective, has affinities with the 'basic techniques of capitalist life' developed in the Renaissance, such as book-keeping, quantity surveying, and the surveying and mapping of land (Cosgrove 1984, pp. 70, 86). And yet, built into the idea of landscape is a conservative resistance to capitalism, especially financial capitalism. This is manifest in cities but most powerfully in the 'villa culture' of the countryside. As a 'way of seeing' landscape developed strongly in the Venetian territories during the phase of so-called 'arrested' capitalism in the 16th century when the focus of investment shifted from trade and finance to land. And the power of landscape then as now was to give the impression that far from being implicated in the commodification of land and any attendant class tensions, it represented a world of nature, or a world where land and life were in harmony. In Renaissance Italy or Georgian England, landscape might offer a complex and comprehensive way of seeing but this, for Cosgrove, is ultimately illusory. Landscape is an ideology, a sophisticated 'visual ideology' which obscures not only the forces and relations of production but also more plebeian, less pictorial, experiences of nature.

It is instructive to compare Cosgrove's approach to landscape with that of the Marxist sociologist Inglis. Far from being a ruling-class illusion, Inglis declares 'a landscape [to be] the most solid appearance in which a history can declare itself'. Moreover, a landscape is more than an 'object', 'it is a living process; it makes its men; it is made by them'. And to conceptualize history as landscape 'is to fasten historical Marxism with unusual purchase on its central and dynamic ideas'. Inglis does not deny more patrician meanings of landscape but does deny that these have enclosed it. 'Although derived [in England] from the design language created in that high cultural moment at the end of the eighteenth century . . . the celebration of community to which [landscape] gave rise is more than a ruling class phantom with which to keep society quiet.' Even the rustic landscapes of televised English soap operas and puppet shows are not 'false consciousness' but the 'solid embodiment of . . . popular culture' (Inglis 1977, pp. 489, 495, 511).

Landscape is a concept of high tension, declares Inglis, because it 'stands at the intersection of concepts a social scientist would strain to hold apart: "institution", "product", "process" and "ideology"' (Inglis 1977, p. 489). We might further observe that this tension between landscape as an élitist (and illusory) 'way of seeing' and landscape as a vernacular (and realistic) 'way of life' is rooted in the very complications of the idea of culture I discussed earlier. Landscape may be seen, as Adorno sees culture generally, as a 'dialectical image', an ambiguous synthesis whose redemptive and manipulative aspects cannot finally be disentangled, which can neither be completely reified as an authentic object in the world nor thoroughly dissolved as an ideological mirage (Jay 1984, pp. 111–60). What can be done is to explore this duplicity of landscape – which has been sensed since the first use of the word landscape in English in the 17th century (Daniels & Cosgrove 1987) – in particular contexts.

Adorno himself sketches the notion of landscape as a dialectical image in some remarks on 'the domain of what the Germans call Kulturlandschaft' – old architecture in local materials arranged around a square or church. Because such *Kulturlandschaften* are now encountered 'mainly in ads promoting festivals of organ music and phoney togetherness', are co-opted by market society as the 'ideological complement of itself' because 'they do not visibly bear [its] stigmata', the joy we might feel at seeing them 'is spoiled by a guilty conscience'. Even so 'that joy has survived the objection which tries to make it suspect.' The factor that gives *Kulturlandschaften* 'the most validity, and therefore staying power, is their specific relation to history . . . [they] bear the imprint of history as expression and of historical continuity as form. They dynamically integrate these elements in ways similar to artistic production. What commands aesthetic attention in them is the manner in which they give expression to human suffering. It is only just that the image of a limited world should make us happy, provided we do not forget the repression that went into making it. In this sense, that image is a reminder' (Adorno 1984, pp. 94–7). And, to press the dialectic further, we might add that the image is also a forgetting.

The power of landscape as a concept resides, to use Inglis's phrase, in the 'field of force' which its oppositional meanings generate. In the remainder of this chapter I will explore and exploit this field of force as it is generated in the works of the two British Marxist cultural critics most concerned with geographical experience and imagination, Williams and Berger. This will involve considering the pressure exerted upon landscape in their works by cognate concepts such as country, region, land, and nature and the counterveiling pressure which landscape exerts upon them.

Raymond Williams

'A working country is hardly ever a landscape', declares Williams in *The country and the city* (1973) 'the very idea of landscape implies separation and observation'. Landscape is an 'elevated sensibility' fundamentally one of patrician control, which is reproduced in 'internal histories of landscape painting, landscape writing, landscape gardening and landscape architecture' and sentimentalized as 'part of an elegy for a lost way of life'. Such ideological histories, Williams maintains, should be connected to and exposed by the 'real history . . . the common history of a land and its society' (Williams 1973, pp. 120–1). The defining feature of this real history and one which is masked by the conventions of patrician landscape, is the condition and experience of those who worked the land. When considering 18th-century writers who questioned patrician taste Williams is more sympathetic to a poet like Crabbe who professed to give 'a real picture of the poor' and who for Williams 'restore[d] the facts of labour to the idyllic landscape' than he is to a poet like Goldsmith whose nostalgic evocation of an independent and leisurely peasantry Williams accused of dissolving 'the lives and work of others into an image of the past' (Williams 1987). As Barrell points out, this appraisal misses the repressive force of Crabbe's insistence on toil, and the egalitarian implications of Goldsmith's celebration of leisure: Goldsmith disengaged the labourer from his 'proper', 'natural' identity as a labourer, as a man born to toil, and suggested that he could be 'as free to dispose of his time as other poets insisted only the richman or the shepherd was free to do' (Barrell 1980, p. 78). In recovering the peasant village as an 'image of the past', overlaid by the park of an engrossing landlord, Goldsmith affirmed that the landscape and its poor inhabitants had a past that was qualitatively different from the present and, by implication, a future that might be so too. In the historicity of the landscape lay its potential liberation from, and resistance to, the actuality of present toil and hardship. Williams is aware of this historical dimension to landscape but locates it not in the narratives of 18th-century poetry but in the narratives of Victorian novels.

While acknowledging the Victorian novel's potential for a more adequate representation of land and society, Williams notes the residual power of 18th-century sensibilities. So although George Eliot 'restores the real inhabitants of rural England to their places in what had been a socially selective landscape she does not get further than restoring them *as a landscape*' (Williams 1973, p. 168; emphasis in original). Williams credits Thomas Hardy with more fully developing the novel as a 'knowable community' in a tradition to which Williams himself as a novelist subscribes. Hardy country is not for Williams so much a location as a relationship, a sometimes tense dialectic of dwelling and detachment expressed for many most forcibly in *The return of the native*:

> At the same time the separation of the returned native is not only a separation from the standards of the educated and affluent world 'outside'. It is also, to some degree inevitably, a separation from the people who have not made the journey; or more often a separation which can mask itself as a romantic attachment to a way of life in which people are merely instrumental: figures in a landscape . . . . [So] the real Hardy country, we soon come to see, is that border country so many of us have been living in: between custom and education, between work and ideas, between love of place and experience of change (Williams 1973, pp. 203, 196).

Border country is the title of Williams's first, autobiographically informed novel published in 1960 and I want to consider this book, the first in a trilogy, and *The fight for Manod* (1979a), the final book, because they forcefully disclose the shifting status of landscape in Williams's thought.

In an interview Williams maintained that in *Border country* he was trying to realize the potential of the Victorian realist novel, especially as written by Hardy, in accounting for 'the experience of work' and (against the trend of modern fiction) trying to recover some of its formal devices in giving characters and places 'a whole network of history'. So a place becomes 'not just the site of an event . . . but the materialization of a history which is often quite extensively retracted'. This, for Williams , 'was the highest moment of bourgeois cultural engagement: a moment from which historical materialism is itself a development' (Williams 1979a, pp. 276, 275). At the beginning of *Border country* Matthew Price, son of a railway signalman, and a university lecturer in economic history, returns, when his father suffers a stroke, to the Welsh valley of his upbringing:

> He had felt empty and tired, but the familiar shape of the valley and the mountains held and replaced him. It was one thing to carry its image in his mind, as he did, everywhere, never a day passing but he closed his eyes and saw it again, his only landscape. But it was different to stand and look at the reality. It was not less beautiful; every detail of the land came up with its old excitement. But it was not still, as the image had been. It was no longer a landscape or a view, but a valley that people were using. He realized, as he

> watched, what had happened in going away. The valley as landscape had been taken, but its work forgotten (Williams 1960, p. 75).

Price finds this narrowly scenic notion of landscape as blunt an instrument for understanding the valley as the demographic concepts and techniques he deploys in his research on 19th-century migration into the industrial valleys of South Wales. But by the end of the novel, after its narration of the social experience of his father's generation (centrally the General Strike and its connections well beyond the valley), Price enjoys a more comprehensive view of the valley, sitting above a black rock, the Kestrel:

> It was strange to be up there alone, with the valley so quiet. The people he had lived with, the voices he had listened to, were all there under his eyes, in the valley. But all he could see from this height were the fields and orchards; the houses white under the sun; the grey farm buildings; the occasional train, very small under its plume of smoke. . . . The station was out of sight, hidden in its cutting. Work went on there, in the ordinary routine, but from here it might not have existed, and the trains might have been moving themselves, with everyone gone from the valley (Williams 1960, pp. 290–1).

But on remaining there, sitting 'very still' Price looks below the Kestrel, 'in legend . . . the guardian, the silent watcher over this meeting of the valleys', and he focuses in:

> He could see the detail of oak and elm, in the full hedgerows. There, in an orchard, was an old tedder turned on its side, and the littered straw round a half-eaten potato clamp. He saw the long white wall of a farmhouse, and the end wall, at a curious angle, was crossed by a melting violet shadow. And here, on the high-banked road, was a cart moving, and a dog barked somewhere, insistently, not too far away. This, seen close, was his actual country. But lift with the line of the Kestrel, and look far out. Now it was not just the valley and the village, but the meeting of valleys, and England blue in the distance. In its history the country took on a different shape (Williams 1960, pp. 290–1).

Price looks in each direction, turning from 'the decayed shape of violence' of Norman dominion 'confused in legend with the rockfall of the Holy mountain, where the devil's heel had slipped as he strode westward into our mountains' to a limestone scarp where 'along the outcrop had stood the ironmasters, Guest, Crawshay, Bailey, Homfray, and this history had stayed'. His witness over, Price descends to the valley:

> On the way down the shapes faded and the ordinary identities returned. The voice in his mind faded, and the ordinary voice came back. Like old Bailey asking, digging his stick in the turf. What will you be reading Will [Price's local nickname]. Books, sir? No, better not. History, sir. History from the Kestrel, where you sit and watch memory move, across the wide valley. That was the sense of it: to watch, to interpret, to try to get clear (Williams 1960, pp. 292–3).

This profound conception of landscape and observation is then characterized by a deep sense of the past – historic, prehistoric, geological, mythical – a sense which diminishes the lives of those in the present, and enlarges them too. This conception of landscape is characteristic of Hardy. In novels like *Tess of the D'Urbervilles* and *Jude the Obscure* the landscape as a densely textured record of history, community, and personal experience is counterposed to those forces of progress and mobility which threaten to erase it, to reduce it to a scenic or utilitarian resource. It is a conception of landscape which has this century powerfully informed many popular and academic narratives of the English countryside, notably Hoskins's *The making of the English landscape* (1970). It subverts narrowly scenic conceptions of landscape both in its detailed focus and in its scope which can best be called regional. For Williams, it is the regional dimension of the English novel which provides the potential to resist patrician or bourgeois forms of rural description. While the notion of region still carries its original implication of subordination and underdevelopment, it can also, in a more modern derivation, denote a counter-movement to a centralized, sophisticated notion of culture, carrying the 'implications of a valuably distinctive way of life'. Williams admits that some of Hardy's novels like *Under the greenwood tree* are regional 'in an encapsulating and enclosing sense' – they contribute to his reputation 'as a regional novelist because he wrote about Wessex – that strange, particular place – rather than about London or the Home Counties' yet other novels like *Tess* 'are set even more deeply in their region but . . . are not in any limiting sense "regional": what happens in them, internally and externally – those two abstractions in a connected process – involves a very wide and complex, a fully extended and extensive, set of relationships' (Williams 1983, pp. 265–6; 1984b, pp. 231–2). Williams wrote *Border country* seven times to find this regional form and in the *The fight for Manod*, the final novel of the trilogy, he 'found that form very consciously and explicitly' (Williams 1984b, p. 231).

Whereas *Border country* was for Williams 'the present, including and trying to focus an immediate past', *The fight for Manod* was 'a present trying to include and focus a future' (Williams 1979a, p. 292). In the language of landscape it is a novel of prospect. Matthew Price, still an industrial historian, is employed as a consultant on a proposal to build a city in mid-Wales. Price's brief is to produce a 'lived inquiry', to dwell in the locality of the proposal for as long as it takes to understand the feelings and needs of local people. The city is conceived similarly, 'in post-industrial terms . . . a city of small towns, a city of villages almost. A city settling into its country' (Williams 1979b, pp. 15, 13). The fight for Manod is a fight for the different meanings of Manod although, as some leftist critics have pointed out, there is 'surprisingly little actual fight demonstrated in the narrative itself . . . the absence of collective scenes is striking. There is no equivalent of the solidarity [in *Border Country*] of the railway line' (Williams 1979a, p. 294). This 'thinning out' (as these critics called it) of the experience of labour is accompanied by a 'thickening' of Price's own experience of Manod, notably his experience of the natural world.

As in *Border country*, Price's sensibilities are initially imprisoned within élitist conventions. Looking at the plans at a ministerial office

> he had caught the familiar smell of a world of arrangements beyond him; of things happening, planned, brought about, without people even being told.

> He stared through the window at the railway yard, the office tower, the streets climbing the hill. By narrowing his eyes he could see the window and the city beyond it as a framed picture. It reminded him of the map on the opposite wall (Williams 1979b, p. 17).

Approaching the valley by car he sensed the physical shape of the land and in 'the long folds, the sudden scarps, the deep watercourses' he found it 'so much his own country: solid, remote, self contained'. Then looking down from a ridge upon the town to be transformed, 'its history stood out very sharply, as if on a slide.' He recalls 'The Kestrel, the watcher' and 'seeing the history of his country in the shapes of the land', but tries now to imagine it inscribed with a future, 'scrawls and projections, lines on glass, over known places, the embedded lives'. 'It is easy to reject it for the warmth, the heaviness, of a known past: a green past, in which lives have been lived and completed . . . this projection and outline of a future. Seen from above, from the height of the Kestrel. From this ridge below the Daren, it must be like this always. It is history past and future, an extended landscape' (Williams 1979b, pp. 29, 36, 38).

As it turns out the 'extended landscape' remains barely realized, torn between, on one hand, a calculating conception of location and, on the other, a sensual conception of land. For the post-industrial city turns out, upon investigation by Price's fellow consultant Peter Owen to be a corrupt speculation by a multinational corporation, 'its local reproduction'. And while Owen, a late 1960s radical sociologist, surgically delineates the circuits of capital, Price fights for words to express the sensory pull of the land. Together they look down at the rush of water under a bridge:

> "You feel it here, do you?", Matthew asked.
> "Feel what?"
> "Something different. Something other. Some altered physical sense."
> "You grew up in this sort of country, Matthew. You're remembering that."
> "Yes, I know when I remember it. And I try to understand it as memory. But sometimes I think it's a different experience. Something that actually alters me."
> "No", Peter said, "I don't get that."
> "It only comes occasionally. Some particular shape: the line of a hedge, the turn of a path round a wood, or in movement sometimes, the shadow of a cloud that bends in a watercourse, or then again a sound, the wind in wires, wind tearing at a chimney."
> "Moments of heightened attention."
> "Yes, but attention to what? What I really seem to feel is these things as my body. As my own physical existence, a material continuity in which there are no breaks. As if I was feeling through them, not feeling about them."
> Peter stared down at the river.
> "I can hear that you feel it. But . . . ."
> "Yes. But. There's usually nothing to be said" (Williams 1979b, pp. 97–8).

This dialogue specifies Williams's concept of 'structure of feeling', the 'area of tension' between articulation and experience, and an area in which Williams often finds himself. He can feel the 'sensory pull' of some monumental buildings like cathedrals and yet be aware of the way they inscribe social dominion in their physical fabric, in their reproduction as 'heritage', and in precisely the kind of powerful feelings which they were designed to induce. 'I find it even more difficult when I take the problem across to the very strong feelings which I have about land . . . I don't fully understand them, partly because I've never had any training in this kind of visual discrimination' (Williams 1979a, p. 348). Berger, primarily concerned with the visual arts not literature, has developed this kind of visual discrimination to a remarkably high degree and it is to his interrogation of landscape that I now turn.

John Berger

If for Williams 'we learn to see a thing by learning to describe it' (Williams 1965, p. 39), for Berger 'seeing comes before words' (Berger 1972, p. 7). Berger has probed the varieties of visual experience – looking, scanning, ogling, imagining – in a variety of visual media, notably painting and photography. Moreover, Berger is a highly visual writer, in his figures of speech and in his tendency to display arguments (in epigrammatic sentences, sometimes with pictures) rather than develop them prosaically. It is then not surprising that landscape with its irreducibly visual dimension is more central to Berger's work than it is to Williams's. It is more troubling for Berger too. For putting pressure on his visual sensibility is Berger's increasing commitment to narrative. Indeed, Berger's work can be seen to explore the territory between image and narrative, between, in his phrases, 'ways of seeing' and 'ways of telling' (Dyer 1986, pp. 118–42). At the beginning of his 1967 documentary essay on a country doctor, *A fortunate man* (Berger & Mohr 1967), and over a photograph of the doctor's district by his collaborator Jean Mohr, Berger declared 'Landscapes can be deceptive. Sometimes a landscape seems to be less a setting for the life of its inhabitants than a curtain behind which their struggles, achievements and accidents take place.' In his endeavour to get 'behind the curtain' Berger next describes an incident mixing qualities that contradict the tranquil view Mohr had photographed – the doctor rushing to minister to an agonized woodcutter crushed beneath a tree (Berger & Mohr 1967, pp. 12–13, 17–19). Nevertheless, many of Mohr's subsequent photographs still show views of the countryside and Berger continues to describe how it looks, but now as part of an attempt to represent landscape as something lived in not just looked at, and to make it express the experiences and knowledge of those who live in it. Berger & Mohr focus again on the drama of woodcutting in a more recent attempt to realize a more adequate notion of landscape, *Another way of telling* (1982), a book documenting life in the peasant village in the Haute Savoie (where Berger now lives) and intending to represent 'the way peasants look at themselves'. One sequence of photographs is commissioned by a woodcutter's wife to commemorate her husband if he should be killed in the forest and is permitted by the husband 'on condition that they show what the *work* is like', centrally the critical and dangerous moment when the tree falls. The sequence shows, from many angles and from many focal lengths, the process of felling a tree, culminating with a landscape, a view of the woodcutter in a clearing burying his axe in the log, posed against the background of the wooded slopes. In opposition to picturesque renderings of the peasantry and their countryside, it is an image that attempts to make 'appearances become the language of a lived life' (Berger & Mohr 1982, pp. 84, 59–70, 288). But it is not an image that is independent of a tradition of rural painting. Indeed, the photograph of the woodcutter burying his axe in a log recalls a painting of woodcutting by Millet, the *Woodsawyers* (1848). And the recollection is surely not accidental, for, according to Berger, it was Millet who 'destroy[ed] the traditional language for depicting scenic landscape' by 'representing the close, harsh, patient physicality of a peasant's labour *on*, instead of *in front of*, the land' (Berger 1980, p. 77).

Much of Berger's analysis of the media of landscape representation has been concerned with their potential both to obscure and to articulate lived experience. As an existentialist Marxist, whose interpretations of the experience of woodcutting owe as much to Heidegger as to Marx or to the woodcutters themselves (Berger 1980, pp. 79–86), Berger is as sensitive to the self-consciousness of those who make landscapes on canvas or paper as he is to that of those who make them on the ground. Not surprisingly, this double sense of lived experience can be a source of tension. For example, in an essay of 1966 'Painting a landscape', he questions why he should feel both innocent pleasure and philosophical significance in 'making lustreless marks upon a tiny canvas in front of a landscape that has been cultivated for centuries and in view of the fact that such cultivation has and still demands lives of the hardest labour' especially when 'any peasant born in the village knows it a thousand times better'. Reasoning that since Cubism has 'broken' the genre of landscape as the replication of appearances he feels free to explore a psychoanalytical dimension of painting. In painting this landscape his sense of mastery is in imagining a world from above and he likens this to 'the experience of every child who imagines he is able to fly', for 'the sexual content assumed to exist in dreams about flying is closely connected to the sexual content of the landscape' (Berger 1972b, pp. 172–7).

In *Ways of seeing* (1972a), a television series and a book, Berger locates the repressive aspect of this pictorial mastery in realistic conventions of landscape painting, as part of his thesis on the complicity of the realistic tradition of Western painting with capitalist strategies of economic, social, and sexual appropriation and control. Linear perspective, centring 'everything on the eye of the beholder', and oil paint, 'reduc[ing] everything to the equality of objects' made the model of this tradition 'not so much a framed window open on to the world as a safe let into the wall, a safe in which the visible has been deposited'. So the point of Gainsborough portraying Mr and Mrs Andrews in the setting of their estate was to give them 'the pleasure of seeing themselves depicted as landowners and this pleasure was enhanced by the ability of oil paint to render their land in all its substantiality' (Berger 1972a, pp. 16, 87, 109, 108). Yet even in *Ways of seeing* the notion of landscape escapes bourgeois appropriation. At the beginning of the book Berger uses it as a metaphor for the kind of social-historical knowledge he himself uses to contextualize and criticize a painting like *Mr and Mrs Andrews* and indulgent interpretation of it; 'when we "see" a landscape, we situate ourselves in it. If we "saw" the art of the past, we would situate ourselves in history.' Indeed, Berger maintains that of all the categories of oil painting, landscape is the one to which his thesis of a proprietorial way of seeing applies the least, for 'landscape painting begins with the problems of painting sky and distance [and] the sky has no surface and is intangible; the sky cannot be turned into a thing or given a quantity.' Furthermore, it was innovative landscape painters like Turner and Monet who subverted the tradition of oil painting by leading it 'progressively away from the substantial and tangible towards the indeterminate and intangible' (Berger 1972a, pp. 11, 105). Nevertheless, this celebration of artists who dematerialize may be seen to be consistent with another thesis of *Ways of seeing*, that derived from Walter Benjamin, that the transformation of painting by modern means of reproduction into a more ephemeral 'language of images' is potentially liberating and consistent also with his photomontagist strategy of placing on a print of *Mr and Mrs Andrews* a sign saying 'Trespassers keep out'.

Subsequent to *Ways of seeing* Berger has further questioned the complicity of oil painting with bourgeois attitudes towards property; indeed he has argued that the very substance of oil paint offers a potential space in which to resist these attitudes, not in dissolving the material world but in reconstituting it in ways inimical to bourgeois versions of materialism. So to explain the subversive power of Courbet's very substantive oil paintings, Berger offers what he calls a 'geographical interpretation' to 'ground, give material, visual substance to, the social-historical one'. This is based on the 'need to interrogate the landscape' Courbet grew up in and painted throughout his life – the valley of the Loue on the western side of the Jura mountains – to 'visualize the mode of appearances in such a landscape in order to discover the perceptual habits it might encourage'. These appearances include dense forests, steep slopes and waterfalls, and fundamentally the outcrops of limestone 'which create the presence of the landscape'; where so much is in shadow and 'the visible . . . has to be grasped when it does make its appearance', the perceptual habits encouraged are 'the eyes of a hunter'. Painted on dark ground and with little atmospheric perspective there is 'no hierarchy of appearances' in Courbet's landscapes: 'Courbet painted everything – snow, flesh, hair, fur, clothes, bark – as he would have painted it had it been a rock face.' And it is in giving life to this landscape 'in which the visible is both lawless and irreduceably real' (as much as in articulating the hard lives of those work in such a landscape) that Berger sees the subversive power of Courbet's art. Proudhon, who came from the same area, wrote: 'I am pure Jurassic limestone.' Courbet, boastful as always, said that in his paintings, 'I even make stones think.' Berger suggests this physiographic interpretation of painting might be extended: 'The Thames developed Turner. The cliffs around Le Havre were formative in the case of Monet' (Berger 1980, pp. 134–41).

In signifying the realistic, historical, and material interconnections they made between land and life, Berger's admiration for Courbet's and Millet's sense of landscape (as against Gainsborough's) corresponds to Williams's admiration for Hardy's sense of landscape (as against Goldsmith's and Jane Austen's). And the correspondence goes further than this. Like Williams, Berger in his recent writings distances himself from semiological varieties of Marxist cultural theory in emphasizing a materialism which is biophysical as well as historical (Berger & Mohr 1982, pp. 111–16). Referring both to bourgeois and to revolutionary theory Berger declares that 'the nineteenth-century discovery of history as the terrain of human freedom inevitably led to an underestimation of the ineluctable and continuous.' Also underestimating it has been 'the new reductionism of revolutionary theory'. So he attacks *Art history and class struggle* (1978) by Hadjinicolaou for its reduction of all paintings to ideology, likening it to the way the culture of capitalism reduces them to commodities or to advertisements for other commodities: 'Both eliminate art as a potential model of freedom.' Resisting Hadjinicolaou's conception of painting as the reproduction of 'visual ideology' and the pleasure of looking at painting to a further reproduction of that visual ideology, Berger emphasizes the energy in the 'work' of art, the energy used by 'every painter since palaeolithic times onwards' to 'push' against the limits of his materials. This 'will to push' is for Berger 'intrinsic to the activity, to rendering the absent present, of cheating the visible, of making images'.

> Ideology partly determines the finished result but it does not determine the energy flowing through the current. And it is with this energy that the spectator identifies. Every image used by a spectator is a *going further* than he could have achieved alone, towards a prey, a Madonna, a sexual pleasure, a landscape, a face, a different world (Berger 1985, pp. 197–204).

Berger's notion of the imagination as a kind of craft is here deployed against the reproductive ideology of image-making in the mass media. In a recent essay, Berger upholds Van Gogh as an artist who closed the space between the acts of painting things and of making them, in his pictures of chairs, beds, and boots, but also in his landscapes, if 'before a landscape the process required was far more complicated and mysterious'. 'If one imagines God creating the world from earth and water, from clay, his way of handling it to make a tree or a cornfield might well resemble the way Van Gogh handled paint when he painted that tree or cornfield.' The religious metaphor is not meant here, or elsewhere in Berger's recent work, to question the materialism of nature or of painting but to restore the spiritual implications of creation to the material world. When Van Gogh 'painted a road, the roadmaker was there in his imagination. When he painted the turned earth of a ploughed field, the gesture of the blade turning the earth was included in his own act . . . when he painted a small peartree in flower, the act of the sap rising, of the bud forming, the bud breaking, the flower forming, the styles thrusting out, the stigmas becoming sticky, these acts were present for him in the act of painting.' The artist whose paintings are perhaps most subject to capitalist commodification – in the number of photographic reproductions, in the enormous prices for his works – Van Gogh, in the original, 'takes us as close as any man can . . . to that permanent process by which reality is produced' (Berger 1985, pp. 176–281).

Berger then closes the space between the aesthetics of art and the aesthetics of nature. While 'the social use to which an aesthetic emotion may be put changes according to the historical moment', he acknoweldges, 'yet there seem to be certain constants which all cultures have found "beautiful": among them – certain flowers, trees, forms of rock, birds, animals, the moon, running water . . .'. And the 'aesthetic moment' of finding such things beautiful 'offers hope', it means 'that we are less alone, that we are more deeply inserted into existence than the course of a single life would lead us to believe'. 'Several years ago', recalls Berger, 'when considering the historical face of art, I wrote that I judged

Figure 9.1 *The Haywain*, Constable (1821) *Cruise Missiles*, USA (1983). Photomontage by Peter Kennard, reprinted with permission.

a work according to whether or not it helped men in the modern world to claim their social rights. I hold to that. Art's other, transcendental face raises the question of man's ontological rights' (Berger 1985, pp. 8–9).

Landscape with missiles

I will conclude this chapter by considering one powerful landscape image which seems to me to be packed with the complications I have identified, Peter Kennard's 1981 photomontage *The Haywain – Cruise missiles* (Fig. 9.1). It shows the wain of Constable's painting laden with nuclear missiles, the carters decked out in military helmets, and the landscape bleached with an incandescent orange light. Some of its power resides in making the wain a parody of one of the American controlled Cruise missile transporters presently based in south-east England, ready to launch an attack from, and in turn to trigger an attack upon, countryside that is quintessentially English – in the words of another anti-nuclear image of an American airbase, 'England's Greenham Pleasant Land'. The power of Kennard's montage also resides in its pedigree in English culture, indeed in its rôle in defining English culture. During this century images of the English countryside (often represented in terms of a Constable) desecrated by un-English (and often American) technological aggression have become commonplace. In 1955 Hoskins concluded *The making of the English landscape* by describing the long deposition of history in the view from his window and the forces which threatened its meaning – American-style harvesters and lorries and 'the obscene shape of the atom-bomber, laying a trail like a filthy slug upon Constable's and Gainsborough's sky', (Hoskins 1970, p. 299). Reproduced in a variety of forms from gilt-framed prints to tea towels to advertisements, Constable's *Haywain* is now a powerful and popular image of England and Kennard has deliberately harnessed its populist nostalgia. Asked why in his images of nuclear horror he did not use 'familiar scenes . . . high streets and suburban estates, for example', Kennard replied that he 'tried to make the work relate to simple memories in people's minds, iconic in a sense' (Berger 1981, p. 9). In Berger's terms from *Ways of seeing* he tried to use the 'language of images' to 'confer a new kind of power'.

The power of Kennard's montage can also be seen to reside in bringing out qualities of Constable's painting often lost in reproduction. Now, the thesis of *Ways of seeing* on the relation of a famous work of art to its mechanical reproduction is that the original becomes an enigma, drained of all meaning other than that of being 'the original of a reproduction' and 'enveloped in an atmosphere of entirely bogus religiosity' which largely reflects (or rather refracts) its market price (Berger 1972, p. 21). We need not quarrel with that. Looking at the original *Haywain* it does seem almost opaque, filmed over with all those reproductions, and, in the setting of the National Gallery and a roomful of famous English paintings, enshrined in Englishness. But the glaring violence of Kennard's montage, its reduction, or more precisely its intensification of visual reproduction to the absurd, seems to strip away much of this perceptual varnish.

The montage brings out, by contrast, the organic qualities of Constable's original, both in the luxuriance of the trees, meadow, and reeds and in the absorption of human actions and artefacts into their surroundings, at one, we might now say, with nature. Such naturalism is no less manipulative than any other style of painting and the entire landscape Constable painted was an artefact, at the time one of the most intensely managed in England. Barrell has emphasized the repressive aspects of *The Haywain*, notably Constable's reduction of labour in the landscape to a few distant and indistinct figures, a prerequisite of that harmony between land and life celebrated in much Romantic art and literature at the time, and also since (Barrell 1980, pp. 146–9). In response to Kennard's montage a more indulgent, but still avowedly Marxist view of Constable's romanticism is taken by Fuller. He maintains that 'through the material process of painting itself' Constable achieves a 'convincing realization of aesthetic wholeness', 'the possibility of a world transformed for the better' in contrast to Cruise missiles, 'a symbol of absolute despair' (Fuller 1983, pp. 38–9). The continuing tradition of English Romantic landscape painting is still for Fuller an 'illusion' but a positive one, 'rupturing the surfaces of the given with imaginative transformations'; landscape provided 'an arena in which the subjective and the objective, the deeply personal and richly traditional, could be mingled in new and previously unseen ways' (Fuller 1985, p. 84).

In the context of Kennard's montage, I think it is both possible and desirable to conserve both an ideological and an ontological interpretation of Constable's landscape and to bring each critically to bear upon the other. In Adorno's terms the 'redemptive' and 'manipulative' dimensions of landscape should be kept in dialectical tension. Landscape is becoming a key word and key image in debates on conservation (Penning-Rowsell & Lowenthal 1987) and important (as the popularity of Kennard's montage attests) to the discourse of green socialism (Mabey 1984). We should beware of attempts to define landscape, to resolve its contradictions; rather we should abide in its duplicity.

Acknowledgement

For their helpful comments on an earlier draft of this chapter I wish to thank Denis Cosgrove, Felix Driver, Nick Entrikin, Mike Heffernan, Dave Matless, and Bob Sack. For her work on a number of drafts I am grateful to Karen Korzeniewski.

References

Adorno, T. 1984. *Aesthetic theory*. London: Routledge & Kegan Paul.
Anderson, P. 1976. *Considerations on western Marxism*. London: New Left Books.
Anderson, P. 1984. Modernity and revolution. *New Left Review* **144**, 96–113.
Barrell, J. 1980. *The dark side of the landscape: the rural poor in English painting 1730–1840*. Cambridge: Cambridge University Press.
Baudrillard, J. 1983. *Simulations*. New York: Semiotext.
Benjamin, W. 1973. The work of art in the age of mechanical reproduction. In *Illuminations*, W. Benjamin, 219–53. London: Routledge & Kegan Paul.
Berger, J. 1972a. *Ways of seeing*. Harmondsworth: Penguin.
Berger, J. 1972b. *Selected essays and articles: the look of things*. Harmondsworth: Penguin.
Berger, J. 1980. *About looking*. London: Writers & Readers Publishing Co-operative.
Berger, J. 1981. Interview. *Leveller*, 7–20 August, 9.
Berger, J. 1985. *The sense of sight*. New York: Pantheon Books.
Berger, J. & J. Mohr 1969. *A fortunate man*. Harmondsworth: Penguin.
Berger, J. & J. Mohr 1982. *Another way of telling*. London: Writers & Readers Publishing Co-operative.
Berman, M. 1983. *All that is solid melts into air*. London: Verso.
Berman, M. 1984. The signs in the street: a response to Perry Anderson. *New Left Review* **144**, 114–23.
Clark, T. J. 1984. *The painting of modern life: Paris in the art of Manet and his followers*. London: Thames & Hudson.
Cosgrove, D. E. 1983. Towards a radical cultural geography. *Antipode* **15**, 1–11.
Cosgrove, D. E. 1984. *Social formation and symbolic landscape*. London: Croom Helm.
Cosgrove, D. E. 1985. Prospect, perspective and the evolution of the landscape idea. *Transactions of the Institute of British Geographers* **10**, 45–62.
Cosgrove, D. E. & S. Daniels (eds) 1987. *The iconography of landscape*. Cambridge: Cambridge University Press.
Cosgrove, D. E. & P. Jackson 1987. New directions in cultural geography. *Area* **19**, 95–101.
Daniels, S. 1981. Landscaping for a manufacturer: Humphry Repton's commission for Benjamin Gott at Armley in 1809–10. *Journal of Historical Geography* **7**, 379–96.
Daniels, S. 1982a. Humphry Repton and the morality of landscape. In *Valued environments*, J. R. Gold & J. Burgess (eds), 124–44. London: Allen & Unwin.
Daniels, S. 1982b. Ideology and English landscape art. In *Geography and the humanities*, D. E. Cosgrove (ed.), 6–13. Loughborough University of Technology, Department of Geography, Occasional Paper 5.
Daniels, S. 1983. The political landscape. In *Humphry Repton landscape gardener*, G. Carter (ed.), 110–21. Norwich: Sainsbury Centre.

Daniels, S. 1985. Arguments for a humanistic geography. In *The future of geography*, R. J. Johnston (ed.), 143–58. London: Methuen.
Daniels, S. 1986. 'Cankerous blossom': troubles in the later career of Humphry Repton documented in the Repton correspondence in the Huntington Library, *Journal of Garden History* **6**, 146–61.
Daniels, S. & D. E. Cosgrove 1987. Iconography and landscape. In *The iconography of landscape*, D. E. Cosgrove & S. Daniels (eds). Cambridge: Cambridge University Press.
Debord, G. 1983. *Society of the spectacle*. Detroit: Black & Red.
Driver, F. 1985. Power, space and the body: a critical assessment of Foucault's *Discipline and punish*. *Environment and Planning D, Society and Space* **3**, 425–46.
Dyer, G. 1986. *Ways of telling: the work of John Berger*. London: Pluto Press.
Fuller, P. 1980. *Seeing Berger*. London: Writers & Readers Publishing Co-operative.
Fuller, P. 1983. *The naked artist*. London: Writers & Readers Publishing Co-operative.
Fuller, P. 1985. *Images of God: the consolations of lost illusions*. London: Chatto & Windus.
Galois, B. 1977. Ideology and the idea of nature: the case of Peter Kropotkin. In *Radical geography*, R. Peet (ed.), 66–94. Chicago: Maaroufa Press.
Gregory, D. 1982. *Regional transformation and the Industrial Revolution*. London: Macmillan.
Hadjinicolaou, N. 1978. *Art history and class struggle*. London: Pluto Press.
Harvey, D. 1979. Monument and myth. *Annals of the Association of American Geographers* **69**, 362–81.
Harvey, D. 1985. *Consciousness and the urban experience*. Oxford: Basil Blackwell.
Harvey, D. 1987. The representation of urban life. *Journal of Historical Geography* **13**, 317–21.
Hill, C. 1977. *Milton and the English Revolution*. London: Faber.
Hobsbawm, E. 1983. Introduction: inventing traditions. In *The invention of tradition*, E. Hobsbawm & T. Ranger (eds), 1–14. Cambridge: Cambridge University Press.
Hoskins, W. G. 1970. *The making of the English landscape*. Harmondsworth: Pelican.
Inglis, F. 1977. Nation and community: a landscape and its morality. *Sociological Review* NS **25**, 489–514.
Jay, M. 1984. *Adorno*. London: Fontana.
Jameson, F. 1984. Post-modernism, or the cultural logic of late capitalism. *New Left Review* **146**, 53–92.
Jarvis, B. 1985. The truth is only known by guttersnipes. In *Geography, the media and popular culture*, J. Burgess & J. R. Gold (eds), 96–122. London: Croom Helm.
Jencks, C. 1986. *What is post-modernism?* New York: St. Martin's Press.
Johnson, R. 1979a. Culture and the historians. In *Working-class culture*, C. Critcher & R. Johnson (eds), 41–74. London: Hutchinson.
Johnson, R. 1979b. Three problematics: elements of a theory of working-class culture. In *Working-class culture*, J. Clarke, C. Critcher & R. Johnson (eds), 201–37. London: Hutchinson.
Johnston, R. J., D. Gregory & D. M. Smith 1986. *The dictionary of human geography*, 86–8. Oxford: Basil Blackwell.
La Capra, D. 1983. *Rethinking intellectual history: texts, contexts, language*. Ithaca: Cornell University Press.
Latimer, D. 1984. Jameson and Post-Modernism. *New Left Review* **148**, 116–28.
Ley, D. 1985. Cultural/humanistic geography. *Progress in Human Geography* **9**, 415–23.
Ley, D. 1987. Styles of the times: liberal and neo-conservative landscapes in inner Vancouver, 1968–86. *Journal of Historical Geography* **13**, 40–56.
Ley, D. & R. Cybriwsky 1974. Urban graffiti as territorial markers. *Annals of the Association of American Geographers* **64**, 491–505.
Mabey, R. (ed.) 1984. *Second Nature*. London: Jonathan Cape.
Massey, D. 1984. *Spatial divisions of labour*. London: Macmillan.

Mitchell, W. J. T. 1980. The language of images. *Critical Inquiry* **6**, 359–62.

Mitchell, W. J. T. 1986. *Iconology: images, text, ideology*. Chicago: University of Chicago Press.

Olwig, K. 1984. *Nature's ideological landscape*. London: Allen & Unwin.

Penning-Rowsell, E. & D. Lowenthal 1987. *Landscape meanings and values*. London: Allen & Unwin.

Punter, J. V. 1982a. Landscape aesthetics: a synthesis and critique. In *Valued environments*, J. R. Gold & J. Burgess (eds), 100–23. London: Allen & Unwin.

Punter, J. V. 1982b. English landscape tastes revisited. In *Geography and the humanities*, D. Cosgrove (ed.), 14–28. Loughborough University of Technology, Department of Geography, Occasional Paper 5.

Reclus, E. 1977. The influence of man on the beauty of the earth. In *Radical geography*, R. Peet (ed.), 59–65. Chicago: Maaroufa Press.

Sayer, A. 1979. Epistemology and conceptions of people and nature in geography. *Geoforum* **10**, 19–43.

Stoddart, D. R. 1986. *On geography*. Oxford: Basil Blackwell.

Thompson, E. P. 1963. *The making of the English working class*. New York: Vintage Books.

Thompson, E. P. 1975. *Whigs and hunters*. London: Allen Lane.

Thompson, E. P. 1978a. *The poverty of theory*. London: Merlin.

Thompson, E. P. 1978b. Folklore, anthropology and social history. *Indian Historical Review* **3**.

Thrift, N. 1983. Literature, the production of culture and the politics of place. *Antipode* **15**, 12–24.

White, H. 1973. *Metahistory: the historical imagination in nineteenth century Europe*. Baltimore: Johns Hopkins University Press.

Williams, R. 1958. *Culture and society (1780–1950)*. London: Chatto & Windus.

Williams, R. 1960. *Border country*. London: Chatto & Windus.

Williams, R. 1965. *The long revolution*. Harmondsworth: Penguin.

Williams, R. 1967. Literature and rural society. *Listener*, 16 November.

Williams, R. 1973. *The country and the city*. London: Chatto & Windus.

Williams, R. 1977. *Marxism and literature*. Oxford: Oxford University Press.

Williams, R. 1979a. *Politics and letters*. London: New Left Books.

Williams, R. 1979b. *The fight for Manod*. London: Chatto & Windus.

Williams, R. 1980. *Problems in materialism and culture*. London: New Left Books.

Williams, R. 1981. *Culture*. London: Fontana.

Williams, R. 1983. *Keywords*. London: Fontana.

Williams, R. 1984a. Between country and city. In *Second nature*, R. Mabey (ed.), 209–19. London: Jonathan Cape.

Williams, R. 1984b. *Writing in society*. London: Verso.

Williams, R. 1987. The practice of possibility. *New Statesman*, 10–15, 7 August.

10 *What is locality?*

Simon Duncan

Introduction: locality – fashion or reality?

Locality has suddenly emerged as one of the more popular ideas in social science, especially in sociology, geography, urban and regional studies, and political science. A content analysis might reveal the term vying for place with structure, and cause, if not yet approaching the use levels of class, status or gender. But it is an infuriating idea. It is one that seems to signify something important, and indeed most people seem to know – roughly – what it signifies for them. Yet few would care to explain what locality (or is it a locality or even the locality) actually is. Even fewer, I suspect, would agree on the result – even if there was one.

This confusion and lack of clarity may be a consequence of the unexamined social assumptions contained within the term. Locality seems first of all to indicate that spatial variations make a difference to how social relations work – although without saying how and why. But even more than this, the term seems to suggest that particular places somehow capture a degree of social autonomy. Again, how and why this happens is rarely discussed and hence there is more than a whiff of spatial determinism – where spatial patterns cause social behaviour – about the term locality. But it is this very spatial determinism which has so marred earlier social concepts like community, the local social system, the urban way of life, region, *pays de vie* and indeed led to their rejection as adequate analytical concepts. Is locality just the latest means of resurrecting a comfortable – but in fact untenable – spatial determinism, handily sidestepping trenchant criticisms by inventing a new term? Why not, after all, simply refer to towns, villages, labourmarket areas, local authorities, case study areas or even – if we want to be completely general – to places? This chapter will attempt to sort out what the term locality is trying to circumscribe and, in so doing, make some judgements about how far and in what circumstances use of the term is justified.

The current popularity of locality can be seen from any cursory glance through the recent literature in the urban and regional studies field. It is sometimes used, in similar but different ways, as a shorthand for spatially affected social process – for 'place based relations between work and community' (Massey 1984), 'local stratification systems' (Urry 1983), or the 'key basis of collective identification in contemporary capitalism' (Cooke 1985; in the end Cooke prefers the older idea of region). Locality also seems to fit the bill by suggesting generality at the same time as it signifies specificity. A book I helped to write is called *Housing, states and localities* (Dickens *et al.* 1985) not – as in our original plan – 'Housing and the state in social democracy: Britain and Sweden compared'. Similarly, a recent volume on political, economic, and social changes in different places is entitled *Localities, class and gender* (Murgatroyd *et al.* 1985). Note how locality is given equal status with housing, state, class and gender; no subtitles here. And so it goes on. A book reporting a mixed-theory findings conference in rural sociology and anthroplogy is entitled *Locality and rurality* (Bradley & Lowe 1984). A paper about gender relations and the division of labour, given at the International Sociological Association conference on Industrial Restructuring, Social Change and the Locality, calls for locality studies to give 'a new unity of approach' (Bowlby *et al.* 1986). The whole locality idea is buttressed by epistemological and theoretical developments stressing the importance of time–space edges, distanciation and the locale in creating human action (Giddens 1981), while contingent relations operating in particular times and at particular places and often specific to them are crucial in establishing causality (Urry 1981a, Sayer 1984).

So the term locality is both useful to people's research and, it appears, significant in the real world. One big problem remains, however – what is it? For the concept remains vague and underdeveloped, even at times contradictory in various forms. The social reasons for using it – to signify one's concern for the empirical and concrete without throwing the structural baby out with the structuralist bathwater – seem quite as important as its scientific use. Indeed, has locality a scientific use over and above its use as a signifier for a research approach and/or social group? Is it just a synonym for the more conceptually neutral local study area, town, local government unit, statistical area, case study areas, etc.? If it is, we should say so, use the precise term, and then go on to describe our research approach without the camouflage of pseudo-scientific jargon.

A good illustration of the social importance but simultaneous intellectual confusion of the locality concept is its major rôle in funded research, at least in Britain. Three major programmes funded by the Economic and Social Research Council (ESRC), which between them account for over £500 000 (quite a substantial sum in the straitened circumstances of British social science), have been centred around locality. Two of these programmes consisted of several research groups, run by a central co-ordinator on behalf of the ESRC, the Changing Urban and Regional System programme and the Social Life and the Economy initiative. The third, for which this chapter was first written, was the Economic Restructuring, Social Change and the Locality programme based at Sussex University. Altogether, these three programmes were made up of about 20 individual projects in the first instance together with central co-ordinators and support services. About 30 localities were taken for intensive case-study analysis studding the map of Great Britain. To judge from the initial briefs, locality played two functions, although these were never properly distinguished (ESRC 1985a, 1985b, Sussex Programme 1984). One was to signify the importance of the link between general or perhaps abstract processes (e.g. the new international division of labour) and specific events (e.g. social changes in particular areas, towns, and cities within Britain). The latter was not to be directly read off the former and locality seems to signify this autonomy and explain why it exists. Second, locality was used to refer directly to particular case-study areas to be examined in the research. On the face of it, these two usages appear to fit uncomfortably with one another. One refers, very vaguely, to an abstracted social process and the other to precisely defined statistical units.

One possible response to this criticism is the claim that locality is merely useful shorthand for case study area, local area, etc. But this is unsatisfactory, for inevitably (in my experience) people are using locality to signify something more than spatial unit; they employ the term to incorporate some notion of local social specificity or even autonomy. That being the case, what about the old concepts of region, *pays de vie*, community, the urban (and rural) way of life, local social system? Why not use these, or are they just unfashionable at the moment? In fact all contain a useful kernel in referring to real social bases for behaviour, some sort of locally referent system of action. Similarly, all have been heavily criticized. These terms very often elevate the local basis of social structure above all others by theoretical fiat, or depend upon spatial determinism, or are used as vehicles to maintain ideologies. None of these terms retains much credibility. So is locality just the latest means of resurrection? The idea can be used to preserve the notion that there is a local social basis to behaviour while at the same time escaping all the overwhelming criticisms of such approaches in the past by simply dispensing with them. Clearly, neither the locality means local area + ingredient = approach nor the locality means local social system – criticisms approach is adequate. But just as clearly, as the studies I quoted above and many more suggest, there *is* something there. How can we make better sense of locality?

At first sight the concern with locality – whatever this is – may seem paradoxical to outsiders, a case of academic navel contemplation, for it is surely the dramatic changes in the world economy and world politics that are most striking. Why concentrate so much on studying small areas when the really important events appear to be linked to the restructuring of multinational enterprises, the development of new forms of technology, the emergence of the Far East as a major economic force, the relations between the super powers, and the decisions of supra-national bodies like the EEC? We are all aware, of course, that there are considerable local variations in the way these wider changes operate, both at national and subnational levels. But, although deep differences remain and new ones continually emerge, the long-term trend is surely towards greater standardization at both these scales. For instance, much has been made of the re-emergence of a North–South (or is it urban–rural?) divide in the economic geography of Britain, with official unemployment rates (for men) of over 20 per cent in some of the older industrial areas and large cities, but below 8 per cent in most small towns and suburbs of the South East. But in the 1930s the scale of these differences ranged between below 5 per cent to over 80 per cent. It would appear that neither the scale of local differences, nor their overall importance, would necessitate a turn to locality research.

In fact, the main reasons for the interest in this sort of research lie, paradoxically, in social science theory – in questions of how the world should be explained rather than in real world changes themselves. Locality itself – or at least a spatial dimension to society – has always been there, and hence the importance of previous concepts like community or the urban/rural way of life attempting to deal with this dimension. That this dimension should re-emerge in the guise of locality depends upon a new appreciation of its rôle in social behaviour which in turn depends upon how behaviour is conceptualized. Very briefly, there has been a growing distaste for functionalist explanations of social behaviour. These functionalist explanations, sometimes connected to forms of structuralist Marxism, attempted to explain social phenomena by reference to the functions they performed for society, or for capitalism. The early work of Castells (e.g. 1977, 1978, Castells & Godard 1974) is perhaps the best known example in urban and regional studies. In this case cities were seen as the site of consumption processes involved in the reproduction of labour power, and this perspective was buttressed by a form of structuralist Marxism following Poulantzas and Althusser. But, as critics pointed out, while this rôle might be the case it did little to explain how people acted in specific cities, still less could it account for variations in what they did (e.g. Foster 1979, Saunders 1979, Sayer 1979a, Duncan 1981). This sort of reaction was codified in the work of Giddens (1979, 1981, 1984) on social theory. He criticized functionalist accounts as having no proper appreciation of human agency, where people's actions were reduced to little more than the performance of functional imperatives laid down by society. Further, the idea that society has distinct needs is itself problematical, even more that this society can simply impose these needs. Instead, the concept of an overreaching and overbearing society with its own coherence and structure has been replaced by a greater emphasis on the specificities of different social structures and entities. There has thus been a move away from theoreticism (where empirical research on specific situations was disparaged) and overgeneralization (where specific variations between places were ignored), and a reaction against structural determinism (in which human agency was dispelled from social explanation).

This theoretical change was reinforced by the social and political context of the late 1970s and early 1980s. Understanding how capitalist society worked at a general level seemed to have little connection with how people acted; ignoring the abstract exposé they voted in large numbers for right-wing governments both in Europe and North America. It would be necessary to work on the level of people's immediate – usually local – concerns both in terms of political action and the research informing it. Hence, in Britain, the very concrete if politically challenging policies promoted by oppositional left-wing city authorities like the former Greater London Council (Boddy & Fudge 1984, Duncan & Goodwin 1988). Questions of specificity and spatially uneven development were very much back on the agenda (e.g. Massey & Catalano (1977) on the land question, Massey (1984) on employment).

Given these linked theoretical and contextual changes, the emergent focus on locality as the substantive basis of a new research programme becomes easier to understand. The locality seemed to offer something concrete rather than abstract, something specific rather than general, and a sphere where human agents mobilized and lived their daily lives. Unfortunately, the danger of the pendulum swinging too far in the other direction seems to have been forgotten.

Having examined the topicality of the term locality, and having briefly discussed reasons for this popularity as well as pointing to possible limitations and criticisms, I will now go on to analyze what locality might actually mean and how the term might best be used. My analysis proceeds in the following way. First of all, the concept of locality clearly rests on the idea that space makes a difference (to quote a current clarion call in the field). But how, if at all, is locality different from the spatial. The first task then, is to sort out exactly *how* space makes a difference to social process (discussed under the heading, The difference that space makes). I then go on to discuss what locality can offer over and above the spatial (under the heading of Local processes). These sections are, then, essentially exercises in setting limits and clearing the ground. Once this is done, is anything left? I follow up this question by referring locality to the history of previous socio-spatial concepts (in a discussion of Spatial differences and social theory) and to current, if only half-articulated, conceptual supports for the term (under the heading, Localities, locales or 'the local'). My general conclusion is that not much is left. But before understanding this primarily abstract argument it will make things clearer to start off with some empirical examples of what people seem to mean when they talk about locality; this now follows.

Locality: some empirical examples

At a global level, the variations and opposing trajectories between First and Third World, and between countries within these worlds is so obvious as hardly to need comment. Think, for instance, of the physical, social, and economic differences between Sudan, Mexico, Bangladesh, Sweden, and Japan. The fact that both capitalism as a world system and the natural world are so clearly uneven directly supports the contention that 'space makes a difference'. Looked at this way, the spatial amnesia common in sociology, economics, and other social sciences, which sees social processes existing on the head of a pin, appears rather misleading. This is why, in the context of empirical research, these disciplines often (and usually implicitly) misidentify society or economy with the nation–state – as in 'British society' etc. Empirical research must recognize spatial and temporal specificity. But this is equally misleading for there is no necessary reason why uneven development should be completely encapsulated by nation–states, however important these may still be in institutionalizing social practices. Super-national differences (e.g. EEC versus Comecon) and subnational differences (Wallonia versus Flanders, Brixton versus Sutton) may also be significant – perhaps increasingly so as the nation–state becomes more and more obsolete as a unit of economic management and even as a unit of political organization.

We are also very well aware, on a day-to-day level if not theoretically, of the importance of subnational variations. Access to jobs and housing is perhaps the major material condition of existence in Britain, yet how such access is obtained, with what success, and what happens even if it is achieved varies widely between different regions and areas. In the Rhondda Valley of South Wales 13 per cent of households lack all three basic amenities (hot water, bath or shower, and internal lavatory), compared to only 1 percent or less in most suburban areas in south-east England. Over 60 per cent of housing in much of urban Scotland is council housing, compared with less than 15 per cent in many small towns in the south of England. As many as 25 per cent of economically active men are registered unemployed in London Docklands, compared to only 5 per cent in Winchester. Similarly, and indicatively with partially correlating patterns, phenomena perhaps more open to individual control vary just as much. In Camden 10.3 per cent of households are single-parent, and only 48.6 per cent follow the supposed normal structure of two parents and children, while in Bracknell these figures are 5.8 per cent and 77.6 per cent respectively. Likewise, rates of participation in industrial disputes vary widely across the country – they can be up to three times higher in Scotland than in the south west of England.

To some researchers, however, spatial differences like these are little more than deviations from the norm and as such of little interest. Some even treat these deviations as somehow less real than the norm. Yet it is, of course, the theoretical norm or the national average which is the statistical fiction – the average family size of 2.56 people can never exist. Nor do these researchers explain why they take national figures as the norm. Are these not merely local deviations in an international socio-economic system, as multinational corporations well know. In fact, these researchers are using an implicit theory which gives some sort of national state social system considerable autonomy, and this may be mistaken. The national level differences may just be local variations on another scale, or they may just be fictional aggregates of local differences.

So local and national variations from statistical averages or theoretical norms can be significant. But the importance of spatial difference goes much further. Not only are they co-variant but, as this suggests, they often seem influential in a causative sense. For instance, the percentage of council housing in any local authority is not just a reflection of variations in income distribution or earlier inadequate housing. It is also a reflection of the strength of local political action to provide such housing, and sometimes this factor may well be crucial. What is more, this strength will co-vary with other social factors like strength of union organization, or the need by dominant bourgeois interests to maintain a low-waged labourforce. Thus abnormally high proportions of council housing in Labour Sheffield and Conservative Norfolk alike could be traced to locally generated action developing out of locally specific social structures (Dickens *et al.* 1985). Other work comes to similar conclusions. There is one, quite instructive example, from research on the variations in municipal maternity and childcare provision in some interwar Lancashire towns (Mark-Lawson *et al.* 1985). These variations were not just some pre-ordained, passive mapping of national policy. Rather, local policies were absolutely crucial. These were consciously and actively formed at the local level, but in turn they did not simply reflect the different political balances in the various towns. It was more a matter of what these balances meant in relation to local consciousness and possibilities. People's experiences of work and gender relations in Nelson meant that Labour Party policy there was different from Labour Party policy in Preston.

Even more suggestive are the few records of those times when masses of people made their own choices. Miller (1977, 1978), Johnston (1983, 1986), and Warde (1985a) have shown that, even if class identification is weakening as a basis for individual voting behaviour in Britain, paradoxically class identification is becoming stronger at a constituency level. They postulate some sort of neighbourhood effect of local interaction dominated by specific political cultures in different places to account for this. Voting variations are not simply reducible to variations in social characteristics, but also depend on how these variations add up to something more, and this more seems to be a locally based

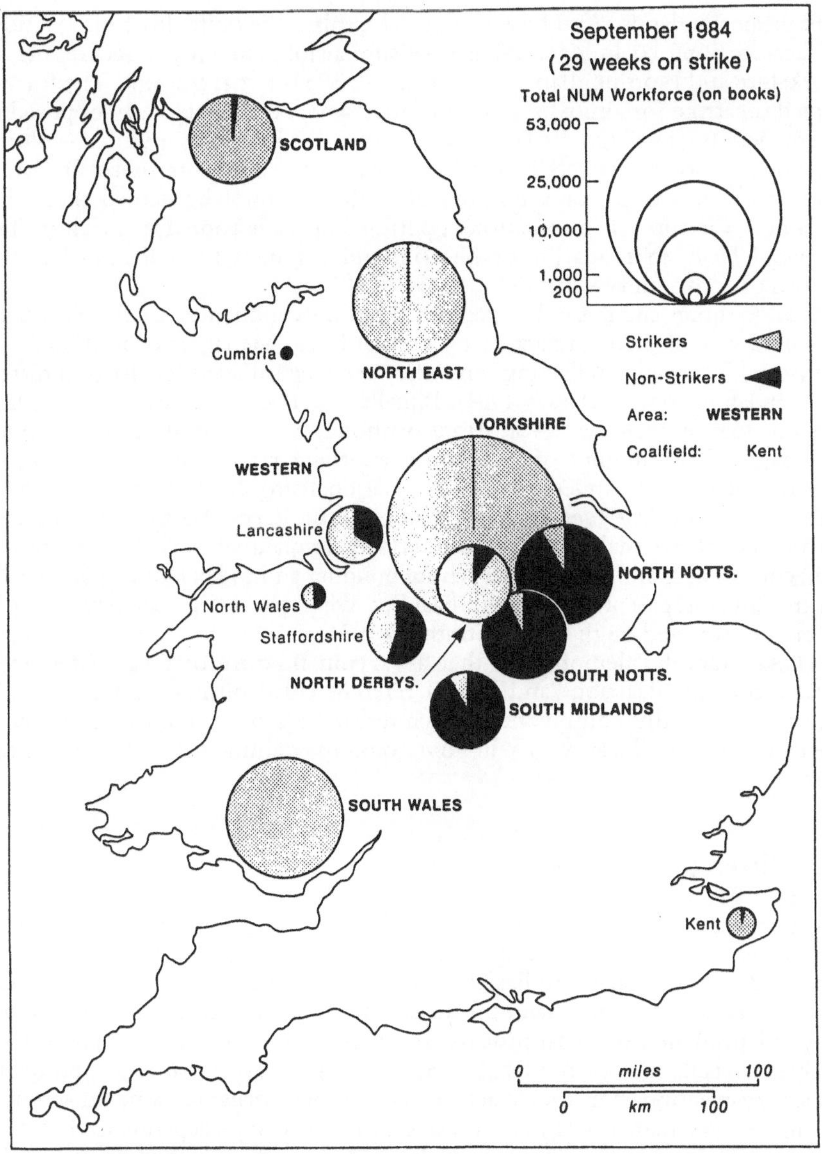

Figure 10.1 Striking and non-striking coalminers in Britain by area, 1984.

neighbourhood effect that is socially active. Variations in strike polls and strike action, which by their nature often minimize class and occupational differences, tell a similar story. In Tyneside and Liverpool as many as 95 per cent of teachers belonging to the National Union of Teachers voted to go on strike in March 1985, compared with a low of 59 per cent in Surrey. Similarly, in July 1985, 84 per cent of APEX members (a white-collar union) in Wales and 83 per cent in Scotland voted to continue with their political fund, compared to only 66 per cent in the Midlands. And how should we explain the quite different reactions of face-working coalminers doing very similar jobs, in equally secure pits, in Yorkshire and Notthinghamshire during the 1984–5 coal dispute? The former stayed on strike for almost a year, the latter were equally strongly opposed to strike action (see Fig. 10.1). Whether this reflects long political–cultural histories (as some analysts believe), or just on the spot differences in group pressures and actions, or a mixture of both, it cannot be denied that what became a variation of momentous political impact is rooted in locality. The *Financial Times*, in its analysis of the strike and its longevity, made much of this revival of community spirit (23, 24 July 1984).

Nor do things end there. For people are not only influenced by their contexts but are aware of them and act partly on that basis. Managers of multinational firms make decisions reflecting what they see as significant local-based differences in labourforces (Massey 1984). Equally 'local people' (!) can act according to some idea of what their local area is or should be, even if this is just as much an imagined community of localism as anything real in material, social, or economic terms (cf. Anderson (1983) on nationalism). In effect, they see 'their' local area as locality even if they do not use the term. So the few hundred inhabitants of Allerdales in Northumberland campaigned successfully both to gain control of their own defunct school building and to keep out a planned ski centre, all – apparently – on the basis of what sort of place they thought Allerdales was and ought to be (Strathern 1984).

These examples demonstrate that a varying mixture of local differences, locally contextual action, and locally referent consciousness can be socially significant. It is this combination which researchers are trying to capture with the catch-all term 'locality'. I will now go on to examine these elements more closely.

The difference that space makes

The debate concerning the significance of space *vis-à-vis* society has important implications for the question of locality. Locality, as well as the related terms local and locale, clearly implies some notion of spatial specificity and spatial bounds. Yet on the other hand the ideas that spatial processes operate in some way independently of social process, and that spatial patterns and areas can be analyzed in isolation, are not tenable. Spatial patterns are not autonomous from social organization, nor does space determine social organization and activity. Similarly, it is incorrect to talk as though there was interdependence between different spatial objects in themselves. It is social objects which interact, not the spatial patterns they form, and presumably this stands as much for localities as for regions or for centre and periphery. This at least is the conclusion to a thoroughgoing debate in both geography and sociology (see Sack 1980, Saunders 1981, Sayer 1982, Urry 1981a).

None the less, despite this conclusion, research can all too easily slip into this spatial determinism or spatial fetishism. This is because spatial arrangements do clearly make a difference (as we saw in the last section) and because 'space' or 'the spatial' is also an appealing commonsense or shorthand way of signifying what we might otherwise have to call 'the contingent effect of the uneven development of social process'. Quite often, however, the shorthand can take over, as is notorious in ecological work and in the whole concept of the urban. This is a potential Achilles Heel of locality. Is this concept just a means of allowing spatial determinism to creep in unnoticed, or even more, is it just a semi-conscious method of allowing the comfortable, if theoretically untenable, autonomy of the spatial to re-emerge in a new form?

It is for this reason that we must sort out how, if at all, locality differs from the spatial. We will first, in this section, review the conclusions of the debate on the social and the spatial, drawing largely on Urry (1981a, 1985) and Sayer (1985a, 1985b). In the next section, we will go on to discuss what locality can offer over and above the spatial.

One way of dealing with the differences that space makes is simply to deny that spatial arrangements have any significance at all. Hence, presumably, there would also be no room for locality. This has traditionally been the view of most of sociology and economics. Societies and economies have been treated as though they exist on the head of a pin. In so far as space was admitted, this was usually just a matter of a minor and limited deviation from spaceless (and usually also timeless) social phenomena. This view was handily confirmed by the opposite extreme of geography and urban sociology, where the autonomy of the spatial was, equally incorrectly, of prime explanatory importance. This situation led to an easy all-or-nothing academic division of labour where neither approach could properly capture the spatial dimensions of social relationships.

For traditional economics and sociology, this fallacy of composition operates in three ways. First of all, societies and human activity necessarily involve passing through – and hence changing in – time as space. Both are temporally and spatially specific. Neglect of this fact has serious, quite possibly crippling implications, if we seek to produce explanations of any concrete, real world problems (which is, after all, the supposed function of social science!). The inner-city problem in Britain is a good example. At one level this is clearly a result of changes in the international division of labour and labour processes. But national (British) and local (city-level) histories, cultures, and social forms equally clearly have their own significance in specifying such processes and mediating them in particular ways (e.g. Massey 1984).

This first problem of ignoring spatial variation is in some ways akin to the problem of the average which never exists in reality. Ironically enough, when researchers go on to look at whatever economic or social abstraction they are using in a concrete way for specific places, they find all sorts of qualifications and particularities. Perhaps, some researchers conclude, class or the new international division of labour do not exist, or may do so only as abstract concepts that cannot by themselves say much abour real world situations. This is the second problem of approaches that deny spatial differences – how can they account for all the detailed real world differences that exist? Finally, if the effects of spatial differences were to be specified, we find that theories developed in their absence are sometimes inapplicable even at this abstract level. The classic example is of those economic models which imply perfect competition where this is quite impossible in a spatial world like the one in which we live.

Interestingly, some apparently non-traditional theories have ended up in a similar position, again partly because of an overreaction against spatial determinism. The new urban sociology of Castells is one example. Criticizing the spatial fetishism of urban ecology, in *The urban question* (1977) Castells and others concluded that space is not something existing outside society, rather it is something produced *by* society. So far so good but then, as Sayer (1985b) points out, comes the *non sequitur* that the spatial is therefore social. While the spatial is indeed *constituted* partly by the social (for we should not forget the natural) this does not mean it can be *reduced* to its constituents.

This brief discussion of disciplinary approaches confirms our previous conclusion that both extremes of space *vis-à-vis* society should be rejected. First of all, space is not an independent entity with its own effects, as the spatial determinism of much previous work in geography and urban sociology had assumed. But on the other hand, space does make a difference to how social processes work, even though much work in mainstream economics and sociology has ignored this. This conclusion leaves us in a less comfortable middle position. It also still leaves us with the question of *how*, precisely, space makes a difference.

The logic of my argument so far is that of the relativist position on space. This view denies that space is an object or substance, in the sense that objects are made up of various substances. Rather, space can only exist as a relation between objects (such as planets, cities, or people) which do have substance. Without these objects, there is no spatial relation, and hence no independent space can exist. (This relativist view is therefore opposed to the concept of absolute space, according to which space does exist independently of objects and where it can have its own effects – like the friction of distance in quantitative geography.)

However, my argument so far would also modify this relativist position in one important way. For, even though space is created by natural and social objects, it does *not* follow that the effects of these spatial relations can simply be reduced to the causal effects that these objects have. This was the mistake Castells made, referred to above. Having been constituted by objects, spatial relations may then go on to affect how, if, and in what ways these objects then relate (see Sayer (1985a) and Urry (1985) for reviews).

Further specification of how space actually makes a difference is allowed by making a distinction between causal processes and contingent effects, a distinction which has been of particular importance to the realist account of scientific explanation (Sayer 1984). Urry summarizes this position. Space in itself has no general effect, and hence no autonomous causal powers. Rather space

> only has effect because the social objects in question possess particular characteristics, namely, different causal powers. Such powers may or may not manifest themselves in empirical events – whether they do or not depends upon the relationship in time–space established with other objects (Urry 1981a, p. 458).

Spatial relations are, therefore, contingent where effects stem not from these relations but from the internal structure of social objects possessing causal powers. However, contingent spatial relations are crucial (among other contingencies) to whether and how such powers are realized.

Sayer (1985a, 1985b) takes this further, again basing his argument on the realist philosophy of science. Rejecting the concept of an autonomous absolute space, but also the complete reduction of spatial effects to its social (or natural) constituents, he similarly concludes that

> space makes a difference, but only in terms of the particular causal powers and liabilities constituting it. Conversely, what kind of effects are produced by causal mechanisms depends *inter alia* on the form of the conditions in which they are situated (Sayer 1985a, p. 52).

However, Sayer takes the implications of this conclusion further by drawing a distinction between abstract and concrete research. Abstract research, according to Sayer, 'is concerned with structures, (sets of internally or necessarily related objects or practices) and with the causal powers and liabilities necessarily possessed by objects in virtue of their nature' (1985a, p. 53). But because it is contingent whether and how these powers are exercised, concrete research is needed to determine the actual effects of the causal powers.

Now, because spatial effects are contingent, abstract social theory need pay little attention to them (and indeed Marx, Weber, Durkheim, etc. did not do so in this theoretical sense; see Saunders 1981, 1985a). This is not to say that such theories can be developed without recognizing this contingency effect, however. They must, says Sayer, recognize that all matter has spatial extension and hence that processes do not take place on the head of the pin and that no two objects can occupy the same relative place at the same time. Hence our earlier comments on the inapplicability of economic models that, in neglecting this fact, are able to assume perfect competition. But while social theory must be spatial in this basic sense, such theories neither have to concern themselves overmuch with spatial effects nor, indeed, can they say much about them.

The situation is quite different with regard to concrete research, where it is the interrelation between structural mechanisms and contingent effects which is crucial. This is certainly the case for the research objects of social science – people and society – for it is rarely, if ever, possible to remove or standardize contingent effects as in the natural science experiment. In natural science, the researcher can physically control and manipulate objects in isolation; this is not possible in social science, where the research object must perforce remain in 'open systems' (see Sayer 1984, Ch. 4). This stays the case despite the existence of various quasi-experimental methods which attempt to circumnavigate openness and contingency, notably comparative analysis and statistical standardization. While useful in particular circumstances, none can create closed systems out of open ones and misapprehension on this point has been a cause of low explanatory power in social science research. Spatial patterns form one such contingent effect and this is why space matters.

The fact that social processes are constituted, contingently, in space and time is something we can often delude ourselves does not happen (at least in the context of social theory – we should be most unlikely to assume this in everyday life). This theoretical blind spot can develop from a misleading definition of general processes. For example, capital accumulation, class conflict, gender relations, or deindustrialization are commonly, but incorrectly, seen as aspatial. It is as though these processes somehow floated above the real world in some space-less realm. Rather, they are from the start constituted in countless places, and are only general in the sense that their spatial dimensions are not those of a single place. It is in this spatial constitution, I should add, that spatial contingency effects take effect. In area X, for instance, the general processes of deindustrialization and gender rôle definition might interact in a specific way, different from that in which they might interact – if at all – in areas A and B.

This artificial distinction between general and local, or macro and micro, results from a confusion of our mental processes of abstraction with what actually happens in reality. Thus we may mentally create abstract theories; these may well be useful enough for theoretical analysis which is, after all, a vital part of any research. For instance, we may mentally abstract theories of the state, say, from the places where real states are created and reproduced. We may then more easily talk about social relations of citizenship or class without becoming over-involved in the very particular histories and contexts of, say, Swedish local authorities. But we should not be fooled by abstractions like this into imagining that processes can actually exist in an abstract way, and hence without spatial relations. States, for instance, can only exist concretely and hence enmeshed in all manner of contingent relationships including spatial ones.

This distinction between abstract and concrete research brings us straight back to locality. For the idea of locality developed precisely in reaction to over-abstract accounts of social forms which appeared to leave no place for variation over space and time or for human agency in creating this variation. The trouble is, however, locality as currently used implies much more than the effects of spatial contingency we have been discussing so far. It implies the existence of some autonomous, locally derived causal effect. An interim conclusion to this discussion would be that if we mean spatial contingency, or simply spatial variation, we should say so. For example, the processes of deindustrialization vary over Britain and in particular labourmarket areas this significantly interacts with gender rôle definitions. There is no need to mystify – and to introduce unexamined and quite possibly false ideas of local autonomy – by using the term locality.

Is there anything left for locality, then, or does spatial contingency completely define its remit? Sayer (1985a) gives some pointers which can be interpreted to suggest that there may, indeed, be a wider remit for spatial relations and hence for the locality concept. People, he says, as self-interpreting beings able to monitor their situations and learn from them, have an exceptionally wide and volatile range of causal possibilities. They learn from their contexts. Furthermore, these contexts will always have particular spatial forms, and hence self-interpretation and actions made as a result also vary over space and time. In this way local variations – the result of actions – will not just reflect spatial contingency effects on general social processes, but will also reflect locally derived social processes. Can this provide a basis for a concept of locality? I will go on to examine this thesis in the next section.

Local processes – derivative or causal?

The conclusion so far is that space does not exist in itself as some autonomous object, and spatial processes do not operate independently of social processes. It is social and physical objects which interact, not spatial patterns. None the less, spatial patterns cannot be wholly reduced to the objects that produce them, and spatial contingency will be influential to how social and physical processes work and what forms result. Similarly, general processes will always be constituted in particular places and at particular times where they will inevitably be affected by this contingency effect. In this way nearly all processes and forms will be spatially differentiated. This also means that most processes and forms will have some sort of local expression, and very often this local expression will be on a subnational scale.

This conclusion also implies, however, that these local processes are completely derivative or responsive. The crucial generative mechanisms that *cause* them to happen – rather than effect *how* they happen – would not be spatially constituted according to this view. On the face of it, this sharply contradicts what we intuitively know about real social changes. It is the activities of people living in particular places and at particular times, whether seen as individuals or as social groups, which actually create change. This is also, of course, the outcome of the debate over structure and agency (see e.g. Giddens 1981, 1984). But, according to our discussion so far, agency would be reduced to a responsive rôle only and structure would dominate in terms of causal influence.

To some extent, this conclusion is unwarranted, resulting from a misleading elision of 'local' and 'agency', and of 'general' and 'structure'. For something which is local may well also be general. Similarly, structures may well be very specific and may well operate locally. We should also remember that 'local' is in fact just as much a statistical aggregate as 'national', although perhaps at a more sensitive spatial scale. For there is no local social situation or system as such – rather there are people, households, firms, state institutions, and so on. It is people, acting individually or in collective situations, that carry out social acts and so respond to and reproduce structures through their agency. Local descriptions of these activities will aggregate and conflate in just the same way as a national description does – but arguably for most processes at a less aggregated level.

However, this very recognition that structures are local just as much as general reintroduces the dilemma. For, according to the discussion in the last section, local processes and actions are merely spatially constituted, and spatially contingent, versions of general processes.

One way out of this dilemma is to make a distinction between contingent local variations and causal local processes. The former refers to the contingent effects of spatial patterns; the latter to the local specificity of generative social relations. The social basis for the former should by now be clear enough, following our discussion of the difference that space makes. However, we can also imagine that some social relations which are generative, that is they cause changes to happen rather than merely contingently affect how change occurs, may be locally specific. This seems logical enough, because as we know societies are unevenly developed. This uneven development is not just a matter of forms and outlines (for example, certain industries locate in certain areas) but also of social mechanisms themselves. In some places, for instance, capitalist social relations will not be fully developed, or may be developed in some very particular ways, or may be partly decomposed.

Two contrasting examples of the development of spatial variations in

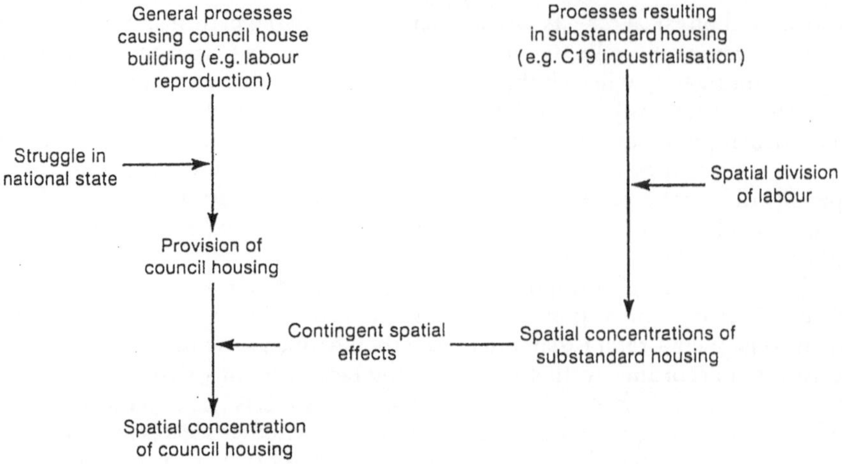

Figure 10.2 Contingent local variations: the case of council housing.

contemporary Britain may make this distinction clearer. Our first example, of contingent local variation, is the relationship between the M4 Corridor, a region of relative prosperity founded on high-tech industries strung out along the M4 motorway between London and Bristol, close to London's Heathrow Airport. The emergence of the M4 Corridor owes something to the fact that Heathrow lies at one end of it. But Heathrow did not in itself cause the development of high-tech industry in Britain and its particular requirements for various sorts of labour supply as well as infrastructural backing like airports, motorways, and government defence laboratories. Nor did the social relations of creating and manning an airport underlie the develoment of high-tech industry. Without Heathrow, or if London's major airport had been sited elsewhere, this high-tech growth would most probably still have happened – although perhaps in that case the corridor would have been somewhere else, up the M11 motorway to Cambridge and Stanstead Airport, for instance. The importance of Heathrow was to affect, contingently, where high-tech growth took place; it did not cause it.

In contrast, causal local processes are those locally specific relations which are socially generative. Our example is the building of council housing in 'red islands' of labour movement local political hegemony. Sheffield is the sort of place we have in mind here, where for a long period much more council housing was built than the average for British local authorities or even the average for large northern towns. Exceptionally, much of this housing was put up using local authority direct labour and hence in a non-commodity form (see Dickens *et al.* 1985, Ch. 5). This local outcome was not, however, just a result of spatial contingency effects – where, for example, more council housing was built because more substandard housing already existed there. Rather, more was built because of the specific social relations existing in Sheffield. A class conscious and organized local political movement dominated the major local organ of state power – local government – and consciously used this power to build council housing which they saw as important to their political project. For

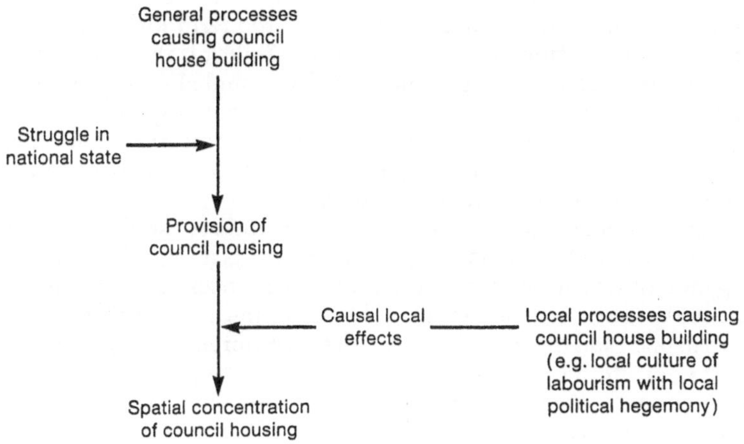

Figure 10.3 Causal local variations: the case of council housing.

the same reason, direct labour was often used to do the building. In other words the causal social mechanisms were locally specific. Variation was not just a matter of the contingent effects of spatial patterns on some invariant and universal (or at least national) mechanism. Many other similar examples can be found (see, for instance, Murgatroyd *et al.* (1985) on the significance of localities for social and economic processes). Figures 10.2 and 10.3 summarize the distinction between contingent local variation and causal local processes for the example of subnational differences in the provision of council housing.

Note that we do *not* claim, by virtue of this argument, that these locally produced mechanisms were the only causal mechanisms involved or even the most important. For example, national legislation and national politics were crucial in allowing Sheffield's labour movement to act the way it did – as the collapse of council house building in the city since 1980 testifies only too well.

Local variation is not only, therefore, a matter of the contingent effects of spatial arrangements. There may also be local causal mechanisms. Social mechanisms are not necessarily universal, but can be derived locally.

This conclusion is not really very surprising. There is after all no necessary reason why causal mechanisms should only operate on a world scale or even a national scale. We know that social mechanisms are historically specific as people create particular societies – for instance, the law of value cannot exist without capitalist society and this society has existed only at particular times. But by the same token we must admit spatial specificity. Capitalism has never been a total world system and the depth of its social penetration varies from place to place. Hence the implicit recognition of spatial bounds in traditional economics and sociology, where society and economy are so often confused with nation–state, for example 'British society'. But if we are to admit the national scale – based on an appreciation of national specificities – then there is no good reason to reject, *a priori*, the local scale. Of course, this logical argument does not in itself prove the existence or importance of local causal mechanisms in any situation – this still has to be demonstrated empirically.

However, the possible existence of locally-derived causal processes in any area still does not, to my mind, justify use of the term locality. For locality implies a spatially specific system of causality. While it is the case that all social systems will be spatially situated and spatially bounded – just because of the spatial constitution of processes and the effect on them of spatial contingency – this does *not* mean that social systems are created *by* local areas. This, however, is what locality implies. And this also looks like a version of that spatial determinism now seen as quite unwarranted.

How far, without re-establishing spatial determinism, can there be some sort of causal locality effect specific to a particular area? Is it logically possible that a whole gamut of spatial contingencies, local causal processes, and locally specific effects could create a spatially structural system, in the sense that what people expect to happen and therefore how they act is comprehensively moulded cross many areas of life. For example, locally specific class relations might act in combination with other locally specific processes and outcomes (say local labour movement hegemony over the local state) to produce a distinctive local political culture. People monitor, learn, and react to their context and so how they act would be partly shaped by their experience of this locally specific political culture. The agenda for action would be changed; *what* actions happen, not just how, would be partly dependent on this locally produced structural context. It is tempting to christen this a locality effect. Two provisos follow, however. First of all, this criterion for the existence of locality effects is a fairly demanding one. It is not something we would expect to happen often. For instance, the evidence for autonomous local political cultures in Britain (as opposed to local variations in political culture) is fairly weak (see Savage 1987). And second, the existence of such locality effects should never be assumed simply because significant local variations exist. This, however, is just what locality studies have assumed.

I have now developed a three-stage hierarchy of how space makes a difference to social process. The first two elements in this hierarchy are based on the distinction between contingent and necessary relations, namely: (a) contingent local variation (where spatial contingency affects how social mechanisms operate in practice); and (b) causal local variation (where the social mechanisms themselves are locally derived). Finally, I argue that there is a third level, of locality effects, where a bundle of complementary and locally derived processes and outcomes produce some sort of local social system. It is only the last which approaches the implied content of locality and hence the rest of the chapter will consider this third category. But before this, one conceptual clearing up task remains. Have we not heard all this before, if in different terms? What of the concepts of community and local social system or even of region and *pays de vie*? These have also attempted to nail the idea of spatially specific social causation, but equally all have come unstuck. Is locality just a case of old wine in a new bottle, and a singularly unsatisfactory wine at that? This section will briefly consider how, if at all, locality might differ from these older concepts.

Old wine in a new bottle? Spatial differences and social theory

The recognition that places are different from one another, and that this difference can have socially meaningful effects, is hardly new. Hence the long sequence of conceptual devices in social science which have attempted to deal with this recognition and to use it analytically – *pays de vie*, region, ecological area, rural and urban ways of life, community, urban social system, local social system – to name only some of the most influential. Locality is only the last in a long list. More disturbingly, however, this is more or less a list of failures. Of course, we should expect concepts to be revised, qualified, and eventually rendered obsolete or axiomatic – otherwise there would have been no explanatory advance. In this sense all concepts, with the exception of those that become axioms, fail. But this list is different. It is a list of false turnings and dead ends. So is locality just going to add one more?

At first reading this may seem too harsh a judgement, and much useful empirical work as well as conceptual development has resulted from these socio-spatial theories. But the literature is more or less agreed that if it was not for these concepts and the way they were used, then more and better work would have resulted (see Dunleavy (1980), Saunders (1979, 1981, 1985a) for reviews). On the other hand, however, the alternative position that space makes no difference is also untenable, as we have seen. This is of course why, as one concept in this list bites the dust, a successor soon emerges only to fall in its turn. Can locality do any better?

This chapter is no place to review these concepts of space *vis-à-vis* society at any length. In any case, this has been done quite adequately elsewhere (e.g. Bell & Newby 1971, 1976, Sack 1980, Saunders 1979, 1981). Rather, we will examine some of the more influential of these concepts in relation to the previous discussion of the possible social bases of locality. How far have these concepts gone along the same path?

The last section ended with a three-fold hierarchy of the spatial specificity of social process, a hierarchy based on the notion of causal and contingent relations. Using this scheme we distinguished between contingent local variations, causal local variations and locality effects. However, some – actually most – of these concepts have attempted to go much further. They have imagined a fourth stage in the hierarchy – the local, or at least the spatially determined, society. Theories of the urban, urban and rural social systems or ways of life, communities and neighbourhood, *pays de vie* and region have all been used in this way. That these concepts have been taken so far (and hence fatally overreached themselves) partly depends on the social need to find a distinctive theoretical base for explanatory theory. Threatened by the intellectual and social dominance of basic economic and social theory, both geography and urban sociology have clutched at spatial specificity in this unwarranted way.

In geography the idea of distinctive regions as both object and method of analysis was not long sustainable in terms of the causal analysis required for this fourth stage in the hierarchy. This depended on a socially inadequate physical determinism, where social behaviour was caused by climatic or geological conditions. This is quite apart from all the practical difficulties of finding regions, although we will return to the various lower-level uses of the concept later. Hence the quip that regional geographers were 'trying to put boundaries that do not exist around regions that do not matter'. More interesting is Vidal de la Blache's notion of distinctive *pays de vie*, which partly emerged as a reaction against the physical determinism necessary to give regional geography causal status. Local areas could be distinguished in terms of a distinctive local social life based on their own local histories and cultures (see Wrigley 1965).

The major subsequent criticism was that the *pays de vie* then discerned were based on a pre-industrial folk life already disappearing. The railway and the telegraph, the saying went, had made the approach obsolete. Although true enough, from our point of view this criticism is misplaced. There is no reason, logically, why pre-industrial local cultures and economies might not be replaced by industrial or even post-industrial ones. Indeed, work like Massey's *Spatial division of labour* (1984) shows very well how current economic processes are spatially specific, while ethnographic and community studies continue to document the importance of social variation even in the midst of industrial capitalist Britain (e.g. Cohen 1982). Indeed, the current spate of locality studies (e.g. Cooke 1985, Murgatroyd *et al.* 1985) suggests that such differences may be *increasing* not least in the more capitalist, or at least more industrial, regions. Ironically enough, the researcher who is credited with showing that the distinction between rural and urban society does not hold (Pahl 1968) has recently published a work on divisions of labour which is explicitly centred on a case study of a very specific place (Pahl 1984, 1985). This case study, or 'local milieu', the Isle of Sheppey in Kent, is even credited with having a 'personality'.

Rather than criticizing the idea that places differ in significant social ways, our objection lies elsewhere. What is not attainable is the fourth level of spatial specificity added on to the *pays* concept – and perhaps added onto locality as well – the idea that spatially specific differences necessarily mean local social systems. Even for Blache's pre-industrial *pays*, there seems little doubt of the importance of wider cultural, economic, and demographic processes (e.g. Watts 1985, Lis & Soly 1982). Hence also the possible treatment of European history as the ebb and flow of long-term, wide scale and almost impersonal social forces (e.g. see the work of Braudel).

Regional geography, although prolonged by the more convincing social causality of the *pays* concept, was of course eventually succeeded by spatial science – the quantitative description of spatial patterns supported by an assumption that spatial arrangements in themselves independently created social behaviour. Despite the internal discordance between the regional and spatial analysis approaches, not least socially (see Duncan 1974, Johnston 1982), in retrospect it is clear that both are similar in the way in which space is viewed *vis-à-vis* society. It is more that the spatial science was emptied of social content and sprayed with the honorific aerosol 'science' (see Sayer (1979b) for one critique).

In urban and rural sociology the fourth-level stage of spatial specificity goes one step further than spatial science. It was not merely any space that was determinate for social behaviour but particular sorts of space. Drawing originally on the urban ecology of the Chicago School and Wirth's 1938 paper 'Urbanism as a way of life', the assumption was that particular spatial arrangements, for example the urban, were fundamental to social organization and behaviour – hence the *urban* way of life, *urban* society, *urban* managers, and so on (see Saunders 1981). That this view was essentially spatially deterministic probably helps explain why these theories were so attractive to geographers looking for replacements for the region and for the socially contentless spatial science. Community studies was a less theoretically assertive form of this fourth-level use of spatial specificity. The notion was that particular social communities existed as spatially defined and autonomous social units and, with the demise of urban ecology as a presentable theory, urban sociology could thus coalesce around community studies.

We are in some danger of repeating ourselves in dismissing these fourth-level approaches. Unfortunately, for these attempts to create essentially spatial theories and disciplines, space is not autonomous from society nor does it determine social behaviour, even if it does make a difference. This is quite apart from other severe theoretical limitations which many of these concepts embody (for which see the reviews cited earlier). But on the other hand we are still being told that 'geography matters' (the title of a book edited by Massey & Allen (1984)). Similarly, as our introduction pointed out the notion of locality is a major way of signifying this importance. (For example the ESRC brief for the Economic Life and Social Change initiative is quite maddening in the way it exchanges the terms communities and localities without explaining what these are or how they relate.) None the less, the resolution of the debate on these fourth-level socio-spatial theories forcefully points out that while space may matter it is not determinate or central. Indeed, space is only relative to objects and does not exist in this absolute way. This fourth level must remain taboo. Locality or, even worse, locality studies should not be allowed to wander over this dividing line.

This conclusion is now generally accepted, at least on the surface. Interestingly, however, much the same sort of work, which one would now have thought inadmissible, still takes place. This is because practitioners still need to relate spatial variation to social processes, they do not know how to do so more adequately and, furthermore, they have found more comfortable and defensible laagers into which to retreat. Their epistemological necks are not so exposed. There have been three chief ways in which this holding operation has been justified and it is worth reviewing these here – for this could be the fate of locality.

A first way of continuing to do so, even when the same tactic is acknowledged as inadequate, is to agree that space is not a basic source of social behaviour. The argument then points out that even if space is contingent, it still makes a difference and so is worthy of analysis. Then urban ecology or spatial analysis can carry on regardless. The fault lies, of course, in the *non sequitur* between the two last statements, for analysis proceeds by using concepts which assume that space is central or at least that spatial processes are separable from social ones. While the assertion of spatial primacy is relaxed, for practical analysis it is confirmed (see Duncan (1980) for a review of this in geography).

The second major resurrection method for these supposedly abandoned theories is less conscious. Practitioners assimilate alternative approaches and relegate these once so powerful theories to the status of mere research techniques. Urban ecology, for instance, becomes just a handy way of presenting census data or community studies is reduced to a convenient way of setting up a research case (cf. Saunders 1981, p. 257). However, presenting data and choosing cases are conceptual activities; inevitably, therefore, urban ecology or community studies will be reintroduced through this back door of research procedure. Rather data presentation or case study selection should be related to those theories or theory which researchers claim inform their analysis.

Finally, and perhaps ironically, these alternative theories have themselves been given a spatial dimension. There are two versions. One depends on a relativism. Thus, for instance, it is sometimes claimed that political economy, urban managerialism, and urban ecology are all appropriate in analyzing British cities because each is seen as specific to a spatially defined level of analysis (see Herbert 1979). In this way all approaches have a role. Now political economies, urban managers (or at least managers), and spatial patterns in urban areas all clearly exist and need to be examined, but each theory approaches these realities in a quite different and often contradictory way. The objects of analysis are incorrectly conflated with the theories used to study them.

A more sophisticated version of the same maneouvre appeals to the reaction against structuralist determinism. Structures are incorrectly identified with national or international scales and agency with a subnational, local scale. Fortified by this reaction, the plea is made for re-erecting regional studies – or even studies of *pays* – as the centrepiece of geography (cf. Gregory 1978, Buttimer 1978). But, of course, the same logic applies as in the last case. Rejecting structuralism does not mean that structures do not exist, and rejecting determinism does not mean that structures do not determine. Also, structure and agency are conceptual abstractions of process, not things which can only exist at particular spatial scales. Structures can be conceived as operating at the local level and just as much agency operates at wider levels.

Our point in drawing attention to these forms of slippage is that the current use of locality also shows them at various times. The briefs for the two ESRC initiatives were a good example. It is no good accepting, in principle or when proved, that localities are not autonomous and then going on to undertake research as though they were. On the other side of the coin, it is just as bad to relegate locality to a definition of case study area and then import unexamined or implicit ideas of social cohesion and autonomy in the case study area. Finally, locality is not a synonym for agency where the structural can thereby be left as some passive framework.

Hitherto we have concentrated on the fourth level of spatial specificity. This reflects the influence of these socio-spatial theories. Most have also been used to refer to the other three levels, but normally without distinction from the fourth level of spatial society and so subsumed within it. However, it is worth spending a little time on two less subsumed lower-level usages, those of region as a classificatory device and of local social system.

Perhaps because the fourth-level use of region and *pays* was so quickly abandoned by the majority, region has a relatively long and clear history as a classificatory device (see Grigg 1969). No claims of causality or social specificity need be associated with this use of region. For instance, there is an olive-growing region where olives are grown. None the less, such claims can easily creep in especially when several criteria are combined. For instance, does the olive growing region in France (which shows a very rough correspondence with the Languedoc region) also delineate some sort of social region? Quite similar in principle are the current classifications of standard metropolitan areas based on definitions of journey to work or to shopping zones. These can delineate spatial patterns and so be very useful for some purposes but have no necessary or even likely correspondence with locality. Other social interactions may take place on quite different scales, so that typical journey-to-work areas, for instance, include rather different suburban and inner-city areas.

Hence the temptation to find a constant spatial container for locality, such as local labourmarkets, may well be mistaken (cf. Warde 1985b). Delineating locality – when we accept that we are not talking about an autonomous social system – may then be a matter of what sort of locality effects we are interested in and drawing boundaries appropriately. With another sort of substantive interest, the locality could well be a different one. Indicatively, some of the work on regions has come to much the same conclusion. Trudgill (1975) gives some interesting examples from linguistic geography. For instance, although we recognize differences between Northern English and Southern English (e.g. in the long and short pronunciation of words), in fact no constant boundary can be found between the two. Not only is the isogloss for short/long 'u' different from that for short/long 'a', but isoglosses differ even for different uses of each vowel (e.g. dance/past). Finally, some usages pay little reference at all to these supposedly crucial dialect differences (e.g. the full use of 'r'). Linguistic geographers consequently distinguish focal areas from transition zones (where the latter are often larger than the former) and create particular zones for particular pieces of analysis (e.g. North Central Bavarian Dialect) even though no such unit really exists in autonomous linguistic use. In other words, these regions are relative units defined theoretically. Perhaps the same is true for locality.

Our last example of 'old wine' is still relatively freshly bottled, although in some ways strangely neglected since. By this we refer to the notion of local social system proposed by Stacey (1969). After the convoluted and over-ambitious attempts to define spatially distinct societies, as with the urban/rural way of life dichotomy, this notion seems refreshingly simple; indeed the paper is entitled 'The myth of community studies.' However, despite assimilating the criticisms of this early work, Stacey still preserves the idea that social systems vary over space. This preservation no doubt seemed quite essential, given her earlier empirical studies of Banbury. No given Banbury community existed, and social interrelations in the town split on a number of dimensions (the newcomer–'local' split seemed particularly important). On the other hand these varying local networks were constituted by specifically local consciousness, institutions, and actors even though at the same time they overlapped with a supposedly national social system.

This sort of research experience had the result of encouraging a conceptual formalization of what these local effects might be. Social systems – patterns of family structure, kinship, belief, etc. – could be constituted locally when 'a set of interrelations exists in a geographically defined locality' (Stacey 1969, p. 140) (locality presumably meaning study area!) None the less, in the face of the structuralist onslaught of the 1970s, this insight into the local specificity of social structure was neglected, partly indeed because Stacey's work was heavily influenced by systems theory and was also deeply functionalist. So the rôle of class or production in creating these networks is neglected (cf. Bradley & Lowe 1984). However, for this discussion these deficiencies are not the crucial ones. What is more limiting is a (connected) inability to locate social causation within local and non-local processes or, to put it another way, an inability to distinguish between the levels of spatial specificity. These leave Stacey in the position of positing local social system as an ideal type which in practice does not exist. For, as in Banbury, social interrelations are most unlikely to be complete or bounded within one geographical area. Hence, says Stacey, actual localities will display partial or even non-existent local social systems.

In part this is a useful result. All too often it has been assumed that communities, rurality, localities and the like must necessarily exist. As with regions, this is not the case and they may be as much analytical constructs as real ones. Ultimately, however, Stacey is left in an unsatisfactory position. On the one hand she dismisses the fourth-level attempts to define spatial societies as myth, but on the other hand she has no way of distinguishing between different types and levels of spatial specificity. She must perforce resurrect the fourth level, if demoting its status to that of an ideal type which does not really exist. What is left in reality is incomplete ideal types. I would claim that the conceptualization developed here of: (a) contingent local variation; (b) causal local variation; and (c) locality effects gives better purchase on what is left after the myths have been dismissed.

Localities, locales, or the local

The increasing popularity of locality as a concept has been paralleled by the rise of the locale. At times the two are even used interchangeably, although at others they seem to be used to denote some sort of distinction between general and particular, or abstract and concrete. Yet again, and quite typically, this distinction is rarely if ever explained. Both locality and locale are at times even interchanged with the local. In this section we will try to establish to what each term does – or should – refer and so escape from this most confusing elision. In doing so we can also establish how, if at all, the local and locale can contribute to our understanding of locality.

We can most easily and quickly deal with the local. This could be used to refer to local variations pure and simple (where local means subnational) irrespective of how these are caused. However, the term might refer to local processes, that is something like the second level in the hierarchy of socio-spatial variations developed under the discussion of local processes, where local variations were more than simply a matter of spatial contingency. But the local could also refer to locality effects (the third level in our hierarchy) or even to the inadmissible fourth level of a spatially determined social unit. The problem is, of course, that it is never clear to which of these levels authors are referring. (One good example is Urry's (1981a) seminal article on the importance of spatial variation for social research in the context of de-industrializing Britain. The paper is entitled 'Localities, regions and social class' yet the key section is headed Class, regions and the 'local'.) Probably authors are not clear what they mean either, and are really covering themselves by trying to include everything. The truncated grammar of the term supports this view; authors do not know whether they mean local variations, local process, local society, or locality, and so end up leaving an adjective standing without a noun. We would back the grammarians in this case. Use of the term the local should be avoided, and including a noun makes it much clearer what the author really means – or at least forces her/him to think about it.

The elision between locale and locality is more complex than this. Examining this elision does, however, allow us to throw more light on the idea of locality. For locale is already a developed conceptual term, although as yet only in the context of abstract social theory (especially in Giddens (1979, 1981)). Can this abstract development, a consequence of lengthy agonizing over the relationship between structure and agency, be married to the empirical development of locality in urban and regional studies, where the reality of everyday research has made it quite plain that local variations are socially meaningful? The synonymous use of the terms in some texts would seem to show that some researchers, at least, think so. To answer this question we must spend a little time following Giddens's formulation of locale and also some of the criticisms it has attracted. Fortunately, this is a task largely completed by others and we need only relate this work to our purposes here.

It appears that Giddens's formulation of locale as a theoretical concept has emerged from an interest in spatial and temporal specificity, which in turn developed from his longstanding concern with structure and agency (Saunders & Williams 1984, Saunders 1985b). The ultimate point of departure can even be traced back to Giddens's earliest aims of relating class structures to social action (Giddens 1973) and in so doing to provide a replacement for both determinism and functionalism. The answer was the theory of structuration. Structure and agency should be seen as a duality where neither is subordinate to the other. Furthermore, both create the other. In particular structure is not some pre-ordained given, rather it is both the medium through which action is produced (enabling action as well as constraining it) and also the product of that action – structure is created, reproduced, and indeed changed through the meaningful activity of agents.

Critics have of course made much of a major limitation of this solution (e.g. Saunders 1985b). It remains almost entirely formal and abstract, with little theoretical substance on, for instance, how class structure does in practice interact with social action. However, the notion of structuration does remain as a useful epistemological starting point, even if the waters inevitably become muddier when empirical work begins. So where Saunders claims that any good empirical sociologist was, by definition, a structurationist of necessity (even if they did not need Giddens to tell them so), I would see Giddens's theorizing as a means of widening the pool of this good empirical sociology. So much, after all, has been structuralist, functionalist, or voluntaristic; that is bad sociology!

It is partly because of this limitation that Giddens became concerned with space (and, before this, with time). How action actually does relate to structure, and how the two can be separated in practical analysis, clearly becomes a considerable problem. Giddens uses spatial and temporal specificity as a major escape route from this difficulty (although, as Saunders & Williams (1984) note, the interest in spatial specificity developed very much as an afterthought to temporal specificity). Time and space are not just contextual backdrops against which actions unfold and hence where structure could more easily rule. Rather, time and space are integral aspects of action. Space helps define how elements are mobilized for interaction, where particular physical settings, specific in space and time, are associated with typical interactions. These settings are then actively organized and used by participants in the production and the reproduction of interaction. The ability to control and manipulate these settings will then become a crucial feature of power in society.

It is this notion of activity and mediation which gives Giddens's use of locale something more than a spatial contingency effect pure and simple (our first level). So it is not just that we cannot explain events without also understanding how they are constituted in time and space. Locale also implies that how social relations are produced depends on the active use of particular spatially and temporally specific physical settings (our second, third, and fourth levels). Unfortunately, however, Giddens has not gone very far in sorting out these levels and so often confuses them. For much of the time he seems to be speaking about spatial contingency effects alone – yet investing these with an unwarranted explanatory or theoretical weight. Hence his interest in Hägerstrand's time–geography, which similarly over-invests specificity with causal weight (see Pred (1977, 1984) for reviews). Compare, for instance, two passages from Giddens:

> (a) the social management of space is . . . in definite ways a feature of all societies. Virtually all collectives have a *locale* of operation, spatially distinct from that associated with others. 'Locale' is in some respects a preferable term to that of 'place', more commonly employed in social geography: for it carries something of the connotations of space as a *setting* for interaction. A setting is not just a spatial parameter, and physical environment, in which interaction occurs: it is these elements mobilised as part of the interaction. Features of the setting of interaction, including its spatial and physical aspects . . . are routinely drawn upon by social actors in the sustaining of communication (Giddens 1979, p. 206).
>
> (b) Locale – a physical region involved as part of the setting for interaction, having definite boundaries which help to concentrate interaction one way or another (Giddens 1984, p. 315).

In the first, earlier passage, locale combines elements from our various levels of socio-spatial specificity. In the second, taken from the 'Glossary of terminology of structuration theory' (in Giddens 1984), locale seems very close to a spatial contingency effect alone. As we have seen under our discussion of the difference space makes, this is important enough for empirical research. But this view of locale only supports our first level of spatial contingency effects. The theoretical apparatus drawn up by Giddens can, in this case, do little for locality as anything more than locale variation.

The idea of locale has nevertheless been appealing to those researchers who were forced to realize that 'geography matters' in the context of empirical work. Partly, this is simply a case of being thankful for support from an unexpected quarter. But despite our critique, all is not necessarily lost for locale as support for locality. For the concept, as developed by Giddens, does at times refer to more than simple spatial contingency. It is here, however, that several more problems arise.

A first problem is the danger of reintroducing spatial fetishism into social research. According to Saunders (1989) this danger is pronounced because of the contradictions in Giddens's development of locale, and a consequent tendency to over-invest spatial contingency with causal status. First of all, Giddens makes a belated discovery that space can be important and has an effect. On this basis, he goes on to accuse social research of space blindness (ironically enough, at about the same time that both geography and urban sociology were escaping from the spatial fetishism of urban ecology or spatial science, and putting space in a more suitable rôle). But in working through how space may really matter (for instance in a discussion of the city and social action), Giddens is forced to the conclusion that space is at best secondary, and does not actually matter very much in terms of causal, social relations. This result is unpalatable, because of the theory of structuration and its need for support. Hence, suggests Saunders, Giddens develops locale as a device with which 'to save space as central to social theory'. As Saunders puts it, space is clearly important in so far as there cannot be an aspatial social science because of contingent effects; 'But to fetishise space is as dangerous as ignoring it' (Saunders 1989, p. 23). The uncritical use of a locale–locality–local combination can easily do just that.

This potential for spatial fetishism is reinforced by another limitation of locale as a concept. Like structuration theory in general, it is so general as to be almost anodyne. It is true enough that all social interactions take place in 'defined locales of operation', and it is useful to point out that these locales are not just environment, but 'are integrally involved in the structural constitution of social system' (Giddens 1981, p. 39). Space matters, and most places will be locales of one sort of another. But to go further requires some operationalization of how locale works in practice and in particular situations. In this connection it is not particularly helpful that Giddens generally reduces social practice to communication. There are many relevant contexts which often remain unacknowledged by actors, for example the world market or their own psychological make-up, but which are nevertheless crucial. To admit practice without communication will, however, reduce the importance of locale as a mediator between structure and action.

Finally, locale is in any case not synonymous with locality. Perhaps localities may be locales (i.e. typical physical settings for interaction), but clearly not all locales are localities; as Giddens point out, the former exist at any number of spatial scales from desk-corner to nation–state. The terms are asymmetrically related. There are also two other important discontinuities between locale and locality. First of all, locale is not a dimension of social organization in the same way that locality (as our third level of locality effects) may be. Rather, locale refers to the mediation of social relations in specific places where, for instance, locality effects may be producing these relations. Second, Giddens defines locale as 'typical interactions of collectivities' (Giddens 1981, p. 39). In contrast, localities are a mixture of different elements. Where locale is typical, locality is unique and heterogeneous.

To conclude, the temptation to jump from locale to locality should be avoided. These are not synonyms, and locale is in any case a limited and sometimes problematic concept. All it really does for locality is (a) reinforce the starting point that space matters, and (b) suggests that this is not merely a

REGIONAL VARIATIONS REVEALED BY NEW REPORT

by Our Boring Staff
Yawn Wells and Princess Zzza-Zzza Gaboroftheyear

Some astonishing regional differences in the way Mr and Mrs Average Briton live today are revealed in a shock new report published by the Department of Lists and Figures, under the title *Some Very Boring Lists and Figures About Modern Britain.*

Neasden emerges from the survey as the most dangerous place in the country.

Those who live in the famous North London suburb, the report shows:

- Eat 1000% more Yorkie Bars than the national average.
- Use local laudromat facilities on average 25 times as often as the inhabitants of the island of Muckay off the West coast of Scotland.
- Are 150% more likely to die from static electricitv caused by wearing anoraks purchased from Millets.

That's enough boring facts. Ed.

Figure 10.4 Locality: academic fashion or social signifier? (source: Private Eye, 8 August 1986).

passive contingency effect but that there may be more – locality, perhaps. But of course we knew this much already (Fig. 10.4).

Both Sayer (1985b) and Saunders (1989) conclude by recommending a rejection of locale in the context of urban and regional studies. Noting that locale is in explanatory terms very much like the old functional region in geography (see under the discussion of spatial differences and social theory), Sayer recommends that instead we should concentrate on setting up each practice of interest with its relevant causal group, including the spatial dimension as necessary. We will go on to discuss this further in the next, concluding section.

Conclusions: using locality

My conclusions so far are as follows:

(a) Locality is currently a fashionable term, but it is used in a variety of unclear and sometimes contradictory ways.
(b) Locality does seem to refer to something important – that spatial variation and specificity are socially significant. However, current use of locality does not help in assessing how, and how far, this spatial variation and specificity are socially significant. Furthermore, the term often seems to incorporate notions of spatial determinism which have, elsewhere, been dismissed as incorrect.
(c) There are three ways in which 'space makes a difference':
(i) Spatial contingency effects. Processes must be constituted in particular places, and as they do so they are influenced by the pre-existing nature of those places (caused by earlier process constitutions) and other process constitutions taking place at that time;
(ii) Local causal processes. Spatial contingency, while affecting *how* processes work, cannot create causal processes (unless we admit spatial determi-

nism). However, such causal processes may be locally derived. This is because determining social systems are spatially variant, and because actors monitor and respond to their variable contexts;

(iii) Locality effects. The contextual effects of local causal powers and spatial contingency may be so significant as qualitatively to alter the nature of social structures in a particular place and hence social action.

(d) Only the third of these three ways in which 'space makes a difference' approximates to current use of the term locality. However, an important proviso affects even this limited rôle for the term. Only spatial contingency is both inevitable and pervasive. Local causal processes and locality effects may not be significant and, if they are, this remains to be demonstrated. All too often locality is simply assumed.

(e) There is a long list of concepts which, like locality, have tried to deal with the difference that space makes. Community is the classic example. This list is also a list of failures. For, unable to distinguish between contingent and causal effects, they ended up with spatial determinism.

(f) The idea of locale should not be equated with locality. The terms are asymetrically related (localities may be locales, few locales will be localities), locale is not a dimension of social organization in the way that locality should be – rather it is a mediation of social relations, and where locale is typical locality is unique.

The overall conclusion, then, is that the locality concept is misleading and unsupported. Localities in the sense of autonomous subnational social units rarely exist, and in any case their existence needs to be demonstrated. But it is also misleading to use locality as a synonym for place or spatial variation. This is because the term locality inevitably smuggles in notions of social autonomy and spatial determinism, and this smuggling in excludes examination of these assumptions. It is surely far better to use terms like town, village, local authority area, local labourmarket or, for more general uses, place, area, or spatial variation. These very useable terms do not rely so heavily on conceptual assumptions about space *vis-à-vis* society.

One major problem remains, with this conclusion, however: what do we do now in empirical research? I will conclude by considering just that.

Sayer (1985b) gives two starting points. First of all, in his discussion of locale he questions why analysis should begin with such a general and loose concept which, in the end, refers only to a secondary dimension in causal terms. It would be better, instead, to determine the relevant causal group for any outcome to be explained, including the spatial dimensions as and if appropriate. The same recommendation can be made for locality. This is not the place to start. And, in building up causal explanation of any outcome we can now be more specific about the relative causal effects of the spatial dimension in referring to our three-fold level of spatial effects. This, of course, means that if there is a locality effect, then it will be included in the analysis as appropriate. But we certainly should not start on the assumption that there is and that this is some prime causal object.

Second, as in our discussion of the difference space makes, Sayer refers to the false distinction often made between general processes and local processes. General processes (for example, capital accumulation or patriarchal gender relations) are commonly but incorrectly, seen as aspatial. For general processes

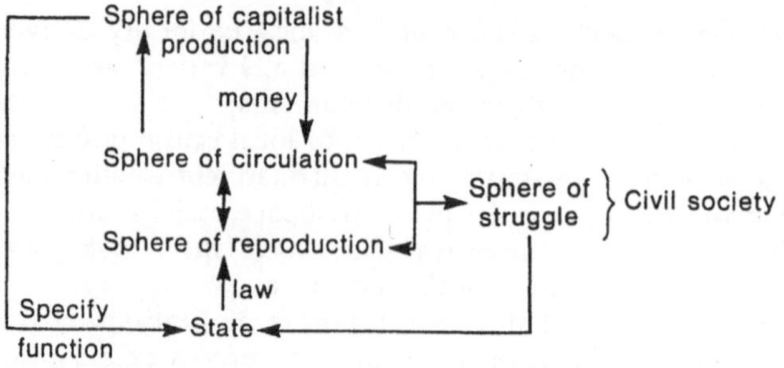

Figure 10.5 The basic structure of capitalist social formations (source: adapted from Urry 1981).

are never aspatial; 'the capital relation does not just float above localities [sic] in some space-less realm but is from the start constituted in countless localities' (Sayer 1985a).

Sayer's observations help us in two ways, one negative and one positive. An overall problem with locality is that the term has been used in an *a priori* manner, both conceptually and empirically. It has been assumed that localities universally exist empirically and that the concept of locality is in itself sufficient to capture how this existence affects social action. Both assumptions are, as we have seen, unwarranted. But why make these misleading assumptions in the first place? One reason is the inadequacy of distinction made between local processes and general processes. If the latter are seen as floating in a space-less realm, but clearly space does exist, then all the more need for a concept like locality to fill the gap. But if we see general processes as processes which are constituted in numerous places, then the need for such a conceptual gap-filler disappears. Locality is then only important if and when locality effects are part of the causal group explaining any event. And locality may well not be important.

On the more positive side, Sayer's observation begs the question of how these general processes are constituted in local areas. One answer is to say that this depends entirely on the specific characteristics of particular places, and hence the importance of local histories and local cultures. This is true enough, but we can improve upon this in analytical terms by providing some more general origins of local histories and cultures. This will then give us a means of analytically relating local characteristics to the concrete construction of general processes.

Our starting point is Urry's (1981b) anatomy of capitalist societies. He identifies three spheres of social relations in capitalist society: capitalist production, state institutions, and civil society (e.g. relations in the home or the community). These spheres of social relations should not be imagined as separate levels or instances, nor as economic base and social superstructure. Rather, these spheres are conceptualizations of sets of social actions, and it is most appropriate to visualize interrelating circuits of social relations. Furthermore, because of the way production is organized, capitalist societies are subject to continual change and the relation between these spheres as well as their extent is constantly changing. Urry (1981b) goes on to describe how these three spheres typically operate and what their links are. Figure 10.5 summarizes this discussion.

Now, the important point for our purposes here is that this anatomy of capitalist society is developed as though it existed on the head of a pin. We know that this is not the case, and that the processes going on within and between the spheres are in fact spatially constituted. It is this spatial constitution, with all the contingent effects involved, which in the first place creates local variations. How can these spatial constitutions be conceptualized?

The same essential problem emerged earlier in the context of Massey's Marxist reformulation of industrial location theory and the changing industrial geography of Britain (Massey 1978, 1979, 1984). The crucial causal dynamic was the law of value and consequently pressures for industrial restructuring. This restructuring was subsequently mapped out concretely as locational changes in industry. Massey formalized this spatial constitution of a general economic process as the spatial division of labour. The economic and occupational structure of any local area can be seen as the product of a whole succession of rôles played by that area within the spatial division of labour. In this way different occupational groups and economic activities will have particular geographical distributions (see Massey (1984), Warde (1985b) for reviews).

The same treatment can be applied to the other spheres of capitalist society. Work on the local state is in fact distinguishing a spatial division of the state (see Duncan & Goodwin 1982, Cooke 1983). Many of the same arguments apply for civil society. The concept starts from the observation that there is, in capitalist societies, a diverse realm of social practices outside the realm of relations of capitalist production and also outside the realm of the state form. But this civil society is highly differentiated and this differentiation is spatially constituted. Hence we can distinguish spatial divisions of civil society. Although clearly interacting with spatial divisions of labour and the state, the spatial divisions of civil society partly find their origin elsewhere and are in this way autonomously constituted. Indeed, local civil society developed in association with a particular spatial division of labour can outlive the latter's dissolution. This is what Cooke (1985) claims has happened in South Wales, for instance. There are very few coalminers and steelworkers left, but we can still speak of a radical region based around the organizations coalminers and steelworkers built in the past.

In this way, perhaps, we should add a fourth sphere to capitalist societies, an imagined sphere of social relations. So where Anderson (1983) talks about the imagined communities of nation–states (where there is no real community of social interaction – the world market is far more concrete), we can see a spatial division of imagined community; that is people's own belief in locality even where actual local social interactions are unimportant. Urry's distinction between the vertical organization of civil society, where the diverse social relations and groupings of civil society are predominantly class specific, and horizontal organization, where the opposite is the case, provides something of a starting point for how this spatial division proceeds.

This solution does not, of course, answer the problem of actually working out in practice how general processes are concretely constituted, and how these then combine in various contingent ways to create the particular outcomes we want to explain. This is, of course, what research should establish. But what this scheme does do is to return us to Sayer's prescription – the explanatory task is one of establishing the causal group for any outcome including the spatial dimensions if appropriate. The notion of a spatial division of labour, state, civil society, and imagined community allows us better to conceptualize in what ways this spatial dimension may exist. Note that locality is no longer central, except perhaps (and probably unfortunately) as a general prescriptive term signifying the importance of spatial constitutions. If real, concrete locality effects do exist and are important to any outcome, then this importance remains to be established.

Acknowledgements

This chapter results from work carried out for the Sussex University Research Programme, 'Economic Restructuring, Social Change and the Locality' financed by the Economic and Social Research Council (grant no. D0023 2093). Thanks to colleagues for their support, and especially to Peter Saunders, Mike Savage, and Rob Shields who made comments on an earlier draft.

References

Anderson, B. 1983. *Imagined communities: reflections on the origin and spread of nationalism*. London: Verso.

Bell, C. & H. Newby 1971. *Community studies*. London: Allen & Unwin.

Bell, C. & H. Newby 1976. Community, communism, class and community action: the social source of the new urban politics. In *Spatial perspectives on problems and policies*, D. Herbert & R. Johnston (eds). Chichester: Wiley.

Boddy, M. & C. Fudge 1984. *Local socialism?* London: Macmillan.

Bowlby, S., J. Ford & L. McDowell 1986. The place of gender in locality studies. *Area* **18**, 327–31.

Bradley, T. & P. Lowe (eds) 1984. *Locality and rurality: economy and society in rural regions*. Norwich: GeoBooks.

Buttimer, A. 1978. Charisma and context: the challenge of la géographie humaine. In *Humanistic geography*, D. Leys & M. Samuels (eds). London: Croom Helm.

Castells, M. 1977. *The urban question*. London: Edward Arnold.

Castells, M. 1978. *City, class and power*. London: Macmillan.

Castells, M. & F. Godard 1974. *Monopolville*. Paris: Mouton.

Cohen, R. (ed.) 1982. *Belonging. Identity and social organisation in British rural cultures*. Manchester: Manchester University Press.

Cooke, P. 1983. Regional restructuring: class politics and popular protest in South Wales. *Environment and Planning D, Society and Space* **1**, 265–8.

Cooke, P. 1985. Radical regions: a comparison of South Wales, Emilia and Provence. In *Political action and social identity: class, locality and culture*, G. Rees (ed.). London: Macmillan.

Dickens, P., S. S. Duncan, M. Goodwin & F. Gray 1985. *Housing, states and localities*. London: Methuen.

Duncan, S. S. 1974. The isolation of scientific discovery: indifferences and resistance to a new scientific idea. *Science Studies* **4**, 109–34.

Duncan, S. S. 1980. Review of Herbert and Johnston *Social problems and the city*. In *Environment and Planning A* **12**, 12.

Duncan, S. S. 1981. Housing policy, the methodology of levels, and urban research: the case of Castells. *International Journal of Urban and Regional Research* **5**, 2.

Duncan, S. S. & M. Goodwin 1982. The local state and restructuring social relations. *International Journal of Urban and Regional Research* **6**, 157–86.
Duncan, S. S. & M. Goodwin 1988. *Uneven development and the local state: behind the local government crisis in Britain.* Cambridge: Polity Press.
Dunleavy, P. 1980. *Urban political analysis.* London: Macmillan.
Economic and Social Research Council (ESRC) 1985a. Research initiative on social change and economic life. London: Social Affairs Committee, Economic and Social Reseach Council.
Economic and Social Research Council (ESRC) 1985b. The changing urban and regional system in the UK. London: Environment and Planning Committee, Economic and Social Reseach Council.
Foster, J. 1979. How imperial London presented its slums. *International Journal of Urban and Regional Research* **3**, 93–113.
Giddens, A. 1973. *The class structure of the advanced societies.* London: Hutchinson.
Giddens, A. 1979. *Central problems in social theory.* London: Macmillan.
Giddens, A. 1981. *A contemporary critique of historical materialism.* London: Macmillan.
Giddens, A. 1984. *The constitution of society.* Cambridge: Polity Press.
Gregory, D. 1978. *Ideology, science and human geography.* London: Macmillan.
Gregory, D. & J. Urry (eds) 1985. *Social relations and spatial structures.* London: Macmillan.
Grigg, D. 1969. Regions, models and classes. In *Integrated models in geography*, R. Chorley & P. Haggett (eds). London: Methuen.
Herbert, D. 1979. Introduction. In *Social problems and the city: geographical perspectives*, D. Herbert & D. Smith. Oxford: Oxford University Press.
Johnston, R. J. 1982. *Geography and geographers.* London: Edward Arnold.
Johnston, R. J. 1983. The neighbourhood effect won't go away. *Geoforum* **14**, 161–8.
Johnston, R. J. 1986. *The geography of English politics: the 1983 general election.* London: Croom Helm.
Lis, C. & H. Soly 1982. *Poverty and capitalism in pre-industrial Europe.* Brighton: Harvester.
Mark-Lawson, J., M. Savage & A. Warde 1985. Gender and local politics: struggles over welfare 1918–1934. In L. Murgatroyd, M. Savage, D. Shapiro *et al. Localities, class and gender.* London: Pion.
Massey, D. 1978. Regionalism: some current issues. *Capital and Class*, no. 6, 106–25.
Massey, D. 1979. In what sense a regional problem? *Regional Studies* **13**, 233–43.
Massey, D. 1984. *Spatial divisions of labour.* London: Macmillan.
Massey, D. & J. Allen (eds) 1984. *Geography matters!* Cambridge: Cambridge University Press.
Massey, D. & A. Catalano 1977. *Capital and land: landownership by capital in Great Britain.* London: Edward Arnold.
Miller, W. L. 1977. *Electoral dynamics in Britain since 1918.* London: Macmillan.
Miller, W. L. 1978. Social class and party choice in England: a new analysis. *British Journal of Political Science* **8**, 257–84.
Murgatroyd, L., M. Savage, D. Shapiro, J. Urry, S. Walby & A. Warde 1985. *Localities, class and gender.* London: Pion.
Pahl, R. E. 1968. The rural–urban continuum. In *Readings in urban sociology*, R. Pahl (ed.). Oxford: Pergamon.
Pahl, R. E. 1984. *Divisions of labour.* Oxford: Basil Blackwell.
Pahl, R. E. 1985. The restructuring of capital, the local political economy, and household work strategies. In *Social relations and spatial structures*, D. Gregory & J. Urry (eds). London: Macmillan.
Pred, A. 1977. The choreography of existence: comments on Hägerstrand's time-geography and its usefulness. *Economic Geography* **53**, 207–21.

Pred, A. 1984. Place as a historically contingent process: structuration and the time-geography of becoming places. *Annals of the Association of American Geographers* **74**, 279–97.

Sack, D. 1980. *Conceptions of space in social thought.* London: Macmillan.

Saunders, P. 1979. *Urban politics: a sociological approach.* London: Hutchinson.

Saunders, P. 1980. Towards a non-spatial urban sociology. Urban and Regional Studies Working Paper 21, University of Sussex.

Saunders, P. 1981. *Social theory and the urban question.* London: Hutchinson.

Saunders, P. 1985. Space, the city and urban sociology. In *Social relations and spatial structures*, D. Gregory & J. Urry (eds). London: Macmillan.

Saunders, P. 1989. A. Giddens: urban space man. In *Critical theory of the industrial societies*, D. Held & J. Thompson (eds). Cambridge: Cambridge University Press.

Saunders, P. & P. Williams 1984. Charity begins at home: some thoughts on recent and future developments in urban studies. Unpublished discussion paper. BSA Sociology and Environment Studies Group Seminar, London School of Economics.

Savage, M. 1987. Understanding political alignments in contemporary Britain: do localities matter? *Political Geography Quarterly* **6**, 53–76.

Sayer, A. 1980a. Theory and empirical research in urban and regional political economy: a sympathetic critique. Urban and Regional Studies Working Paper 14, University of Sussex.

Sayer, A. 1979b. Epistemology and conceptions of people and nature in geography. *Geoforum* **10**, 1.

Sayer, A. 1982. Misconceptions of space in social thought. *Transactions of the Institute of British Geographers* **NS 7**, 392–406.

Sayer, A. 1984. *Explanation in social science: a realist approach.* London: Hutchinson.

Sayer, A. 1985a. The difference that space makes. In *Social relations and spatial structures*, D. Gregory & J. Urry (eds). London: Macmillan.

Sayer, A. 1985b. Locales, localities and why we want to study them. Unpublished working note. University of Sussex, June.

Stacey, M. 1969. The myth of community studies. *British Journal of Sociology* **20**, 134–45.

Strathern, M. 1984. The social meaning of localism. In *Locality and rurality: economy and society in rural regions*, T. Bradley & P. Lowe (eds). Norwich: GeoBooks.

Sussex Programme 1984. Industrial restructuring, social change and the locality. Unpublished programme statement. Urban and Regional Studies Working Paper 14, University of Sussex.

Trudgill, P. 1975. Linguistic geography and geographical linguistics. *Progress in Geography* **7**, 23–48.

Urry, J. 1981a. Localities, regions and social class. *International Journal of Urban and Regional Research* **5**, 455–74.

Urry, J. 1981b. *The anatomy of capitalist societies: the economy, civil society and the state.* London: Macmillan.

Urry, J. 1983. De-industrialisation, classes and politics. In *Capital and politics*, R. King (ed.). London: Routledge & Kegan Paul.

Urry, J. 1985. Social relations, space and time. In *Social relations and spatial structures*, D. Gregory & J. Urry (eds). London: Macmillan.

Warde, A. 1985a. Space, class and voting in Britain. In K. Hoggart & E. Kofman (eds). *Politics, geography and social stratification.* London: Croom Helm.

Warde, A. 1985b. Comparable localities: some problems of method. In *Localities, class and gender*, L. Murgatroyd, M. Savage, D. Shapiro *et al.* London: Pion.

Watts, S. 1984. *A social history of Western Europe 1450–1720.* London: Hutchinson.

Wrigley, E. 1965. Changes in the philosophy of geography. In *Frontiers in geographical teaching*, R. Chorley & P. Haggett (eds). London: Methuen.

Part IV
NEW MODELS OF SOCIAL THEORY

Introduction

Nigel Thrift

Social theory is in a state of uncertainty at present. To begin with, it has a past but its practitioners seem unsure as to whether that past is a dead hand or an opportunity. Then again, social theory now consists of 'a varied, often confusing, array of approaches' (Giddens & Turner 1987, p. 3), but its practitioners seem uncertain whether this diversity of approaches is a good or a bad thing. Finally, the practitioners of social theory seem uncertain whether to lay claim to constructing the base of a social 'science', or of a 'critical reason' which does without some of the claims of science.

Of course, this state of uncertainty in social theory only mirrors the state of uncertainty of philosophy as a whole for, 'notwithstanding the repeated denunciation of philosophers who believe in *a priori* knowledge and self-evident givens, in necessity and certainty, in totality and ultimate foundations, there are relatively few of them around today' (Baynes *et al.* 1987, p. 7). Suspicion is the watchword.

Such a state of uncertainty hits particularly hard in geography where many were socialized into social theory via Marxism. Having slain the dragon of positivism they now find themselves with more trials ahead. As Urry points out, this position is to a degree the result of the relatively ascendant position of Marxism and Marxian theories in human geography compared with other subjects where the dominance of any kind of paradigm is much less apparent. (For example, Wiley (1985, p. 174) noted that American sociology had been 'without a dominant or hegemonic theory for about fifteen years', ever since the functionalism of Talcott Parsons and Merton ceased to hold sway.) The position is also linked to many human geographers' lack of knowledge of social theory, outside a closely drawn range of theories. (This lack of knowledge can sometimes be quite depressing. How often in the past, for example, were positions denounced as neo-Weberian by authors whose reading of Weber was perfunctory, to say the least?)

This introduction is therefore concerned to draw out some of the main themes of modern social theory and how they can be related to developments in the political-economy approach in human geography. Its starting point will be in Marxism but, as will soon become clear, Marxism alone can no longer be the end point. Classical Marxism was a child of the Enlightenment. As such it subscribed to principles which are now under attack: an Hegelian faith in history; a Cartesian faith in reason; and a Kantian faith in humanity (Merquior 1986). For Marxism, these Enlightenment origins do not have to prove a crushing blow, but they certainly demand something of a transformation. That transformation demands three retheorizations. The first of these is of the idea of totality. The issue of totality has, of course, been a long-running one in Marxist thought (Jay 1984). Its specific danger is that in its strong form it can produce a metanarrative in which all phenomena are viewed simply as instances of the workings of capitalism: 'capital . . . remains theoretically constant underneath the froth on the surface' (Storper 1987, p. 419). According to Storper (1987) three effects follow from such a metanarrative. The first is that what is important is selected in advance of considering particular events, so running the risk of missing new developments. The second is that it assumes that modern societies can never escape capitalism. Capitalism is regnant, and there is no prospect of civil society ever changing that fact. Finally, it makes it difficult to envisage capital as a concept in process, that is as a set of social relations that are constantly changing, from within and without, as various forms of civil society intervene. In such a conception the local can never make the global because the global is always already there. In other words, a strong conception of totality can do in theory what capitalism tries to do in actuality: it can stifle certain lines of enquiry, it can blank out true novelty, and it can continually try to colonize civil society.

The second retheorization is of the subject and consciousness. The autonomous, rational subject, to be found in certain varieties of classical Marxism (although, it might be argued, not always in Marx), is flawed:

> 'I think, therefore I am' is disqualified on a number of grounds. The 'I' is not immediately available to itself, deriving its identity as it does from its involvement in a system of signification. The 'I' is not the expression of some core of conscious selfhood that is its basis. The 'being' suggested by the 'I am' is not the expression of some core of continuous selfhood that is its basis (Giddens 1987, p. 206).

Even the idea of 'I think' is seen to be more problematic than it once was, involving concepts of practical reason, accounting, context, and intersubjectivity (dialogic) which make the idea of a subject increasingly slippery (Dennett 1987).

The last retheoretization consists of rethinking the power of transcendental claims. Such claims as are made now concerning science and truth are made with a good degree of caution because it is clear that all knowledge is, at the least, *situated* and therefore vulnerable. As the modern-day example of Habermas shows, this does not mean that transcendental claims cannot be made. Habermas, for example, makes the claim that speech acts routinely raise four universal validity claims of comprehensibility, truth, sincerity, and normative right. But, even he situates these 'gentle but obstinate' claims to reason within 'the general structures of possible communication' (Habermas 1979, p. 97).

The problem, of course, is how to refashion the confidence of the Enlightenment into something more suited to less confident times, something uncertain but not so uncertain that it degenerates into the free play of purely local narratives. We need a social theory that 'feels comfortable with the uncertainties of transition' (Wallerstein 1987, p. 324), but does not revel in them. Modern social theory can provide some clues here but it will provide no ticket of entry to some shining city on the hill which, at the end of an arduous intellectual climb, will provide a revelation. Modern social theory is about capturing human practices which are constantly being invented, changing, being adopted and adapted, so that it will always be an open-ended project: society talking to itself (Castoriadis 1987).

Modern social theory

Modern social theory provides a number of ways in which we can rid social theory of some of its Enlightenment optimisms without sinking into decadent nihilism, built on three different but interrelated orientations. The first of these orientations is the theory of social action, or more generally, of culture. This theory stems from the complex meeting of anthropology, sociology, literary theory, and social psychology with problems of the subject, consciousness, and reason. It is caught up with, and in some cases is indivisible from, the methodological correlate of the hermeneutic turn in social theory, the study of the understanding of understanding with its investment in words like symbol, meaning, text, and so on. The theory of social action has gone ahead in three different areas of the world. In North America, inspired by pragmatism, it has been developed as the traditions of symbolic interactionism (Joas, 1984, Smith 1987), and ethnomethodology (Heritage 1984, 1987). In Germany, the theory of action took other turns. In particular, it developed as an interest in aesthetic methods of interpretation and especially art (in its many forms) as a critical source of knowledge. Here we might consider the work of such Frankfurt School cultural theorists as Adorno and Benjamin as well as, in a slightly different vein, Gadamer's recasting of hermeneutics. In France, meanwhile, we would have to trace out the Saussurean diaspora through structuralism to a Nietzchean inspired post-structuralism, especially to the work on reading and writing (via Heidegger) of Derrida and other deconstructionists (Dews 1987, Norris 1987), and the treatises of Foucault and the early Lyotard with their insistence on the links between knowledge and power to be found in the will to truth (Dreyfus & Rabinow 1982). Here we come across the free play of meaning in discursive formations as the only reality and end a traverse from North America to France, having moved from subject-centred meaning to subjectless meaning.

There is also a second orientation in social theory which cross-cuts the first, namely those 'social scientists who have argued that the proper study of society is one that studies society as a complex system, where there are underlying interacting structures, systematic imperatives, and dynamic forms of systemic integration and/or breakdown' (McCarthy 1987, p. 22). In the United States Wallerstein's world systems theory with its complex historical mosaic of mini-systems, world empires, and world economies is notable. In France there is the work of Braudel (and other Annalistes), which was one of the inspirations for a vision of history as a series of faster moving ripples and deeper, slower moving waves which has a number of modern exponents like Duby and Bois. In Germany, writers like Luhmann and Schluchter have been particularly important, drawing on a mixture of Talcott Parsons and Weber to construct comprehensive systems theories of society (Munch 1987).

These two strands of social theory are not discrete. Like DNA they interlock in a complex spiral of influences. For example, in the strand of theories of social action can be found writers like Foucault who eschew conventional modes of historiography but still sometimes wrote in ways which look suspiciously like a large-scale systemic interpretation of history. As Turner (1987) writes, Foucault's work on biopolitics is an 'implicit history'. Similarly, those who are in the grand systems camp are likely to stray over into social action theory territory, especially in work on the logic of the cultural system.

Some writers have made more thoroughgoing attempts to incorporate both approaches in their work in such a way that it is now possible to identify a third direction within social theory. Writers like Bourdieu, Castoriadis, Elias, Giddens, Habermas, and Mann are particularly notable for their attempts to twist the strands of social action and system into greater concordance. Bourdieu (1977, 1984, 1987) makes the move between what he calls 'subjectivist' and 'objectivist' moments via an expanded theory of capital where the purely economistic form of capital is joined by cultural capital and social/symbolic capital. This system of different kinds of capital is fuelled by the structure of predispositions of actors which links categories and categorization together. Castoriadis (1986, 1987) tries to found a concept of action and system quite similar to that of Bourdieu, but in a much more generalized form, by developing the Marxian concept of praxis to include processes of signification. Elias (1978, 1982, 1987) develops an epic history of the civilizing process which ties the history of state formation to the internalization of restraint by subjects, thus grafting system on to action. Giddens (1984, 1985, 1987) develops a complex model of the constitution of society in which societies wax and wane as a result of structures of signification, domination, and legitimation. He is particularly concerned to explore the importance of agency as a vital part of these systems as a combination of language and praxis. Habermas's (1984, 1987a, 1987b, White 1988) main goal has been to forge a dialectical synthesis of social systems and the life world of action, most especially through his stress on intersubjective communication, buttressed by the philosophy of speech act theory. Modern society is increasingly the plaything of processes of rationalization straying over into the life world and bringing with them multiple crises for state and economy. Finally, Mann (1986) provides a social theory based upon four sources of social power (the ideological, economic, military, and political) which combine to form 'multiple, overlapping socio-spatial networks' (Mann 1986, p. 2).

The approaches of these writers are not identical. They have their differences, both of substantive method and in the degree to which their theoretical schemes have been worked out as full-scale historical accounts (Giddens 1985). But I would claim that their differences are not as great as their similarities. These similarities are *very* instructive for they provide a set of six presuppositions about the current 'best practice' in modern social theory. First of all, each of these writers is avowedly non-theoreticist, that is they are careful to separate out theoretical abstraction and reality. They always try to avoid confusing 'the things of logic' with 'the logic of things' (Bourdieu 1987, p. 7). In an allied vein, they find a continual puzzle for social theory is social creativity, how *new* institutions and practices are formed which continually outrun social theorists' mental reach. Third, and following on from this second point, they are all concerned with the importance of the cultural realm of signification and meanings as a vital area of investigation. Fourth, they all feel that this cultural realm is under threat from processes of rationalization emanating from state and economy. Systems of domination are spreading and becoming more 'systematic'. Fifth, they are all concerned to stake a claim for grand social theory (Skinner 1985), but it is not a theory which is monolithic. Each of these writers is concerned to acknowledge that societies are made up of a series of different institutions, all of which have widely variable and effective causal powers. Society is not a force field but a field of forces. Sixth, and finally, they draw their inspiration from many sources. Marx looms large in many of their accounts. So does Weber (see Collins 1986, Whimster & Lash 1987). But the essence of these writers is that they are not ashamed to be eclectic; they are only ashamed of eclectism that does not work.

Key issues in social theory

Each of these three orientations in modern social theory has had to tackle two important issues as they have explored social action, social system, and the interaction between them. The first of these issues is quite simple. What special claims can be made in these uncertain times which dignify systems of knowledge as science? This is not a comfortable debate for it strikes at the heart of academic self-interest and arrogance. Three main camps can be identified. In one camp are those like Habermas who still want to cleave to the logic of integration and reconstruction of the Enlightenment, especially by asserting that universal (not metaphysical, but metabiological) validity claims are still possible. Habermas takes seriously the idea of the academic as a guardian of reason, operating with the credentials of communicative reason. This is not to say that Habermas naïvely goes on supporting an untenable position. Thus he argues that reason is inescapably situated, concretized in history, society, body, and language.

In the opposing camp are writers who, like Nietzche (one of their chief mentors), want to explode occidental rationalism. Sometimes labelled as post-structuralist, they are those like Foucault, the early Lyotard, and Baudrillard who oppose the very concept of theory and so find 'science' an untenable philosophical position. Here is the logic of disintegration and deconstruction, of 'the modern art of thought' (Merquior 1986, p. 40). As McCarthy (1987) puts it:

> To the necessity that characterises reason in the Cartesian–Kantian view, the radical critics typically oppose the contingency and conventionality of the rules, criteria, and products of what counts as rational speech and action at any given time and place; to its universality, they oppose an irreducible plurality of incommensurable lifeworlds and forms of life, the erremediably 'local' character of all truth argument, and validity; to the *a priori*, the empirical; to certainty, fallibility; to unity, heterogeneity; to homogeneity, the fragmentary; to self-ended givenness ('presence'), universal mediation by differential systems of signs; to the unconditioned, a rejection of ultimate foundations in any form (McCarthy 1987, pp. viii–ix).

Thus, instead of science, genealogy, or deconstruction, or paralogy; the non-methods.

Set between these two camps is the half-way house of realism. Realism is an attempt to hold on to a science, but it is a science whose cloth is cut to the times. Thus realism takes due account of: the specificity of social science, that is the limits on its knowledge (in Bhaskar (1986) activity-dependence, concept dependence, time–space dependence, and social relation dependence respectively); the importance of examining internal agency, beliefs, and mental states in social science; and of the need to incorporate a historically situated hermeneutics into social science. Thus we reach a picture of science as a 'historically situationally specific judgemental rationality':

> We now have a picture of a traveller on a particular epistemic world-line, marshalling, and transforming the historical materials and media at its disposal, aspiring to assess and express claims in the only way open to it, that is, using these historically generated, transmitted and transferable (so transient) materials and media, about how the world is, independently of these claims (and the materials and media and more generally the conditions that make these claims possible), in a continually intricate process of an identification, description, explanation and redescription of deeper strata of reality. In the ongoing process of science, as deeper levels and wider shores of reality come to be known and reknown, historically *situated subjects* make *ontic* (being-expressive) claims about a reality which transcends their situation, in a dialectic which affords *objective grounds* for their inevitably *local choices* (Bhaskar 1986, p. 92).

The second important issue in social theory has been the nature of the experience of modernity. Berman (1982) sums up what is at stake here:

> There is a mode of vital experience – experience of space and time, of the self and others, of life's possibilities and perils – that is shared by men and women all over the world today. I will call this body of experience 'modernity'. To be modern is to find ourselves in an environment that promises us adventure, power, joy, growth, transformations of ourselves and the world – and at the same time, that threatens to destroy everything we have, everything we know, everything we are. Modern environments and experiences cut across all boundaries of geography and ethnicity, of class and nationality, of religion and ideology: in this sense modernity, can be said to unite all mankind. But it is a paradoxical unity, a unity of disunity: it puts us all into a maelstrom of perpetual disintegration and renewal, of struggle and contradiction, of ambiguity and anguish. To be modern is to be part of a universe in which, as Marx said, 'all that is solid melts into air' (Berman 1982, p. 15).

The body of work on the experience of modernity has grown out of the renewed attention being paid to the writings of Baudelaire and to Benjamin's interpretations of Baudelaire (e.g. Benjamin 1973). It stresses the transitory nature of modern life, the headlong rush of change. It is perturbed by modern life's lack of authenticity, its loss of 'aura' or 'halo'. It is puzzled by a world where everything is open-ended and pregnant with its contrary. It is disturbed by a world full of terrible ills which is also very often enjoyable and exciting. It is maddened by the problems of analyzing and representing such a world.

As one might expect, work on modernity divides in much the same way as do modern debates on the existence of science. In each case, some of the same exponents are active and some of the same problems persist. On one side are those like Habermas, who are only too willing to acknowledge the importance of understanding modernity but do not see the evidence that it has outgrown our powers of understanding. Instead what he sees is the result of the organizing logic of modernization, the spill-over of means–end rationalism from the world of production into the life world, systematizing it and bringing with it pathological effects. The structure of everyday life is deformed. New social movements like the women's movement or the ecological movement are defensive reactions to this process.

On the other side are authors like Lyotard and Baudrillard. For Lyotard the modern world has recently undergone a qualitative shift as 'societies enter what is known as the postindustrial age and cultures enter what is known as the postmodern age' (Lyotard 1984, p. 3). The postmodern world consists of a hyperreal society of multinationals, media, and computers in which culture is commodified in an economy of the sign, and the subject becomes surplus to requirements. In Baudrillard (see Poster 1988), the logic of cultural commodification is taken even further, into a postmodern world of pastiche and spectacle and simulation:

> And if reality, under our eyes, suddenly dissolved? Not into nothingness but into a real which was more than real (the triumph of simulation)? If the modern universe of communication, the space of hyper communications through which we are plunging, not in forgetfulness, but with an enormous saturation of our senses, consumed us in its success – without trickery, without secrets, without distance? If all publicity were an apology, not for a product, but for publicity itself? If information did not refer any longer to an event, but to the production of information itself as the event? If our society were not a 'spectacle', as they said in 1968, but more cynically a ceremony? If politics was a confinement more and more irrelevant, replaced by the spectre of terrorism, by a generalised taking of hostages, that is to say by the spectre of an impossible challenge? If all this mutation did not emanate, as some believe, from a manipulation of subjects and opinions, but from a logic without a subject where opinion transited into fascination? If pornography signified the end of sexuality as such, and from now on, under the term of obscenity, sexuality invaded everything? If seduction were to succeed desire and love, that is to say, the reign of the object over that of the subject? If strategic planning were to replace psychology? If it would no longer be correct to oppose truth to illusion, but to perceive generalized illusion itself as truer than truth. If no other behaviour was possible than that of learning, ironically, how to disappear? If there were no longer any fantasies, lines of light or raptures but a surface, full and continuous, without depth, uninterrupted? And if all this were neither the matter of enthusiasm nor despair, but fatal (Baudrillard 1987; cited in Kroker 1988, p. 185).

(As one commentator has remarked (Dews 1987) there is an irony here. For this postmodern world seems very much like Adorno's totally administered world, but with commodification of the sign replacing administration as the totalizer.)

In the half-way house this time between Habermas's gallant, and very germanic, defence of modernity and the postmodernity of Lyotard and Baudillard, with its distinctively French mode of proceeding – 'svelte philosophy' (Sim 1988) as it has been called – lies the work of those like Jameson and Lash and Urry who are willing to take on some of the analysis of Lyotard and Baudillard but relate these changes to the shifting contours of capitalism. For Jameson (1984, 1985) and Lash & Urry (1987), the new postmodern culture is the cultural correlate of a new phase of capitalism: late capitalism, disorganized capitalism, or what have you. This is a capitalist economy which is decentred, flexible, and at the same time monumental in its influence and scale. This is a capitalist culture in which the subject has become decentred, becoming an economy of desire on the wire, in response to continual innovations in the sphere of consumption: the transformation of reality into images and the fragmentation of time into a series of perpetual presents, both leading to historical depthlessness and pastiche.

Marxism and social theory

What then of Marxism, from which this introduction emanated? Where has it been in all these debates? First, within the debates themselves, it has formed an implicit subtext as they have unfolded. It is critical in the work of Giddens, Habermas, and others who want to deconstruct and reconstitute historical materialism. It is influential even in the work of those like Foucault who see Marxism as something of a foil (and even here there is some ambiguity) (Lentricchia 1987). Second, within the Marxian tradition itself, these debates have lead to quite considerable changes of emphasis and procedure. Marxism has changed as the six prescriptions of practice in social theory set out above have filtered through, and as the implications of the debates on science and modernity have become more and more apparent. Currently, it is possible to discern three main reactions. One of these might be called 'the enthusiastic embrace'. The work of the School of Analytical Marxism of Cohen, Elster, Roemer, Wright, and others is probably most representative of this reaction. With its emphasis on delineating modes of exploitation through the ages, this School sometimes seems to be close to a Marxist systems theory approach. It has been criticized as non-Marxist (e.g. Harvey 1987b), yet it retains some links with Marxism, for example in its fascination with game theory as a means of retaining some notion of rationality. A second approach has been to take on some of the implications of the debates in modern social theory and reject others. The work of Jameson and others on postmodern culture is a case in point. The Regulationist School, with its wariness of functionalism and reductionism, is one more. Resnick & Wolff's (1987) nondeterminist approach to political economy counts as yet another case. Finally, there is a group of Marxists who rejects much of the content and all of the style of these debates, who see themselves as besieged keepers of the tradition. For them, Marxism is a separate tradition of intellectual enquiry which has been caricatured in current debates on social theory, devalued and distorted by a flood of inflated rhetoric. Of course, there are still problems with Marxism, as with any intellectual tradition, but these can be solved and Marxists can solve them (see Harvey 1987a, *Society and Space* 1987).

Geography and social theory

Last, but by no means least, where has geography been in all this? The answer has been, with a few exceptions, on the outside looking in. But that position is now changing. Geography is now becoming an increasingly important pivot of debate for at least two reasons. The first reason is because the recent debates on social theory have all placed intensified emphasis on the importance of context. Societies are seen as contextual entities, that is they are seen as locally variable right from the start: they are constituted in time and space. Social structures cannot be separated from spatial structures or, as Urry puts it (1985, p. 23), 'the social world should be seen as comprised of space–time entities having causal powers which may or may not be realised depending on the patterns of spatial/temporal interdependence (between them).' This argument can be further illustrated by means of a metaphor of perspective. Societies are no longer perceived as vertical entities, with the implication

> that concrete realities of social life are the actual result of a series of hierarchical causal interactions from the more general (mode of production) or more powerful forces (forces of production) down to the more particular (late capitalism) or weaker forces (ideologies) (Storper 1987, p. 421).

Rather, they are seen as horizontal where forces are 'not sequential or hierarchical, but simultaneous' (Storper 1987, p. 421). In other words, social structures are geographies – overlapping, partially integrated, and messy geographies – and they have to be not just perceived but theorized and even represented as such (Mann 1986, McHale 1987).

These thoughts have considerable implications for a social theory which, all too often, has failed to conceive terms like local as spatial. For example, it becomes clear that the whole process of local socialization is crucial to understanding the constitution of society; how subjects are produced in different areas, with their different capacities, powers, and local knowledges (Giddens 1987). But, at the same time, it is also clear that considerably more attention must be paid to how subjects' local knowledge and values can become homogenized by the spread of global communications and media (Meyrowitz 1985).

In the postmodern culture debate, this might be phrased as an enquiry into the degree to which the decentring of the subject has lead to a decentring of the local, with resonances concerning the degree to which individualism was a myth and the degree to which it actually existed. Such thoughts also point to a crucial problem, if social structure is to be grasped as a geography, of scale. In such a conception causality will be 'multi-tiered' (Gregory 1986), which points to the important task of refining basic geographical ideas such as region and locality and of developing many more mediating concepts (Taylor 1982, 1987).

To summarize, geography provides context. Whereas prior to the 1970s that might have meant the addition of a little local detail to the grand sweep of social theory, it now means that local detail is coming to be seen as crucial to the constitution of society. It is not a detail!

The second reason why geography is now becoming a pivot of debate concerns the nature of capitalism. For a long period Harvey has argued for a historical-geographical materialism which gives equal weight to how capitalism produces history *and* geography. This interest in the *production of space* by capitalism has become ever more emphatic since the discovery of three literatures by geographers. The first of these was the work of Lefebvre (Harvey 1973, Soja 1985, 1989) on space as a diachronic discourse: space is 'medium, milieu and intermediary, instrument as well as goal' (Lefebvre 1974; cited in Shields 1988, p. 45). The second literature was the work of the situationists, especially Debord and Vaneigem on space as a crucial aspect of the society of the spectacle: 'in all societies where modern considerations of production prevail, all of life presents itself as an intense accumulation of spectacles. Everything that was directly lived has moved away into a representation' (Debord 1987, para. 1). The third literature was the work of Baudrillard (referred to above) on the consumption of the sign and hyperreality. Each of these literatures refers to the production of space as a key element in the production and reproduction of modern capitalism, especially the design of space as a spur to consumption.

Thus social theory and human geography are coming together in ways which would have been undreamt of – by social theorists or human geographers – only a few years ago. The chapters by Slater, Urry, Soja, and Gregory are all concerned to show the consequences of these new theorizations of society and space, space and society. Social theory is moving onwards now, towards post-postmodern theories (see Dews 1987, Fekete 1988, Giddens 1987). In the future, human geography will be moving with, and as part of it. Social theory has a new passenger.

References

Baudrillard, J. 1987. *L'Autre par lui-même*. Paris: Grasset.

Baynes, K., J. Bohman & T. McCarthy (eds) 1987. *After philosophy: end or transformation?* Cambridge, Mass.: MIT Press.

Benjamin, W. 1973. *Charles Baudelaire. A lyric poet in the era of high capitalism*. London: New Left Books.

Berman, M. 1982. *All that is solid melts into air. The experience of modernity*. London: Verso.

Bernstein, R. J. 1985. Introduction. In *Habermas and modernity*, R. J. Bernstein (ed.), 1–32. Cambridge: Polity Press.

Bhaskar, R. 1986. *Scientific realism and human emancipation*. London: Verso.

Bourdieu, P. 1977. *Outline of the theory of practice*. Cambridge: Cambridge University Press.

Bourdieu, P. 1984. *Distinction. A social critique of the judgement of taste*. London: Routledge & Kegan Paul.

Bourdieu, P. 1987. What makes a social class? On the theoretical and practical existence of groups. *Berkeley Journal of Sociology* **3**, 1–20.

Castoriadis, C. 1986. *Crossroads in the labyrinth*. Brighton: Harvester.

Castoriadis, C. 1987. *The imaginary institution of society*. Cambridge: Polity Press.

Collins, R. 1986. *Weberian sociological theory*. Cambridge: Cambridge University Press.

Debord, G. 1987. *Society of the spectacle*. Detroit: Rebel Press.

Dennett, D. 1987. *The intentional stance*. Cambridge, Mass.: MIT Press.

Dews, P. 1987. *Logics of disintegration. Post-structuralist thought and the claims of critical theory*. London: Verso.

Dreyfus, H. C. & P. Rabinow 1982. *Michel Foucault. Beyond structuralism and hermeneutics*. Brighton: Harvester.
Elias, N. 1978. *The civilising process*. Oxford: Basil Blackwell.
Elias, N. 1982. *State formation and civilization*. Oxford: Basil Blackwell.
Elias, N. 1987. *Involvement and detachment*. Oxford: Basil Blackwell.
Fekete, J. (ed.) 1988. *Life after postmodernism. Essays on value and culture*. London: Macmillan.
Giddens, A. 1981. *A contemporary critique of historical materialism*. London: Macmillan.
Giddens, A. 1984. *The constitution of society*. Cambridge: Polity Press.
Giddens, A. 1985. *The nation state and violence*. Cambridge: Polity Press.
Giddens, A. 1987. Structuralism, post-structuralism and the production of 'culture'. In *Social theory today*, A. Giddens & J. Turner (eds), 195–223. Cambridge: Polity Press.
Giddens, A. & J. Turner (eds) 1987. *Social theory today*. Cambridge: Polity Press.
Gregory, D. 1986. Structuration theory. In *A dictionary of human geography*, R. J. Johnston (ed.), 464–9. Oxford: Basil Blackwell.
Habermas, J. 1979. *Communication and the evolution of society*. London: Heinemann.
Habermas, J. 1984. *The theory of communicative action*, Vol. 1: *Reason and the rationalisation of society*. Cambridge: Polity Press.
Habermas, J. 1987a. *The theory of communicative action*, Vol. 2: *Lifeworld and system. A critique of functionalist reason*. Cambridge: Polity Press.
Habermas, J. 1987b. *The philosophical discourse of modernity*. Cambridge: Polity Press.
Harvey, D. 1973. *Social justice and the city*. London: Edward Arnold.
Harvey, D. 1987a. Flexible accumulation through urbanisation: reflections on post-modernism in the American city. *Antipode* **19**, 260–86.
Harvey, D. 1987b. Review of J. Elster, Making sense of Marx. *Political Theory* **14**, 686–90.
Heritage, J. 1984. *Garfinkel and ethnomethodology*. Cambridge: Polity Press.
Heritage, J. 1987. Ethnomethodology. In *Social theory today*, A. Giddens & J. Turner (eds). Cambridge: Polity Press.
Jameson, F. 1984. The cultural logic of late capitalism. *New Left Review* **146**, 53–92.
Jameson, F. 1985. Postmodernism and consumer society. In *Postmodern culture*, H. Foster (ed.), 111–25. London: Pluto Press.
Jay, M. 1984. *Marxism and totality. The adventures of a concept from Lukacs to Habermas*. Cambridge: Polity Press.
Joas, H. 1987. Symbolic interactionism. In *Society theory today*, A Giddens & J. Turner (eds), 82–115. Cambridge: Polity Press.
Kroker, A. 1988. Panic value: Bacon, Colville, Baudrillard and the aesthetics of deprivation. In *Life after postmodernism. Essays on value and culture*, J. Fekete (ed.), 181–90. London: Macmillan.
Kroker, A. & D. Cook 1988. *The postmodern scene. Excremental culture and hyper-aesthetics*. London: Macmillan.
Lash, S. & J. Urry 1987. *The end of organised capitalism*. Cambridge: Polity Press.
Lefebvre, H. 1974. *La Production de l'espace*. Paris: Anthropos.
Lentricchia, F. 1987. *Ariel and the police*. Brighton: Wheatsheaf.
Lyotard, J. F. 1984. *The postmodern condition. A report on knowledge*. Manchester: Manchester University Press.
McCarthy, T. 1987. Introduction. In *The philosophical discourse of modernity*, J. Habermas, vii–xx. Cambridge: Polity Press.
McHale, B. 1987. *Postmodernist fiction*. New York: Methuen.
Mann, M. 1986. *The sources of social power*. Cambridge: Cambridge University Press.
Merquior, J. G. 1986. *From Prague to Paris. A critique of structuralist and post-structuralist thought*. London: Verso.
Meyrowitz, J. 1985. *No sense of place*. New York: Macmillan.

Munch, R. 1987. Parsonian theory today: in search of a new synthesis. In *Social theory today*, A. Giddens & J. Turner (eds), 116–55. Cambridge: Polity Press.
Norris, C. 1987. *Derrida*. London: Fontana.
Poster, M. (ed.) 1988. *Jean Baudrillard. Selected writings*. Cambridge: Polity Press.
Resnick, S. A. & R. D. Wolff 1987. *Knowledge and class. A Marxian critique of political economy*. Chicago: Chicago University Press.
Shields, R. 1988. An English precis of Henri Lefebvre's 'La Production de l'espace. University of Sussex. Urban and Regional Studies. Working Paper 63.
Sim, S. 1988. 'Svelte discourse' and the philosophy of caution. *Radical Philosophy* **49**, 31–6.
Skinner, Q. (ed.) 1985. *The return of grand theory in the social sciences*. Cambridge: Cambridge University Press.
Smith, N. 1987. Dangers of the empirical turn: the CURS initiative. *Antipode* **12**, 30–9.
Society and Space 1987. Reconsidering social theory: a debate. *Environment and Planning D, Society and Space* **5**, 376–76.
Soja, E. W. 1985. The spatiality of social life: towards a transformative retheorisation. In *Social relations and spatial structures*, D. Gregory & J. Urry (eds), 90–107. London: Macmillan.
Soja, E. W. 1989. *Postmodern geographies*. London: Verso.
Storper, M. 1987. The post-Enlightenment challenge to Marxism. *Environment and Planning D, Society and Space* **5**, 418–26.
Taylor, P. J. 1982. A materialist framework for political geography. *Transactions of the Institute of British Geographers* **NS7**, 15–34.
Taylor, P. J. 1987. The paradox of geographical scale in Marx's politics. *Antipode* **19**, 287–306.
Turner, B. S. 1987. The rationalisation of the body: reflections on modernity and discipline. In *Max Weber, rationality and modernity*, S. Whimster & S. Lash (eds), 222–41. London: Allen & Unwin.
Urry, J. 1985. Social relations, space and time. In *Social relations and spatial structures*, D. Gregory & J. Urry (eds), 20–48. London: Macmillan.
Wallerstein, I. 1987. World systems analysis. In *Social theory today*, A. Giddens & J. Turner (eds), 309–24. Cambridge: Polity Press.
Whimster, S. & S. Lash (eds) 1987. *Max Weber, rationality and modernity*. London: Allen & Unwin.
White, S. K. 1988. *The recent work of Jurgen Habermas. Reason, justice and modernity*. Cambridge: Cambridge University Press.
Wiley, E. 1985. *American Journal of Sociology* **56**.

11 *Peripheral capitalism and the regional problematic*

David Slater

Thematic demarcation and point of departure

Reflecting upon much of the recent theoretical literature on urban and regional development, and especially those studies which have applied a broadly conceived Marxist method of enquiry, it seems clear that the advanced capitalist societies have been retained at the thematic centre of investigation. This observation must not be overemphasized since it is also evident that from the early 1980s interest and concern for global trends and processes of internationalization have been increasingly inserted into the contemporary research agenda. Moreover, there definitely does seem to be a greater awareness of the existence and specificities of peripheral societies than a decade ago. *Nevertheless*, there still remains an underlying, perhaps unconscious, tendency towards universalism or Euro-Americanism in the formulation and priority given to topics for theoretical discussion. Even though this tendency is not only to be located in the realms of Marxist urban and regional analysis, its presence in this sphere has received far less interrogation than elsewhere.[1]

My purpose in the following pages will be to set out certain central issues in the development of regional analysis for peripheral societies. In so doing, I shall inevitably touch on a number of problems that require far more elaboration than can be attempted here. In particular, I want to examine some of the difficulties involved in the production of spatial concepts in the context of capitalist development at the periphery. As one illustration, I shall refer to the important conceptual and political boundaries that need to be drawn between terms such as the regional problem, the regional question and the regional problematic. In addition, I shall include some considerations on the way both econocentrism and universalism have moulded and aligned the inner structure of much current debate on the significance of space and its conceptualization.

The thematic focus of the chapter will be demarcated as follows. First, I shall concentrate on the regional or more broadly expressed territorial dimensions of capitalist development in peripheral societies, and the more strictly urban-oriented literature will fall outside my purview (see in this volume Chatterjee, Ch. 6). Second, in order to limit further the potentially extensive scope of the discussion, treatment of the post-revolutionary societies of the Third World will be largely left out of account.[2] Finally, although key elements of the impact of development theory on regional analysis will inevitably surface at different stages of the argument, a comprehensive survey of the interconnections between theories of development and regional research will not be taken up here.[3]

My argument will be structured around three analytical components. Initially, I shall provide an unavoidably brief survey of the emergence and evolution of the main theoretical tendencies of regional development analysis with a Third World orientation. This will lead me into looking at some of the problems arising from the production and application of spatial concepts within a Marxist problematic. Then in the final section of the chapter I shall address myself to one or two central issues concerning the peripheral state and the modalities of regional crises.

Theory, territory, and the Third World – divergent currents of interpretation

When tracing the origins and development of particular social interpretations it is instructive to situate the discussion within some specific historical limits. In this sense, it is pertinent to bear in mind that since the mid-1970s the domain of regional development studies has been markedly influenced by the incursion of Marxist and neo-Marxist ideas, and their linkage with the crisis of capitalism. In general, this crisis and the associated upsurge of critical Marxist analysis seriously undermined the dominance of conventional social science thinking. In the particular sphere of Third World regional development studies, and in the period from the mid-1970s to the present time, I believe it is possible to identify three major responses to the introduction and extension of Marxist theory.

The continuance of traditional conceptions

Although the 1970s witnessed a growing presence and application of critical ideas emanating from Latin America *dependencia* perspectives, radical underdevelopment theory, and more classical Marxist approaches, it would be quite wrong to assume that conformist perspectives etiolated through lack of intellectual nourishment.[4] Moreover, starting in the 1980s with the resurgence of conservative politics, many of the long established positivist notions of modernity and Western progress have been reasserted with noteworthy rigour. It may be recorded that an amalgamated series of constructs such as the dual economy thesis, achievement motivation, and national integration models have permeated much of the traditional work on regional development and spatial organization (Allor 1984, Adarkwa 1981, Dickenson 1980, Drake 1981, Leinbach 1986, Rondinelli 1980, 1983, Rondinelli & Evans 1983, Siddle 1981, and Townroe & Keen 1984). Above all, however, it has been modernization theory that has acted as the central articulating core for these models or constructs, and its influence is still to be found in the recent literature (Chisholm 1982, Gwynne 1985, Potter 1985, Riddell 1985, Scott 1982).[5]

Over ten years ago, I sketched out the main deficiencies of modernization theory, as it was then being deployed within development geography (Slater 1974), and more recently I have restated the central inadequacies of this theory as it has influenced studies of Third World urbanization (Slater 1986a). Two general criticisms can be immediately signalled.

In the first place, modernization theory was grounded on categories invented in response to Western experience and viewed as constants of every possible society. From this starting point, analysis was reduced to a strategy of recognition in which the very manner of posing questions already presupposed the basic essence of the answers. Consequently, Western universalism, rooted in this particular paradigm, was unable to account for the historical specificity of social development in Third World regions.

Second, no effective understanding of the spatial expansion of capitalism and its impact within peripheral societies was possible since modernization theory furnished an idealized and highly partial perspective on world development. In particular, no theorization of capitalism was presented and development was visualized in terms of the diffusion of Western capital, values, institutional arrangements, and social practices, the description of which hardly captured the antagonistic nature of social relations in the heartlands of the already constituted modernized societies, let alone the violent forms of incorporation of peripheral societies into the international capitalist system. But the discourse of modernization, as Escobar (1984–5) reminds us, was effective in providing the ideological foundation for a wide-ranging series of Western and mainly American interventions in the countries of the Third World.

In the current era the discourse of modernization finds expression in official doctrines on Third World development: 'African societies must be encouraged to open their doors to more foreign private investment so as to promote economic development'; 'with the freeing of the market, there will be more creation of wealth, and therefore improved possibilities for the reduction of poverty.' Equally, US definitions of democracy are used as a gauge for the provision or withdrawal of economic aid, and in the more acute cases, the judged absence of democracy, so defined, is used to legitimize military intervention, covert and overt, in recalcitrant peripheral societies such as Nicaragua or Grenada.

Modernization, as an official doctrine, has always denied the societies of the periphery a history of their own, but today there is also a growing inclination to deny some of these societies a future of their own. Visvanathan (1987), for example, trenchantly argues that whilst in the past Western man constructed the savage as the other, in order to impose his own savagery on him, in the contemporary world, societies and cultures are now being destroyed because they are considered refractory to the Western scientific gaze (1987, p. 48). Hence, the Western encounter with the other ends, in its eventual logic, in erasure and the 'abandonment of modernity as a universalizing project' (Visvanathan 1987, p. 48). Although such a view may well find echoes in the labelling of some Third World countries as 'basket-cases' (for instance, Bolivia), irremediably lost to the civilizing cause of modernization, we would be well advised not to extend Visvanathan's argument too far, since there are still a not inconsiderable number of peripheral societies experiencing new waves of modernization (for instance, the so-called newly industrializing countries).[6]

In the academic literature, adherence to the canons of modernity and Western-style capitalist democracy receives a more nuanced articulation. For development geography, the double-track nature of today's discourse of modernization, with encouragement for those countries that have already attained a certain level of industrialization, and annulment for those countries judged to have failed the test of modernity, is still not fully visible. But when its influence does seep through, for that group of peripheral societies deemed to be obsolescent, there will be no more need for a codification of spatial policies; rather these territories of the periphery will be abandoned to their world market fate. Furthermore, perhaps, it might be suggested that in a postmodern First World, these obsolescent territories that were never able to join the modern era may now be portrayed in the postmodern imaginary as a set of romanticized images of past cultures, as well as objects of the present tourist gaze.[7]

Although the reproduction of traditional conceptions, and in particular modernization theory, has sometimes received little if any critical consideration in the literature on regional development and spatial organization, it is certainly not my intention to imply that these conceptions have retained a sovereign position. On the contrary, the 1970s witnessed the rise of a new and increasingly influential research tendency which attempts to combine elements of the traditional with elements of the radical or critical.

The emergence of a syncretic perspective

In its central lineage our second research perspective is embedded in the mainstream of bourgeois liberal thought. Although the writers falling within this tendency have firmly set themselves against the theoretical and political thrust of historical materialism *tout court*, they have not abjured the use of Marxist ideas in particular instances. Rather, *extracted* concepts and ideas originating in the Marxist tradition have been appropriated and reinserted into a newly constituted *mélange*, which has been formed by the amalgam of immiscible and antagonistic concepts. I would argue that within this perspective an attempt is made to employ Marxist categories, uprooted from their defining theoretical terrain, in order to fulfil two main ideological functions:

(a) On the basis of selective appropriations of certain Marxist notions an effort is made to fortify an approach which in some respects is critical of the unalloyed modernization position, but which maintains a significant distance from the central theses of a materialist perspective. In this sense, one has a neutralization of emergent and opposed ideas within a discourse that seeks to rearticulate and redeploy antagonistic concepts inside its own ideological space. As a result, conceptual reconciliation becomes an ideological sublimation of reformism.

(b) In a related fashion, this kind of attempted reconciliation permits the development of new orientations and modes of interpretation, which, while displaying, in varying degrees, some reliance on Marxist thought, are able, through the preservation and renewal of non-Marxist propositions, to avoid becoming parasitic on or subordinate to that same thought.

In giving a title to this subsection of the chapter I have used the term syncretic. Normally, one finds reference to eclecticism, but I prefer to define this particular research current as syncretic because of the assemblage of concepts which are not only incompatible but also *antagonistic*. In other words, syncretism can be viewed as a specific form of eclecticism in which the antagonisms existing among the aggregated concepts and positions prevent the formation of a coherent theoretical method. Sometimes, however, ideas taken from a number of different sources can be and have been effectively merged into a new synthesis containing a coherent set of concepts. But in these cases each concept has to be carefully rearticulated into a newly constituted theoretical mode of analysis.[8] In this context, I want to add that I am not arguing for one preconstituted, all-encompassing theoretical perspective that requires no alteration. Rather the point is that the simple aggregation of constructs which are rooted in antagonistic schools of thought can only vitiate the development of a coherent method of analysis.

A preliminary expression of a syncretic position in development geography may be found in the work of Brookfield (1975) and de Souza & Porter (1974).[9] Subsequently, around the turn of the decade, the studies of Friedmann & Weaver (1979), Gilbert & Gugler (1982), Krebs (1982), Mabogunje (1980), Mehretu & Campbell (1981–2) and Riddell (1981) exemplified the coming of age of syncretism within regional and urban development research. In associated spheres of geographical enquiry the rise of what is customarily referred to as eclecticism occasionally received explicit support. Hall (1982, p. 75), for instance, in offering some thoughts on the Marxist and positivist traditions, wrote that 'one might look hopefully to some mutual understanding and to the development of some eclectic body of theory – otherwise two important traditions will be the poorer.'

More recently too, Corbridge (1986) in surveying some contemporary trends, makes a plea for a sensible dialogue between Marxist and non-Marxist development geographers, adding that 'there are no privileged concepts or "facts", which need to be defended on epistemological grounds' (p. 68). Consequently, according to Corbridge, there is no need for a 'ritual purging of bourgeois ideologies' (1986, p. 68). The reader is further encouraged to believe that if 'radical development geography is to claim the future which is surely its own' (p. 247), it must begin to engage constructively with the arguments of those who oppose the development of Marxist thought, since for Corbridge 'epistemological confrontation' has only weakened radical development geography. There are at least three problems with the Corbridge formulation.

First, it draws a veil over the origin and effects of the fundamental rift between the Marxist and non-Marxist approach to issues of Third World development. Second, by advocating a *rapprochement* between conflicting schools of thought, the incoherence of a syncretic mode of interpretation is ignored. Finally, it is far from certain that radical development geography has a sure future, especially if it resigns its critical posture so as to accommodate antagonistic positions. In addition, it remains incorrect to assume that there is only one analytical pathway within the domain of Marxism or that the Marxist theory already constituted is sufficient and effective enough to permit a transference of emphasis to empirical work alone. I shall return to this aspect of the debate in the next section of the chapter.

I have already discussed, albeit briefly, the importance of keeping in mind the existence of antagonisms between Marxist and non-Marxist positions.

However, the second problem, already considered above in quite general terms, requires some further exemplification. I shall refer to two texts.

Mabogunje (1980), in his attempt to outline certain spatial dimensions of the development process, includes a treatment of state and class structure, and also a brief incursion into the literature on imperialism. However, symptomatically, he concludes his work with a passage that contradicts his previous orientation. He refers, for example, to the recommendations of the 1971 Conference of Ministers of the Economic Commission for Africa. For the member states of Africa, three policy suggestions were recorded, and following Mabogunje, these were

> that they first effectively marshall their national and external development resources; second, mobilize all sectors of their population for participation in activities which should lead to the integration of the traditional sector, . . . with the modern dynamic sector; and third, promote structural changes to reduce the almost exclusive dependence on external factors for the imitation of the processes of transformation and development (Mabogunje 1980, p. 344).

Although one of the key recommendations of the above report refers to the posited need to integrate the so-called traditional and modern sectors, thus reflecting an adherence to the conventional dual-economy model, Mabogunje offers no critical evaluation. Instead, he contends that 'the development effort must be evaluated only in so far as it enhances the capacity of individuals and societies to cope effectively with the changing circumstances of their lives' (p. 345). But naturally one has to ask, which individuals? Do the owners of the means of production in peripheral societies benefit from the development effort in the same way as workers and peasants? It is somewhat surprising that such questions have to be posed in relation to Mabogunje's final few paragraphs, for a few pages earlier he acknowledges the significance of class analysis (1980, pp. 339–40). The difficulty stems from the fact that when one has a straight aggregation of immiscible theoretical positions, contradictions and incoherencies inevitably result.

Similarly, in Gilbert & Gugler's (1982) text on Third World urbanization, comparable incompatibilities are in evidence. Although, for example, the reader is encouraged to believe that the authors intend presenting a political-economy approach, linked to Marxist thought, the assemblage of theoretical terms comes to resemble an extraordinary farrago. While, on the one hand, concepts such as the mode of production and class fractions are brought forward to substantiate claims of a political-economy approach, on the other we find references to subsistence economies, separate informal and formal sectors, and to the old dichotomous notion of the so-called push-and-pull factors of migration.[10] The failure to appreciate the incompatibility of these separate terms is further reflected in the authors' stated belief that the Marxist and non-Marxist literature 'demonstrate many similarities of interpretation about spatial change' (p. 38). As a result, salient disputes over the diagnosis and explanation of urban and regional change are spirited away.

On the basis of the above criticisms it might seem that I am adhering to a rather rigid, uncompromising approach which is intolerant of innovation, but, in fact, what I am highlighting is the defining deficiency of a research current which arbitrarily amalgamates opposed concepts and ideas into a *nouveau mélange*. And it is hard to see how such amalgamations can generate either clarity of explanation or consistency of policy recommendation.

A move towards radical spatialism

The last current to be assessed in our brief overview tends to be both more coherent in its argument and more critical in its stance toward capitalist development. However, we now have a new problem which relates to the place of spatial concepts in a radicalized socio-political analysis. In some cases, we may come across a mystique of spatiality, recalling Luporini's (1975, p. 229) comment that Althusser's conceptualization of time implied a 'mystique of temporality'. Before identifying the characteristic difficulties of this third current, it is necessary to clarify what I mean by spatialism.

In its most undiluted form the spatialist approach may be defined by its pretension to explain the spatial organization of socio-economic phenomena while remaining solely at the spatial level. Socio-economic phenomena are transformed into spatial processes and an attempt is made to delimit an apparently autonomous scientific domain whose object is everything spatial.[11] In the context of Third World studies three forms of spatialism can be delineated:

(a) An orientation within which social relations are interpreted as, or misleadingly transformed into spatial relations; for example, the widespread notion that one area dominates another, or that the city exploits the countryside. In these examples, the reality of socially determined exploitation is displaced on to a reified spatial level.
(b) An inclination to view the realization of spatial planning goals, such as the limitation of the size of large cities, or the development of a spatial planning machinery, in isolation from the social processes and political discourses which largely determine the formulation, essential directions, and effects of the practice of planning.
(c) Third, because, within the purest expressions of spatialism one encounters no notion of how social processes produce particular spatial forms, the ways in which spatial configurations are changed, or, under crisis conditions, transformed, cannot be explained. Instead, one finds a tendency to extrapolate existing spatial patterns so that not only are the underlying causes of those patterns not analyzed, but the social forces which, under certain historical circumstances, may transform the generative conditions of those patterns cannot be identified.

These three forms of spatialism do not surface in a combined fashion in all the three research currents mentioned here. The second and third forms are normally to be found in the traditional and syncretic tendencies, whereas the first form generally inhabits the more radical literature. Soja (1980, pp. 207–25), in a well-known article, develops an argument that can be associated with this radical form of spatialism. He writes, for example, as follows:

> The opposition between dominant centres of production, exploitation and accumulation, and subordinate, dependent, exploited peripheries represents

> the primary horizontal structure arising from the process of geographically uneven development and from the dynamic tension between equalization and differentiation. It is fundamentally homologous to the vertical structure of social class, in that both are rooted in the same contradiction between capital and labour that defines the capitalist mode of production itself. *In this sense, core and periphery are the spatial expressions of the same underlying relations of production which define bourgeoise and proletariat* (pp. 221–2; emphasis added).

For Soja, the spatial structure as well as the social structure forms a part of '*the general relations of production*' (p. 219; my emphasis), and of the two posited structures which determines which, or which structure possesses pre-eminence over the other remain 'empty questions' (p. 208). There are a number of difficulties with Soja's position on the spatial dimension.

Initially, his conception of social determination is weakened by an adherence to the view that the spatial structure or the spatial relations of production carry equal theoretical weight to the social structure or the social relations of production. For this to be accepted, it would have to be maintained that somehow space *qua* space possesses a causative dynamic, set apart from social forces.

Second, the posited equivalence of centre–periphery relations with class relations is untenable. It is not only the case that Third World societies have complex class structures within which all manner of antagonistic social relations and dispositions present themselves, but that within such a vision of a centre–periphery dichotomy the crucial connectivities between international capital and the dominant social classes of the Third World are invariably left out of account.[12]

Lastly, by overprivileging the status of the spatial, Soja's argument can quite easily lead to a counterreaction in which the spatial dimension and the rôle of spatial concepts are inappropriately characterized as irrelevant to critical socio-political analysis.

In a more recent statement, Soja (1985a) has retuned his 1980 argument suggesting that 'social and spatial relations are intercontingent and combined' (p. 177), and that whilst the social and the spatial can be separated 'they cannot be dichotomized into independent realities, each with their own laws of formation and transformation' (p. 177). Referring to the core–periphery issue, Soja contends that bourgeoisie–proletariat and core–periphery are 'interpretable together as social products, intercontingent and combined' (p. 181). Although we are warned that there can be no neat division into bourgeois and proletarian regions, it is maintained that the theoretical and political significance of the core–periphery structure is related to the degree in which it is the 'material geographical expression of the fundamental class structure of capitalist society' (p. 181). And, in the last instance, 'insofar as capitalism continues to exist, there will remain an antagonistic opposition between labour and capital and between core and periphery' (p. 182).

Couched in a broader context of the spatiality of social life, Soja (1985b) reaffirms his earlier view that spatiality is not only a social product but also a producer and reproducer (pp. 98, 110), and that spatial and social relations of production co-exist.[13] However, in a somewhat ambivalent but revealing footnote, Soja adds that 'to speak, for example, of regions affecting (exploiting, politically dominating, influencing the production process in) other regions does not necessarily abrogate the ultimate basis of the relationship in people, human beings' (p. 124). It does, none the less, leave wide open the cardinal question of social determination, as well as the proper place of the spatial dimension.

I have quoted at some length from Soja's work because in my view it represents the clearest and most detailed example of a radical spatialism that has surfaced in the writings of other urban and regional researchers.[14] Keeping in mind our three previous criticisms, outlined in relation to Soja's 1980 article, the following additional and complementary observations are in order at this juncture.

(a) In the first place, it needs to be restated that space *qua* space has no causal power and spatial relations of themselves do not produce effects. As Sayer (1985, p. 52) encapsulates the point, the spatial relation of between-ness cannot, of itself, be said to have any effects or make any difference'. However, *depending on the nature of the constituents*, their spatial relations may make a crucial difference' (emphasis in the original). Hence, for Sayer, space only makes a difference in terms of 'the particular causal powers and liabilities constituting it'.[15]

(b) Furthermore, although Soja does not assign the spatial an independent reality, invested with its own laws of formation and transformation, by stating that the same point holds for the social, the *determining effects* of social processes are occluded.[16] Likewise, with the coupling of social and spatial relations of production the causal effectivity of the former is blurred, whilst the requisite conditions of co-existence of the latter are not identified.

(c) Again with reference to core and periphery, the notion that there exists an antagonistic opposition between them, conflates social relations with spatial relations. Similarly, to suggest that the core–periphery structure can be visualized in terms of a material geographical expression of the fundamental class structure of capitalist society makes it more difficult to see that the spatially peripheral is not necessarily peripheral in a functional sense and conversely the functionally central is not necessarily spatially central.[17]

From the above remarks it ought not to be assumed that I regard the spatial dimension of socio-political analysis as either banal or unproblematic. Its significance is being increasingly recognized, although at the same time it is clear from much of the relevant literature that discussions of spatial concepts and the rôle of the spatial dimension are certainly not free from ambiguities and ambivalences.

Massey (1985, p. 11), for instance, quite correctly warns us that there are not such things as 'purely spatial processes', whilst in the same volume, Cooke (1985, p. 213), refers to just such processes in a way which is somewhat discordant with the thrust of his argument. In addition, a certain ambivalence can be noted in Urry's (1985, p. 39) essay on social relations and space, for in employing the terms central and peripheral economy he sets the two spatial adjectives in inverted commas. Again, Sayer (1985, p. 53) is useful here, reminding us that our descriptions of the world would be clumsy if we always evaded the spatial dimension and the categories which are cognate to that dimension. Terms such as peripheral societies or regional problems act as convenient surrogates for the more awkward but more accurate formulations of societies at the periphery of the world capitalist system, or socio-political problems which are expressed in a regionally specific form etc. As long as we do not attribute any causal or theoretical status to the regional or to the spatial such usages need not pose any problem.

Having outlined some of the main features of our third research current, we are now in a better position to examine further, and within a broadly conceived Marxist perspective,[18] a number of issues concerning the relations between society and space. As indicated at the outset, the guiding theme will be the problems involved in analyzing the spatial dimensions of capitalist development at the periphery.

The deployment of spatial concepts – a political question?

In order to clarify my position on what is becoming an interesting arena of controversy I want to consider three interrelated issues.

Are spatial concepts neutral?

As a preliminary note it is worthwhile indicating that there are at least three kinds of spatial concept; first, concepts such as distance, contiguity, diffusion, and enclosure are employed in and traverse the broadest possible range of scientific discourses; second, concepts such as the spatial division of labour or territorial socialization primarily inhabit the domain of economic analysis – one could refer here to spatio-economic concepts; and finally one can envisage a series of spatio-social or spatio-political concepts such as regional coalitions or territorial breaks in the power bloc.[19]

Much of the recent critical regional analysis has tended to focus on the economy and the concept of, for example, the spatial division of labour has gained a quite central position in much contemporary research, and not only in the advanced capitalist societies. Is this concept neutral? I would argue that it is neutral in the sense that it is not fixed, *a priori*, to any one theoretical position. In other words it has a floating significance until the moment it is actually used within a specific discourse that assigns to it a specific meaning within an ensemble of related concepts, both spatial and aspatial. Hence, as Läpple (1985) correctly argues, potentially, the spatial division of labour can as well be linked to Adam Smith as to Karl Marx. With that very linkage, however, the concept's neutrality is immediately lost. However, contrary to Läpple, the concept of the spatial division of labour does not of itself automatically inscribe a political meaning. On the other hand, a concept such as the territorial socialization of labour is much less neutral since the analysis of the socialization process is closely linked to a Marxist position.[20] Expressed more generally, propositions or concepts change their meaning according to the positions held by those who use them, so that, for instance, the meaning of spatial diffusion will vary in relation to the way in which it is situated within a given discourse.

Econocentrism and the spatial dimension

The recent discussions surrounding the meaning and application of the spatial division of labour reflect what I would call the predominance of econocentrism. By econocentrism I am referring to an analytical orientation within which the study of the economy constitutes the determining focus of investigation. Granting the economy a pivotal position within the structure of analysis is characteristic of both non-Marxist and Marxist literature; in this sense, and in contrast to economism, econocentrism exists in two main and opposed forms.[21]

In the domain of development theory, the Marxist and/or neo-Marxist form of econocentrism is characterized by a concentration on issues such as the dynamics of capital accumulation, the articulation of modes of production, and the international division of labour. Other themes, such as militarism, the state and political regimes, popular mobilization and political ideology are often left out of account or implicitly subsumed under the more familiar themes mentioned above. Critical urban and regional research that has been launched along Marxist lines has, as suggested, tended to follow an econocentric orientation, and it is not coincidental that much theorization concerning space has been concerned with economic restructuring, labourmarkets and changes in the spatial division of labour. Consequently, in the light of the predominance of econocentric tendencies in both Marxist development theory and critical urban and regional analysis, it can hardly be surprising that Third World urban and regional research, when located at the confluence of these two currents, has flowed in the same direction.

There are three problems with Marxist econocentrism. In the first place, it leads to a kind of conceptual centralism which relegates the inner composition of state, civil society, ideology, and discourse to the outer bounds of reflection. Second, the crucial interrelations between these entities and the spatial implications thereof are not examined. This is, without doubt, an exceedingly difficult task. Lastly, there is a deeply rooted assumption that since in the final instance the economy is politically determinant, theoretical analysis on the development of the capitalist economy must be retained at the centre of our research.[22]

Although econocentric tendencies have exerted a moulding influence on the theorization of space under contemporary capitalism, it would not be correct to assume that there have been no alternative or counter-tendencies at work. In particular, Urry (1981a, 1981b, 1985) and Cooke's (1983) work on social relations and space, the contributions of a number of French writers on politics, power, and territory (Alliès 1980, Bataillon 1977, Lacoste 1976), and, in Latin America, the writings of a broad range of social scientists on questions of state power and the territorial dimensions of socio-political conflict and struggle, point in another direction altogether (Calderón & Laserna 1983, Federico 1982, Henríquez 1986, Lungo 1984, Pírez 1984).[23] But what is the relevance of this other direction, especially when it originates from the capitalist periphery?

Conceptualizing social space: against universalism

How do we regard spatial concepts in relation to the passage of time? Are all spatial concepts transhistorical? Poulantzas (1978, pp. 90–107), in his last book, began to address just these sorts of questions and imaginatively reminded us of the importance of connecting our thoughts on social space with the mode of production. As he noted, 'towns, frontiers and territory do not at all possess a single reality and meaning in both capitalism and pre-capitalist modes of production' (p. 100). For Poulantzas pre-capitalist political power and forms of state involved a specific space that is 'continuous, homogeneous, symmetrical, reversible and open' (p. 100); in contrast territory under capitalism is to be characterized in terms of 'the serial, fractured, parcelled, cellular and irreversible space which is peculiar to the Taylorist division of labour on the factory assembly line' (p. 103). However, capitalist social space does become homogeneous in the end but very much in relation to a twofold dimension; 'it is composed of gaps, breaks, successive fracturings, closures and frontiers: it has no end: the capitalist labour process tends towards world-wide application' (pp. 103–4). Further, Poulantzas integrates the rôle of the state, suggesting that the capitalist state tends to monopolize the procedures of the organization of space and comes to play a key rôle in the forging of national unity and the development of homogenization. Given that social space has no intrinsic nature but is determined by the contradictory development of social relations, is it possible to sustain the view that capitalist social space consists of a series of universally relevant characteristics? Before answering this question I need to explain what I mean by universalism or Euro-Americanism. I want to suggest the existence of three modalities of universalism.[24]

First, in many of the critical and theoretical discussions of urban and regional development that take place in the advanced capitalist societies Third World societies are conspicuous by their absence. This particular modality of universalism, which omits to consider non-Western societies, represents an implicit negation of their existence. Such a negation requires far more critical opposition than has so far emerged.

In a second instance, whilst there may well be a passing recognition of the existence of Third World societies, the formulation and deployment of theoretical concepts is assumed to have a wholly generic quality; in other words, concepts which are produced in the context of historically concrete social conditions are invested with a universal significance. In this way the historically specific is conflated with the theoretically generic. Similarly, in the domain of political practice, orientations, actions, and meanings judged to be appropriate in Western Europe and/or North America are not infrequently bestowed a relevance far beyond their initial boundaries of applicability.

A third modality of universalism concerns the failure to recognize that in some areas of research imaginative new conceptualizations or methods of analysis are being pioneered outside the First World. However, a tangible lack of intellectual curiosity, combined perhaps with a certain unconscious condescension towards Third World research, tends to preclude the enrichment of critical thought in the original heartlands of the capitalist world.

In exemplifying these three forms of universalism I want to concentrate on the second instance, since it is the most significant and also the most difficult to substantiate. I shall refer to the third modality in the following section of the chapter.

An immediate difficulty relates to the demarcation of historically specific from theoretically generic concepts. Some theoretical concepts do have a far greater generic value than others. Thus, within the capitalist system, the concepts of, for example, relative and absolute surplus value have an analytical applicability for a wide range of social formations and not just for the advanced capitalist societies. At the same time, the concrete application and meaning of these particular concepts will certainly vary, especially in relation to the historically constrained forms of capitalist development at the periphery. However, the extreme Third Worldist position whereby it is argued that any concept originally formulated in the West must by that very fact be inapproriate in a peripheral capitalist context cannot be seriously maintained. There are concepts which can be deployed across the so-called North–South divide; for example, the spatial displacement of labour, the spatial circuits of power, or the regionalization of the state are spatial concepts that can be generally used. What then is the problem? I shall take two kinds of illustration, relating to non-spatial and spatial concepts:

(a) Urry (1981b, pp. 80–1) in his consistently incisive and stimulating treatment of the capitalist state argues that the general form of the state in capitalist societies 'would appear to be best expressed as representative democracy', and further, that 'democracy seems to be the best possible political shell for capitalism'. This might well be true for the advanced capitalist societies, but it certainly does not hold for the capitalist societies of the periphery where state repression is more overtly coercive and ideological subordination much more fragile. In a similar fashion, with conceptualizations of democracy and socialism the immanent substance of these terms is built around Western historical experience. Bobbio (1987), for instance, develops a consideration of democracy, which while alluding to universalist meaning, is in fact rooted in the particular socio-political circumstances of the West. Equally, Jay (1987) attempts to capture the mood of *fin-de-siècle* socialism in terms of 'left melancholy' and the waning appetite for any 'wholesale repudiation' of contemporary society. Whether or not this is true for Euro-America remains somewhat of an open question, but it certainly does not apply for many peripheral societies where 'wholesale repudiation' is being expressed in armed struggle.

(b) Referring to spatial concepts two examples can be cited. As I argued previously (Slater 1978), theoretical definitions of capitalist urbanization which do not take into account the urban–rural dialectic do not have a ubiquitous applicability. Similarly, the notion of the hypermobility of capital (Urry 1985, p. 33) must be used carefully, since although it corresponds to new trends within the advanced capitalist societies, and is also in evidence with relation to the expansion of transnational capital, it is much less relevant within most peripheral capitalist societies where capital is not as 'spatially indifferent' as elsewhere. This distinction can be partly connected to the much more uneven spatial extension of the general material conditions of production – here, 'space has still not been annihilated by time' – but also to the territorially centripetal functioning of internal power blocs, a point to which I shall return subsequently.

Finally, it is not only important to consider how widely applicable a given spatial concept may be, but to realize that there are always questions of analytical and thematic priority. A research agenda formulated in the United

States, Britain, or the Netherlands, and quite naturally carrying with it a certain set of conceptual and methodological priorities ought not be parachuted into a society like Tanzania or Colombia. More awareness of the significance of historical specificity would not be out of place in much critical Western writing on regional and urban development.[25]

How far is geographical scale a central issue?
In the last few years, the importance of geographical scale has been raised in two connections; first, in terms of the relations between spatial coalitions and spatial policies of the state (Pickvance 1985, pp. 132–4), and second, in relation to the implications for regional geography of Wallerstein's work on the modern world system (Taylor 1986). In the first example, it is observed that coalitions may join together or split in order to alter their spatial scale of policy demands on the state. Potentially, this opens up a useful discussion of the scale of demand articulation and operations of not only spatial coalitions or regional defence movements but also of indigenist or guerrilla movements, a theme to which I shall briefly return below. Taylor's plea for a Wallerstein-based regional geography poses the question of the relative significance of scale within the broader and determining debate on the construction of a theoretical framework. Whilst accepting the need for an imbrication or interpenetration of 'levels of analysis', so that even micro-level research does not remain divorced from a consideration of the effects, within a particular area, of changes at the global, national, and regional levels, geographical scale, as such, ought not to divert our attention away from the critical domain of socio-political theory. Likewise, the necessity of examining the effect of changes in the internationalization of capital and political power (imperialism) within specific territories and zones – the so-called global levels of analysis – can be detrimental if it leads to trends in internationalization being seen as somehow superimposed on peripheral societies rather than as a penetrative process interwoven with the internal specificities of capitalist development and state–society relations within given social formations of the periphery.

I began this section of the chapter with a question. Although it should be clear that my answer is in the affirmative, what kind of political question are we dealing with?

The peripheral state and the regional question: towards a clarification of the issues

It is quite evident that when concepts and ideas, originally formulated in the critical literature on Third World development, are utilized in an attempt to reconstruct a new geography of underdevelopment the potential range of application is daunting. Some research in this direction has already been undertaken (Forbes 1984, Gore 1984, Rakodi 1986, Rauch 1984) and rather than recapitulate some of these positions I intend to select one theme for examination: namely, the constitution of the regional question in relation to state power. First of all, however, I need to clarify what I mean by the regional problematic (see Fig. 11.1).

Regional analysis as a field of enquiry based on traditional conceptions

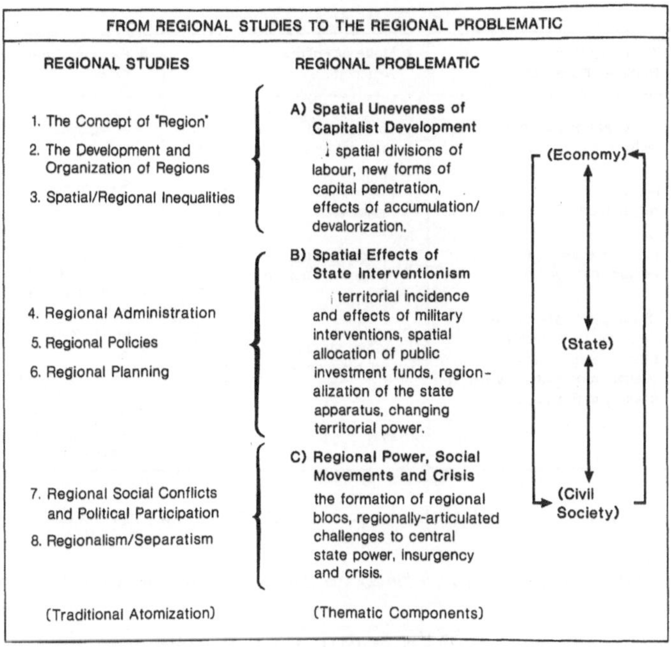

Figure 11.1 From regional studies to the regional problematic.

(dualism, modernization theory, economic growth and integration models) lacks an articulated thematic structure. For example, the eight topics listed under regional studies in Figure 11.1 tend to be treated arbitrarily with little if any attempt being made to establish the interlinkages among these themes as potentially constitutive of an underlying theoretical problematic. Hence, with the term regional problematic I am referring to the possibility of transforming the terrain and horizon of traditional regional studies as a result of a theoretical/conceptual displacement. Within the thematic framework of this new problematic it is possible to deploy notions such as the spatial unevenness of capitalist development or the spatial effects of state interventionism, and these notions acquire an articulatory unity through being anchored in a broadly developed Marxist perspective. Economy, state, and civil society are not seen as separate levels or instances but as interlocking spheres of social relations and political practices. The three thematic nuclei or clusters proposed define the terrain of analysis, but their potential unity or articulation can only be realized through the clear specification of a theoretical problem, which carries with it and usually entails the identification of a political issue. A consideration of the regional question can constitute one such theoretical problem.

In much of the analysis on regional social movements, the regional problem, and regional crises these terms are not always clearly demarcated. Moreover, as noted earlier, under the sway of econocentrism the first thematic nucleus has been given the most obvious priority; thus the spatial dimensions of, for example, the accumulation, valorization, and devalorization of capital, new divisions and socialization of labour, new forms of capital penetration and

I A Typology of Regional Problems	II Mode of Political Articulation		III Regional Question/ Crisis
1. Regionally-based antagonisms within the power bloc ⇒	Formation of regional blocs	IDEOLOGICAL MEDIATION AND STATE INTERVENTION	A regional question may be defined in relation to territorially-specific state intervention, aimed at politically resolving either a) a perceived threat to the political security of the power bloc or b) a perceived threat to the state's own territorial jurisdiction and control within a given region
2. Popular demands for regional power and control ⇒	Regional social movements		
3. Indigenous expression of ethno-regionalist identity ⇒	Movements for indigenous autonomy		
4. Challenges to state power (Regional, Local, National) through operations of guerrilla organisations with specific territorial bases ⇒	Armed struggle and insurgency		

Figure 11.2 From regional problems to regional crises – a proposed schema.

changing labourmarkets have become the focus of considerable investigation and reflection.[26]

Similarly, examination of the spatial effects of state interventionism has not infrequently fallen within the orbit of econocentrism, so that whilst analysis of the spatial allocation of public investment funds and spatial policies on credit and taxation have often been prioritized, enquiries into the territorial incidence and effects of military interventions, the regionalization of the state apparatus, and the changing nature of the state's overall territorial power have been, until quite recently, somewhat neglected.

With the third component of our regional problematic we are presented with a crystallization of several theoretical difficulties. Here it is necessary to treat, *inter alia*, the problems involved in accounting for the emergence and development of regional social movements, the potential durability and seriousness of regionalist challenges to central state power, the conditions affecting the formation of regional crises, and the changing modalities of state intervention. How do we proceed? Can we simply direct our attention towards phenomena such as the contemporary accentuation of uneven spatial development, the territorial effects of new forms of capitalist penetration, and the spatial concentration of state investments? Although it would be unwise to ignore these trends, an effective understanding of the major contours of regional questions requires a perspective that includes but also goes beyond these trends.

In Figure 11.2 I have sketched out a possible typology of regional problems. This typology is largely based on the Latin American experience. In relation to my earlier argument the four regional types represent or express particular socio-political problems with a clear regional inscription or articulation. These regional problems are to be distinguished from the regional question or crisis in the following way. A regional problem may develop into a regional question along two different but not completely unconnected routes. First there can be a conjuncture wherein the central state is forced to intervene, within a specific region, to control and resolve what is perceived to constitute a potential or actual threat to the political security of the power bloc.[27] If this intervention is not successful and the threat not only becomes actual but prolonged, it may be appropriate to refer to the existence of a protracted regional question or crisis. In a second instance, there may not be any immediate regionally based threat to the national dominance of the power bloc, but the central state may judge that its territorial jurisdiction and power are being undermined within one particular region.

Embodied in the above specification are two underlying issues which require further substantiation. In both instances a key factor concerns the state's perception of what might constitute a threat or a challenge either to the political security of the power bloc or to the state's own territorial power. That perception is of course interwoven with the nature of the social conflict surfacing within a specific region and the kind of political regime in power at a given historical moment. The matter of state intervention and the regional question, as a question of the state, has to be merged, analytically, with the territorially based characteristics of social conflict. Also, as Coraggio (1984) rightly argues, this conflict must have deep roots in the structures of civil society; otherwise it would make little sense to talk of a state question. The course taken by a regional crisis will be affected by a number of interlocking factors; by the modalities of state intervention which, in turn, will be moulded by the ideological disposition of the regime in power; by the type, acuteness, and durability of the social conflict in question; by the political direction of the leadership of the given regional social movement, indigenist or guerrilla movement; and by the intention and or relative effectiveness of such movements to extend the territorial scope of their activities and mobilization so as to constitute a political question at the national as well as regional level.

This last factor raises the issue of territoriality since the terms regional problem and regional question invest the category of the regional with a significance that remains unspecified. Ultimately, concepts of the regional are most usefully viewed as unfixed and open. They can assume different meanings in relation to their insertion and deployment in a variety of discourses. Sometimes, for example, an indigenous movement in its opposition to an integrationist state may incorporate regionalist symbols and sentiments to help safeguard its overall identity. In other cases, a guerrilla organization may use images of regional impoverishment and centralist domination to aid in its project of securing a territorial support base from which it may launch wider-reaching operations against the state. In contrast, a given political regime can develop a territorial ideology within which notions of harmonious regional development and spatial equity may be employed to win regional support for its policies.

There are a number of ways in which peripheral state interventionism can be examined in terms of its territorial dimensions. In addition to the creation of a spatial or regional ideology and the above-mentioned rôle of the state in the regional question, there are two other primary modalities of intervention that can be identified. First, the spatial concentration of state resources (general material conditions of production and means for the reproduction of labour power) in areas where predominantly transnational but also internally dominant capital is based, constitutes a generalized territorial effect of state policy in peripheral societies. Second, the changing territorial organization of the state itself is a highly relevant phenomenon in this context. Not only does the establishment of regional development agencies need to be taken into account but also, and more systemically significant, the development and deepening of the territorial hierarchy of state power (e.g. regional governments under central jurisdiction, the regional aggregation of public administrative functions, and the subordination of local state bodies to their regional equivalents) can be seen as a pivotal component of state–space relations. In some examples, the introduction of legislation for the regionalization of the state bureaucracy can be connected to the eruption of regionalist protests and the emergence of anti-centralist politics in the regions (for example, as in Peru in the late 1970s and early 1980s), but in other cases, as in Tanzania for instance, the expansion and entrenchment of the state's organs of territorial power have proceeded during a conjuncture devoid of regionalist struggles and mobilizations. Here, geo-political factors together with the exigencies of territorially extending the state's control over production, primarily agricultural, were far more important. Geopolitical influences, such as border disputes or socio-political instability in areas contiguous with or close to national frontiers frequently have an impact on the territorial organization of the state, but in societies where the military is in power these influences often become paramount.[28]

Returning to Figure 11.2 and the typology of regional problems it is now necessary to sketch out a few specific details. With any such specification it is worthwhile remembering the following:

(a) The scale and persistence of regional social conflict will vary enormously among peripheral societies. The reasons for and implications of these variations still remain largely unexplored and underresearched, even in relation to groups of contiguous societies (e.g. the Andean countries).[29]
(b) Within the same country, regional social conflicts frequently go through cycles of development, evolving, dissolving and re-emerging in different conjunctures, expressing ideological continuities and discontinuities. As such it would be quite wrong to assume that regional tensions and issues are only contemporary phenomena explicable in the context of the crisis and the re-structuring of the world capitalist system. Instead, what needs to be explained is the specificity and novelty of today's regional conflicts.
(c) Due to the fact that the emergence of regionally oriented movements cannot be explained simply in terms of the effects of economic restructuring or new forms of capital penetration, it is not possible to posit a direct relationship between the level of capitalist development in a society and the likely incidence of regional social conflicts.

The notion, for example, that the more advanced a peripheral society may be in terms of its level of capitalist industrialization, the less likely we are to encounter regional movements and the constitution of a regional question, not only assumes a linear pattern of development but also suffers from the usual shortcomings of economic reductionism.

In Figure 11.2 the first example of a regional problem relates to regionally rooted antagonisms within the power bloc. In societies where the territorial structure of capitalist development became highly segmented the national constitution of a power bloc was not infrequently impeded by regionally articulated cleavages, schisms, or even open moves towards territorial secession. Although many such examples can be located in the pre-1940 period, when the centripetal forces of state power had still not been firmly anchored, it ought not to be assumed that regionally based antagonisms of this nature have not developed in the postwar period, as Oliveira (1977) has shown for the case of the Brazilian North-East in the early 1960s.

In general, whenever these antagonisms have been characterized by the struggle of one or more regionally based factions of the power bloc against the political dominance of the centrally established factions, the centralist rôle of the state has always been cardinally important. Equally, however, the frequent failure of 'regional blocs' to develop a nationally oriented discourse of development, combined with the fact that their material bases have almost always been relatively less developed than the corresponding bases of the dominant factions, have always been important factors contributing towards the reabsorption of centrifugal forces within the power bloc. Even in instances where the army has also been internally fissured, with important garrisons allied with regionalist forces, and where regionalist insurrections have occurred, as was the case in Peru at the beginning of the 1930s (Slater 1988), this kind of territorial rupture in the power bloc has never led to a permanent breakdown in state–society relations. Although under some historical circumstances, as in the case of Peru in the period between 1930 and 1933, and perhaps in the Brazilian North-East in the early 1960s, one can posit the development of a type 1 regional problem into a regional question, subsequently to be resolved by central state intervention, no potential for revolutionary change can be discerned. Does this also hold for regional problem, type 2?

In the first place, it is appropriate to note that a regional social movement is not necessarily synonymous with a social movement located in a region and nor does the outbreak of regional protest, of itself, provide sufficient evidence for the existence of a regional social movement. I would suggest that a regional social movement (RSM) is most usefully designated in relation to the following constituent elements.

An RSM must have some sort of political leadership, which can at the very least guarantee a rudimentary form of organized capability. Similarly, and also minimally, this leadership ought to be in a position to adumbrate a political project for regional change. This would include a clear statement of political objectives and the articulation of a series of concepts, images, and values into a discourse that can effectively address or interpellate a fairly wide range of regional social subjects. Third, and inserted within this discourse, there needs to be an identification of those socio-political tendencies and forces that constitute the object of transformative struggle. Furthermore, the activities and political engagement of the movement must generate some concrete effect on the system of power relations and social practices found within its region. Fifth and last, although an RSM ought not be seen as an ephemeral social phenomenon, normally it will be more historically temporal than a political party, and obviously much less administratively structured.

In the context of the above elements is it possible to classify RSMs into different types, based on, for example, the orientation of their demands and the ideological thrust of their project? As far as the primary demands are concerned three can be mentioned: (a) varying degrees of devolution and democratization of decision-making power to and at the regional level; (b) improvement of the material conditions of production, collective means of consumption, and employment opportunities; and (c) a greater democratization of the practices of social and economic life. To these three major demand orientations one can add the accompanying objective of greater social control over the operation of transnational companies within the region – in relation, for instance, to questions of environmental impact, taxation, reinvestment in the region, and labour conditions. Nevertheless, it would not be advisable to label RSMs according to only one of these demands, since frequently, within a given RSM, there tends to be a process of coalescence or amalgamation of demands. Also, the way in which this process emerges will be closely connected to the political leadership of the RSM and the corresponding ideological trajectory of its project. In some cases, as a regional social movement evolves its political leadership becomes gradually radicalized and initial demands for more central state investment can develop into more militant opposition to regional economic spoliation and eventually crystallize around calls for regional autonomy.[30] In other instances, where ideological fractures open up within the more militant, popular sectors of the RSM, the political leadership can be taken over by representatives of the regionally dominant social forces with a consequent realignment of ideological direction.[31] Before answering the question of whether RSMs have any possible revolutionary potential, let us briefly consider regional problem, types 3 and 4 (Figure 11.2).

With respect to the indigenous expression of ethno-regionalist identity, the most common objective is to win regional autonomy or at the very least a meaningful degree of regional self-management. Certainly in the case of the Atlantic coast in Nicaragua it is possible to refer to the existence of an ethno-regionalist question (Slater 1986b). The introduction by the Sandinista Government of legislation providing for the establishment of a new tier of regional government has helped to resolve this political question, thus avoiding the potential deepening and exacerbation of indigenous discontent and opposition to central state rule. Usually with this kind of regional problem the ethnic content of social conflict overdetermines the associated connotations of regionalism. In other words, whilst regionalist symbols may well be inscribed within an *indigenista* discourse, they are essentially auxiliary to the driving force of ethnic validation and autochthonous identity. Although the state may have to intervene to mediate and resolve a potential threat to its territorial power, indigenous movements do not easily combine with other oppositional movements and their demands and revendications tend to remain within the ideological matrix of *indigenismo*.

Conversely, with regional problem type 4 guerrilla organizations such as Sendero Luminoso in Peru or the FARC (Colombian Revolutionary Armed Forces) develop regional bases as both sanctuaries and potential springboards for the spatial extension of their operations against the state. Although a regionalist discourse may be articulated in the early stages of their development, the ultimate goal is the seizure of state power through armed struggle and insurgency. In the Peruvian case, the previous guerrilla movement of the 1960s was severely handicapped by its inability to establish a national network of operations; as a consequence the armed forces were more able to encircle and destroy its original territorial bases. In contradistinction, Sendero Luminoso's territorial strategy has been to extend its operations from its initial Ayacucho base not only to Lima but also to northern, central, and southern zones of the Sierra. In addition, given the peasant background of many of its leading cadres, the organization has been capable of setting up much more effective channels of support in many of the rural areas, certainly in comparison to the movement of the 1960s. Even though organizations like Sendero Luminoso and the FARC have developed a military capacity for limited engagement with the armed forces of the state, such organizations have not constructed a politically effective national project, in contrast, for example, to the FSLN in the period prior to the overthrow of Somoza. One reason for this, certainly in the case of Sendero Luminoso, is its leadership's implacable hostility towards any notion of political alliances or of winning political support through debate and consent. Its discourse is structured according to a system of equivalences within which, in the last instance, political struggle is dichotomized into a contest between protagonists and supporters of reactionary violence on the one hand and organizers of revolutionary violence on the other (i.e. Sendero itself).

Returning to the regional social movement it might well seem that, in comparison, these movements are fundamentally reformist, posing little if any serious threat to the capitalist state. How far is this true? Any answer to this question inevitably involves the issue of the differential conception of reform and revolution, *and* also of socialism and democracy. Set in the framework of the three primary kinds of demands made by RSMs it is possible to envision a rising scale of political engagement along which protests, mobilizations, and continuing social struggles can escalate into passive or active revolt, insurrection, and finally rebellion and chronic insurgency. If, along the various points of condensation of this scale of political intervention, alliances were to be made with other similar movements and initial regionalist goals came to be overshadowed by the outlines of a national project of political transformation, a revolutionary potential could be posited. But the RSMs will then have been superseded by a quite different sort of political movement. As far as I am aware there is still no historical example of such a route to revolutionary transformation, which is perhaps not surprising given the ideological heterogeneity of most RSMs, even under popular-democratic leadership, as well as the well-known difficulties of constructing a workable political unity round the broad constellation of left-wing forces present in such highly fragmented societies. What political potential then do regional social movements actually have?

Although the emergence of RSMs, and similarly ecological, urban, and also new rural movements, has taught the crucial lesson that not all sites of social conflict and struggle can be reduced to a unitary class contradiction, none the less these struggles often seem circumscribed by their localized or regionalized nature, since what they contest is not infrequently overdetermined by the general process of capitalist development.[32] In order to resolve this dilemma some principle of articulation would appear to be required, so that in a 'war of position' these movements, without jettisoning their autonomy, can be linked together in a hegemonic project of socialist democratization and collective control. In this way, a potentially unifying articulation of struggles, no longer pre-given by the sovereignty of class, can itself become a purpose of struggle.

Conclusion

I hope it is by now clear that a theoretically informed approach to problems of regional analysis does not necessarily have to be founded on the customary centralism of the economic. Whilst avoiding a certain mode of theoretical enquiry which proceeds as if there were no economy, it is surely time to reject the pervasive assumption that theory, in either regional or urban research, means economic theory. In the context of the capitalist societies of the periphery, and in relation to our interests in the spatial dimensions of development and change there is a burning need for critical *political* theory, and not just the ritual invocatory flourish, tagged on at the end of a traditional Marxist economic analysis. In this chapter I have worked out a preliminary approach to some aspects of the regional problematic in peripheral societies. In doing so I have used certain elements of Marxist political theory, and it will be readily appreciated that the interpretation here presented is still in process of formation. Of one thing, however, I am convinced and that is this: critical regional analysis of peripheral societies will stagnate unless we recast our perspective, throw off the fetters of econocentrism, and begin developing a substantive political theory of the peripheral state and social conflict.

Notes

1 Over ten years ago, Anderson (1976, pp. 114–21) drew our attention to universalist traits in Marx's thinking on the spread of capitalism, and to Lenin's 'wholly generic' discussion of the bourgeois state in his *State and revolution* (1956, p. 117). More recently, women's studies have been characterized by an emerging debate on ethnocentrism and Euro-Americanist biases (see, for example, Carby (1982) and Barrett & McIntosh (1985)), whilst in the area of Middle-Eastern studies Said (1985) continues to penetrate, with good effect, the inadequate perceptions of many Western scholars.

2 Elsewhere, I have examined some aspects of the territorial dimensions of revolutionary change in Cuba and Nicaragua (see Slater 1982, 1986). For a much wider discussion see Forbes & Thrift (1987).

3 A few reflections on the links between development theory and peripheral urbanization can be found in Slater & Pansters (1986).

4 This point does need to be made, especially in relation to a certain brand of critical writing that often proceeds as if the more conventional paradigms of social science research had all but disappeared. For a recent defence of modernization theory, for example, see Levy (1986).

5 Although, with the exception of Adarkwa (1981), all the authors listed under the traditional paradigm come from North America or Western Europe, it would be unwise to assume that no Third World researchers can be located in this theoretical sphere. In Latin America, Boisier (1981), and in Africa, Abumere (1980), and Sada (1977) provide evidence of the successful intellectual diffusion of modernization ideas.

6 It is sometimes suggested that there now exists a three-tier system of Third World countries (or triage) within which a first group is seen as being able to become modernized through their own efforts, with a minimum of assistance (e.g. Brazil, South Korea, India); a second group of lesser-developed countries needs far more aid and investment to ensure that they can eventually make it into the club of modernized societies (e.g. Zimbabwe, the Philippines, Peru); and a third group of

countries that is obsolescent and beyond help (Bolivia, Upper Volta, Bangladesh). This kind of classification misses out, of course, any geopolitical consideration. For one discussion of social triage, see Visvanathan (1987).

7 And, of course, as previously, whenever necessary these territories may still perform the function of sites of resource extraction, or theatres of war.

8 Rée (1985) discusses some of the problems involved in combining ideas from different traditions, for example, Marxism, structuralism, and psychoanalysis, and he rightly emphasizes the need to examine both the specificity of what is written within each tradition, and the particularity of the origins of each tradition.

9 As a concise illustration of what I am arguing here, de Souza & Porter (1974, p. 15), for instance, called for an alternative approach to research that would be 'historical and dialectical', but then added, two lines later, that, 'exploitation is exploitation in whatever age and place'.

10 Some of these terms also surface in Riddell's (1981) discussion of proletarianization in West Africa.

11 Aspects of this theme will be taken up in subsequent sections of the chapter, but for the moment it is worthwhile recalling that ten years ago, two Latin American writers clearly criticized the spatialist orientation. Coraggio (1977, p. 15) wrote that, 'it would be nonsensical to attempt to explain the organization of cells in a tissue in purely spatial terms without following biological theories. In the same way, it is not possible to advance in the explanation of spatial phenomena regulated by social processes without recurring to theories on society.' Similarly, Santos (1977, p. 3) observed that 'it might be said that geography has been more interested in the *form* of things than in their *formation*.'

12 Within development geography, an earlier expression of this kind of position can be found in Blaut's (1974) discussion of internal colonialism.

13 Hadjimichalis (1987, p. 39) slips into the application of a similar argument, although elsewhere he takes a clear stance against spatialist positions (p. 44).

14 For example, Weaver (1981) berates Marxist theory for having 'no appropriate causal niche for the geographical or "spatial" component of social relationships' (p. 86). Similarly, Friedmann & Weaver's (1979) earlier discussions of the posited need for territorial closure in Third World societies abstract from the determining rôle of socio-political forces and relations.

15 An earlier statement of a related position can be found in Urry (1981a, p. 462) where it is noted that 'spatial relations never have a general effect separate from the constitutive properties of the social objects which are in some determinate spatial relationship with each other'.

16 It can be also added here that as Sayer (1985, p. 59) suggests, 'the spatial is *partly* constituted by the social, but it is reducible neither to natural nor social constituents'.

17 It is also worthwhile recalling that radical spatialism has taken much of its inspiration from earlier expressions of dependency writing. Frank's (1967) *Capitalism and underdevelopment in Latin America*, with its invocation of dominating metropolises and dominated satellites provided an initial and lasting spark. Related notions of internal colonialism and cities exploiting their surrounding countrysides became quite common in the early and mid-1970s.

18 Not only does this mean that I avoid siding with any of the warring factions in contemporary Marxist economic and political thought, but that *additionally*, although I shall be using ideas from writers like Laclau and Mouffe, I do not place myself on a post-Marxian terrain, but rather identify with the tradition going back to Gramsci.

19 Although the split between the economic and the socio-political is in this instance somewhat overdrawn, it is important to be aware of some such distinction as I shall argue subsequently in the text.

20 Nevertheless, even within a Marxist analysis some concepts are more easily transferrable than others, so that, for example, the general material conditions of production can be and have been retranslated into social and economic infrastructure and subsequently neutralized, whereas relative surplus value, or the rate of exploitation retains a greater degree of discursive fixity.

21 In contrast to econocentrism, economism can be most appropriately regarded as a directly political interpretation of the course of social change under capitalism, whereby the logic of economic development is seen as paramount. An economistic position entails the expression of an explicitly political vision or judgement on the course and probable direction of societal change under capitalism (for some further discussion see Slater (1987)).

22 Urry (1981a, p. 462) develops a similar point in relation to a critical examination of Buch-Hansen & Nielsen's (1977) article on spatial structure and Marxism. A recent example of the continuance of this standpoint can be found in Harvey's (1985) discussion of the geopolitics of capitalism.

23 Two points are in order here: (a) I am referring to those studies that have concentrated on regional rather than urban issues, and (b) the research of the cited authors illustrates but does not encompass the existence of an alternative trend.

24 And, as above, I am referring to that corpus of literature which falls under a Marxist or radical/critical rubric. As far as other tendencies are concerned, there are many examples of Eurocentric perspectives; for instance, West European specialists on Latin America who publish books on regional and urban development which include virtually no reference to and certainly no serious discussion of the relevant and comparable research of their social science counterparts in Latin America.

25 The above discussion of universalism has only touched on one or two of the more obvious aspects of the problem and much more reflection is required.

26 Previously, I also tended to orient my attention too strongly in an economic direction (see Slater 1985).

27 In general terms I am following the Poulantzian definition of power bloc whereby emphasis is given to the contradictory unity of the politically dominant social classes and class fractions in relation to particular forms of the capitalist state.

28 In Chile the regionalization of the state under Pinochet and the general importance attached to territory and conceptions of space in the military doctrine of national security provide a clear illustration of the significance of these influences.

29 Although some comparative research on regional social movements has been initiated recently in these countries (see Calderón & Laserna 1983, Henríquez 1986). It may well be worthwhile posing the question of why in Latin America have some countries recently experienced a rapid growth of political interest in regional problems (e.g. Peru, Ecuador, Mexico, and Argentina) whilst others have not (e.g. Brazil and Venezuela)?

30 This seems to have occurred in the Santa Cruz region of Bolivia (see Calderón & Laserna 1983).

31 This occurred in the case of the regional defence front in Cuzco, Peru during the late 1970s.

32 Of these movements I have not mentioned the new forms of the women's movement since it would hardly be accurate to categorize their effects and influences as localized or regionalized. They do, however, share a common border in their struggle for democratization and social empowerment.

References

Abumere, S. I. 1980. Spatial development in an ex-colonial territory. *Singapore Journal of Tropical Geography* **1**, December, 1–10.

Adarkwa, K. 1981. A spatio-temporal study of regional inequalities in Ghana. *African Urban Studies* **11**, Fall, 39–64.
Alliès, P. 1980. *L'Invention du territoire*, Collection Critique du Droit 6. Grenoble: Presses Universitaires de Grenoble.
Allor, D. J. 1984. Venezuela: from doctrine to dialogue to participation in the processes of regional development. *Studies in Comparative International Development* **19**, Spring, 86–97.
Anderson, P. 1976. *Considerations on Western Marxism*. London: New Left Books.
Barrett, M. & M. McIntosh 1985. Ethnocentrism and socialist-feminist theory. *Feminist Review* **20**, June, 23–47.
Bataillon, C. 1977. *Etat, pouvoir et espace dans le tiers-monde*. Paris: Presses Universitaires de France.
Blaut, J. 1974. The ghetto as an internal neo-colony. *Antipode* **6**, 37–41.
Bobbio, N. 1987. *The future of democracy*. Cambridge: Polity Press.
Boisier, S. 1981. Algunas interrogantes sobre la teoría y la práctica de la planificación regional en países de Pequeño Tamaño. *Revista Latinoamericana de Estudios Urbano Regionales. EURE* **7**, May, 9–16.
Brookfield, H. 1975. *Interdependent development*. London: Methuen.
Buch-Hansen, M. & B. Nielsen 1977. Marxist geography and the concept of territorial structure. *Antipode* **9**, September, 1–11.
Calderón, F. & R. Laserna (eds) 1983. *El poder de las regiones*. Cochabamba, Bolivia: Ediciones Ceres-Clacso.
Carby, H. 1982. White woman listen! Black feminism and the boundaries of sisterhood. In *The empire strikes back – race and racism in 70s Britain*, CCCS (ed.), 212–35. London: Hutchinson.
Chisholm, M. 1982. *Modern world development*. London: Hutchinson.
Cooke, P. 1983. Regional restructuring: class politics and popular protest in South Wales. *Environment and Planning D, Society and Space* **1**, 265–80.
Cooke, P. 1985. Class practice as regional markers: a contribution to labour geography. In *Social relations and spatial structures*, D. Gregory & J. Urry (eds), 213–41. London: Macmillan.
Coraggio, J.-L. 1977. Social forms of space organization and their trends in Latin America. *Antipode* **9**, February, 14–27.
Coraggio, J.-L. 1984. *Los terminos de la cuestión regional en América Latina*. Mexico, unpublished manuscript.
Corbridge, S. 1986. *Capitalist world development: a critique of radical development geography*. London: Macmillan.
Corrigan, P. 1974. The local state: the struggle for democracy. *Marxism Today*, July, 203–9.
Dickenson, J. 1980. Innovations for regional development in N.E. Brazil. *Third World Planning Review* **2**, Spring, 57–74.
Drake, C. 1981. The spatial pattern of national integration in Indonesia. *Transactions of the Institute of British Geographers* **6**, 471–90.
Escobar, A. 1984–5. Discourse and power in development: Michel Foucault and the relevance of his work to the Third World. *Alternatives* **10**, Winter, 377–400.
Federico, A. 1982. La espacialidad social de la cuestión etnico–campesina y el desarrollo desigual del territorio en países de América Latina. *Boletín de Antropología Americana* **5**, July, 7–33.
Forbes, D. 1984. *The geography of underdevelopment*. London: Croom Helm.
Forbes, D. & N. Thrift (eds) 1987. *The socialist Third World*. London: Basil Blackwell.
Frank, A. G. 1967. *Capitalism and underdevelopment in Latin America*. New York: Monthly Review Press.
Friedmann, J. & C. Weaver 1979. *Territory and function: the evolution of regional planning*. London: Edward Arnold.

Gilbert, A. & J. Gugler 1982. *Cities, poverty and development: urbanization in the Third World*. Oxford: Oxford University Press.
Gore, C. 1984. *Regions in question: space, development theory and regional policy*. London: Methuen.
Gwynne, R. N. 1985. *Industrialisation and urbanisation in Latin America*. London: Croom Helm.
Hadjimichalis, C. 1987. *Uneven development and regionalism: state, territory and class in southern Europe*. London: Croom Helm.
Hall, P. 1982. The new political geography: seven years on. *Political Geography Quarterly* **1**, January, 65–76.
Harvey, D. 1985. The geopolitics of capitalism. In *Social relations and spatial structures*. D. Gregory & J. Urry (eds), 128–63. London: Macmillan.
Henríquez, N. 1986. Notas y tesís sobre los movimientos regionales en el Perú. In *Movimientos sociales y crísis: el caso Peruano*, E. Ballón (ed.), 165–224. Lima: DESCO.
Jay, M. 1987. Fin-du-siècle socialism. Paper presented to the Conference on Socialism at the End of the Twentieth Century, 26–28 March, Groningen. Mimeo.
Krebs, G. 1982. Regional inequalities during a process of national economic development: a critical approach. *Geoforum* **13**, 71–81.
Lacoste, Y. 1976. *La Géographie, ça sert, d'abord, à faire la guerre*. Paris: Petite Collection Maspero.
Läpple, D. 1985. Internationalization of capital and the regional problem. In *Capital and labour in the urbanized world*, J. Walton (ed.), 43–75. London: Sage Publications.
Leinbach, T. R. 1986. Transport development in Indonesia: progress, problems and policies under the new order. In *Central government and local development in Indonesia*, C. MacAndrews (ed.), 190–220. Oxford: Oxford University Press.
Lenin, V. I. 1956. *State and revolution*. Moscow: Progress Publishers.
Levy, M. J. 1986. Modernization exhumed. *Journal of Developing Societies* **11**, April, 1–11.
Lungo, M. 1984. El Salvador 1979–1983: la guerra revolucionaria y los cambios en la estructura regional. *Tabique* **5**, 1–12.
Luporini, C. 1975. Reality and historicity: economy and dialectics in Marxism. *Economy and Society* **4**, 206–31.
Mabogunje, A. 1980. *The development process: a spatial perspective*. London: Hutchinson.
Massey, D. 1985. New directions in space. In *Social relations and spatial structures*, D. Gregory & J. Urry (eds), 9–19. London: Macmillan.
Mehretu, A. & D. J. Campbell 1981–2. Regional planning for small communities in rural Africa: a critical survey. *Rural Africana* **12–13**, Winter–Spring, 91–110.
Oliveira, F. de 1977. *Elegia para Uma Re(li)giâo – SUDENE, Nordeste, planejamento e conflito de classes*. Rio di Janeiro: Paz e Terra.
Pickvance, C. G. 1985. Spatial policy as territorial politics: the role of spatial coalitions in the articulations of 'spatial' interests and in the demand for spatial policy. In *Political action and social identity: class, locality and ideology*, G. Rees (ed.), 117–42. London: Macmillan.
Pirez, P. 1984. El estado y Lo regional. Un intento de integración conceptual. *Revista Interamericana de Planificación* **18**, June, 30–41.
Potter, R. B. 1985. *Urbanisation and planning in the Third World: spatial perceptions and public participation*. London: Croom Helm.
Poulantzas, N. 1978. *State, power, socialism*. London: New Left Books.
Rakodi, C. 1986. State and class in Africa: a case for extending analyses of the form and functions of the national state to the urban local state. *Environment and Planning D, Society and Space* **4**, 419–46.
Rauch, T. 1984. An accumulation theory approach to the explanation of regional disparities in underdeveloped countries. *Geoforum* **15**, 209–29.
Rée, J. 1985. Marxist modes. In *Radical philosophy reader*, R. Edgley & R. Osborne (eds), 337–60. London: Verso.

Riddell, J. B. 1981. Beyond the description of spatial pattern: the process of proletarianization as a factor in population migration in West Africa. *Progress in Human Geography* **5**, 370–92.

Riddell, R. 1985. *Regional development policy: the struggle for rural progress in low-income nations.* Aldershot: Gower.

Rondinelli, D. A. 1980. Balanced urbanization, regional integration and development planning in Asia. *Ekístics* **47**, September/October, 331–9.

Rondinelli, D. A. 1983. *Secondary cities in developing countries: policies for diffusing urbanization.* London: Sage Publications.

Rondinelli, D. A. & H. Evans 1983. Integrated regional development planning: linking urban centre and rural areas in Bolivia. *World Development* **2**, January, 31–54.

Sada, P. O. 1977. Political structure and economic regionalism in a developing polity: the case of Nigeria. In *Regional planning and national development in tropical Africa*, A. L. Mabogunje & A. Faniran (eds), 265–72. Ibadan: Ibadan University Press.

Said, E. 1985. Orientalism reconsidered. *Race and Class* **27**, Autumn, 1–16.

Santos, M. 1977. Society and space: social formation as theory and method. *Antipode* **9**, February, 3–13.

Sayer, A. 1985. The difference that space makes. In *Social relations and spatial structures*, D. Gregory & J. Urry (eds), 49–66. London: Macmillan.

Scott, I. 1982. *Urban and spatial development in Mexico.* Baltimore: Johns Hopkins University Press.

Siddle, D. J. 1981. Achievement motivation and economic development: farming behaviour in the Zambian Railway Belt 1912–1975. *Third World Planning Review* **3**, August, 259–74.

Slater, D. 1974. Contribution to a critique of development geography. *Canadian Journal of African Studies* **8**, July, 325–54.

Slater, D. 1978. Towards a political economy of urbanization in peripheral capitalist societies: problems of theory and method with illustrations from Latin America. *International Journal of Urban and Regional Research* **2**, 26–52.

Slater, D. 1982. State and territory in post-revolutionary Cuba. *International Journal of Urban and Regional Research* **6**, 1–34.

Slater, D. 1985. The state and issues of regional analysis in Latin America. In *Capital and labour in the urbanized world*, J. Walton (ed.), 76–105. London: Sage Publications.

Slater, D. 1986a. Capitalism and urbanization at the periphery: problems of interpretation and analysis with reference to Latin America. In *Urbanisation in the developing world*, D. Drakakis-Smith (ed.), 7–21. London: Croom Helm.

Slater, D. 1986b. Socialism, democracy and the territorial imperative: elements for a comparison of the Cuban and Nicaraguan experiences. *Antipode* **18**, September, 155–85.

Slater, D. 1987. On development theory and the Warren thesis: arguments against the predominance of economism. *Environment and Planning D, Society and Space* **5**, September, 263–82.

Slater, D. 1988. *Territory and state power in Latin America.* London: Macmillan.

Slater, D. & W. Pansters 1986. City and society in Latin America: introductory reflections. *Boletín de Estudios Latinoamericanos y del Caribe* **41**, December, 3–4.

Soja, E. W. 1980. The socio-spatial dialectic. *Annals of the Associations of American Geographers* **70**, 207–25.

Soja, E. W. 1985a. Regions in context: spatiality, periodicity, and the historical geography of the regional question. *Environment and Planning D, Society and Space* **3**, 175–90.

Soja, E. W. 1985b. The spatiality of social life: towards a transformative retheorisation. In *Social relations and spatial structures*, D. Gregory & J. Urry (eds). London: Macmillan.

Souza, A. de & P. W. Porter 1974. *The underdevelopment and modernization of the Third World.* Commission on College Geography Resource Paper, 28. Washington: Association of American Geographers.

Taylor, P. J. 1986. The world-systems project. In *A world in crisis? Geographical perspectives*, R. J. Johnston & P. J. Taylor (eds), 269–88. Oxford: Basil Blackwell.

Townroe, P. M. & D. Keen 1984. Polarization reversal in the state of Sao Paulo, Brazil. *Regional Studies* **18**, 45–54.

Urry, J. 1981a. Localities, regions and social class. *International Journal of Urban and Regional Research* **5**, 455–74.

Urry, J. 1981b. *The anatomy of capitalist societies: the economy, civil society and the state.* London: Macmillan.

Urry, J. 1985. Social relations, space and time. In *Social relations and spatial structures*, D. Gregory & J. Urry (eds), 20–48. London: Macmillan.

Visvanathan, S. 1987. From the annals of the laboratory state. *Alternatives* **12**, January, 37–59.

Weaver, C. 1981. Development theory and the regional question: a critique of spatial planning and its detractors. In *Development from above or below? The dialectics of regional planning in developing countries*, W. B. Stöhr & D. R. F. Taylor (eds), 73–106. Chichester: Wiley.

12 *Sociology and geography*

John Urry

Introduction

In the last decade or two the disciplines of sociology and geography have come much closer together. If it is the case that sociology studies society or the social, and geography the spatial, then there has been a growing research programme oriented to the analysis of socio-spatial interactions (see Soja 1985). It has come to be accepted both that social life is spatially organized and that this makes a difference, and that space is necessarily the outcome in part at least of social processes. This is not to say that all sociologists and all geographers are aware of this sea-change in modes of thinking and analysis. But there has been a considerable body of writings in both subjects which suggests that there is a major shift occurring in the very relationship between them.

Thus far I have presumed that these two disciplines are discursively organized in roughly the same kind of way. This is, however, not the case and it is therefore inappropriate to maintain that there is a simple process of convergence whereby sociologists 'add' space and geographers 'add' society to their respective analyses. To know how disciplines come together it is necessary to have at least a rudimentary understanding of what they are in fact like *as* disciplines. This involves consideration of how they are organized as academic discourses. And this can roughly be understood as how the relationship between power and knowledge is structured and reproduced within each discipline.

First of all, sociology is organized as a relatively unusual discourse (see Urry 1981b):

(a) There is a multiplicity of perspectives with no common concept of society which unifies them.
(b) Sociological concepts and propositions cannot be easily demarcated from commonsense concepts and propositions.
(c) It is difficult to establish that there is sociological progress and that a progressive research programme has been established. Progress mainly follows from theoretical innovations.
(d) One major form of such innovation results from the *parasitic* nature of sociology, from the fact that innovations sometimes originate in discourses outside sociology itself.

I shall consider certain aspects of point (d) in more detail. What, we might ask, are the circumstances that permit this parasitism to occur? Within sociology's neighbouring disciplines there is a simultaneous process of both presupposing and rejecting what I will loosely call the social, by which I mean the analysis of general social relations which link together individuals and groups. In these disciplines, which include geography, these social relations are presumed to be both of importance, and yet are partly ignored. The social is thus both present and absent simultaneously. Instead, in these disciplines some particular dimension or aspect of social life is abstracted for study, such as the distribution of social activities within space. But this means that such a discipline is discursively unstable. On occasions, certain texts will break through the limitations implied by that discourse. New understandings emerge that will involve more systematic comprehension of the general form of these social relations which will not obscure or neglect the realm of the social.

How, though, does such a development in a neighbouring discipline relate to sociology? First, these other disciplines are in varying degrees discursively unified – which will mean that blocks will be placed upon the new, more 'social' interpretation. Yet, second, because there is no strong essence to sociological discourse, apart from a broad commitment to this idea of the interdependence of individuals and social groups, sociology may attract this new 'social' interpretation. So sociology is important in permitting analysis and elaboration of aspects of the social world which are generally neglected by the other social sciences. It can thus be defined negatively – as a discourse with relatively minimal organization, structure, or unity into which many contending developments from other social sciences become incorporated. So although it is parasitic it enjoys two crucially important features: first, it provides a site within which further elaboration of the original innovation may occur; and second, it provides the context in which a wide variety of contending social theories can be juxtaposed. This has the function of promoting inter-discursive debate and confrontation, something which I shall show developed particularly in the 1970s.

Geography by contrast is discursively more centralized with a considerably higher degree of policing and controlling over what should constitute geographical topics. Roughly speaking, geography is to be viewed as seeking to explain the spatial distribution of one phenomenon in terms of the spatial distribution of one or more other phenomena. The social is present in the sense that much of the subject-matter of geography (that is, its non-physical aspects) concerns the spatial distribution of human or social phenomena. However, there is conflict over the exact degree to which the character of social relations should enter into either the description of what is being explained, or into the processes which are supposedly explaining the phenomena in question. Much of this debate is fought out in terms of contrasting philosophies of science since one of the characteristics of space is that aspects of the spatial can apparently be more easily mapped, measured, and mathematically manipulated than can the social. Hence, part of the resistance to the social is couched in terms of claims about the proper character of geographical *science* which is presumed to be based on precise mapping, measuring, and manipulation of relatively hard spatial data. The main way in which social relations have come to constitute the basis for geographical explanations has been via Marxist political economy which appeared to provide a ready-made theory. Geography has been characteristically atheoretical. In the 1970s, however, the apparent sophistication and growing influence of 'Young Turks' (including here Gregory, Cooke, Massey, Peet, Thrift, and Soja amongst the contributors to these volumes) was sufficient to force Marxism into mainstream geography (see, for example, the journal *Antipode*). This shift in the power/knowledge relationship was facilitated by the fact that perhaps *the* central concern of geography is the relationship between human beings and nature, which is of course the basis of Marxism. As a result geography became more obviously polarized between Marxists and non-Marxists with the latter, not surprisingly, in overwhelming ascendancy. However, as this debate died down one consequence has been that social relations, now often considered in a less orthodox Marxist fashion, became more central to geography and this resulted in an increased openness to the concerns of sociological discourse.

I have so far sketched out some initial features of the discursive organization of sociology and geography. I will now consider briefly certain aspects of the early history of these disciplines and show how it might have been possible for a closer relationship to have emerged. This I will do by considering four of the central figures in classical sociology, Marx, Durkheim, Simmel, and Weber, who were each reasonably aware of the importance of the spatial. I shall concentrate on these sociological contributions since the classical geographers have been dealt with elsewhere in this book. I shall also mention one potentially important trend in British sociology. Thereafter a very brief account will be given of the period between World War I and 1970, a period in which apart from the area of urban and rural studies there was little significant overlap between sociology and geography. In the lengthy following section I show how in selected areas of research extremely constructive overlap occurred during the 1970s and 1980s. A concluding section points to some possible future directions of debate and research.

Classical sociological contributions

Karl Marx

As is well known, Marx argued that crucial to the development of Western societies has been the growth of what he called the capitalist mode of production, which in turn transforms existing patterns of economic and social life (see Quaini (1982), Smith (1984); on these classical contributions see Saunders (1986)). One central feature of these changes consists in the exceedingly rapid growth of industrial towns and cities. Marx considered that the division between town and countryside has been a feature of all societies where there was some development of the division of labour. However, up to the Middle Ages the countryside was the more important aspect of society. Of the ancient city–states of Greece and Rome, Marx noted that they were fundamentally based upon land ownership and agriculture – there was 'ruralization of the city' (1973, p. 479). In the Middle Ages, however, this slowly began to change especially as, first, agriculture itself came to be much more based on producing commodities for sale in the market and much less tied to producing for relatively self-sufficient communities; and, second, small-scale industrial production was gradually established in the countryside. He says that the 'modern (age) is the urbanization of countryside' (Marx 1973, p. 479). The contradiction between town and country, although necessary for the growth of capitalism, is itself minimized as new industrial towns emerged in the countryside. However, these towns then developed exceedingly rapidly and as a result a new conflict arose between town and countryside.

Marx and Engels also wrote more generally of how in capitalist society

> Constant revolutionising of production, uninterrupted disturbance of all social conditions, everlasting uncertainty and agitation distinguish the bourgeois epoch from all earlier ones. All fixed, fast-frozen relations . . . are swept away, all new-formed ones become antiquated before they can ossify. All that is solid melts into air, all that is holy is profaned (1888, pp. 53–4).

Bourgeois or capitalist society, then, is one of intense change, particularly of where people live and how their lives are organized over time. As production is revolutionized in order to bring about massive savings of labour-time, peoples' relationships to each other across *space* are transformed. There are a number of changes, namely: (a) capitalism has 'pitilessly torn asunder the motley feudal ties that bound man [sic] to his "natural superiors"'; (b) the need for a constantly expanding market 'chases the bourgeoisie over the whole surface of the globe and destroys local and regional markets'; (c) the 'immensely facilitated means of communication draws all . . . nations into civilisation' – hence reducing the distance between societies; (d) enormous cities are created and this has 'rescued a considerable part of the population from the idiocy of rural life'; (e) political centralization is generated as independent, loosely connected provinces 'become lumped together into one nation'; (f) masses of labourers 'organised like soldiers' are 'crowded into the factory', the proletariat 'becomes concentrated in greater masses'; and (g) the development of trade unions is 'helped on by the improved means of communication that are created by modern industry and that place the workers of different localities in contact with one another' (Marx & Engels 1888, pp. 53–65). Overall, it is argued that the aim of capitalist production is to annihilate space with time, and hence to overcome all spatial barriers to capitalist industrialization (see Harvey 1982).

Emile Durkheim

In *The Division of labour in society* (1984) Durkheim argued that there are two types of society with associated forms of solidarity, mechanical (based on likeness or similarity) and organic (based on difference and complementarity). It is the growth in the division of labour, of dramatically increased specialization, that brings about the transition from one to the other. Two factors give rise to this heightened division of labour: increases in material density and increases in moral density. By the former he means that the density of population in a given area increases, both because of the development of new forms and speed of communication so that by 'abolishing or lessening the vacuums separating social segments, these means increase the density of society' (1984, p. 203), and because of the growth of towns and cities. Moral density refers to the increased density of interaction and social relationships within a given population.

The increase in the division of labour is due to the fact that different parts of society lose their individuality; the partitions between them become more permeable. This occurs because of heightened material and moral density. There is a drawing together of individuals who once were separated. Social relationships become more numerous and complex. Durkheim says that as the 'division of labour progresses the more individuals there are who are sufficiently in contact with one another to be able mutually to act and react upon one another' (1984, p. 201). However, Durkheim also notes that because cities normally grow through immigration rather than through natural increase, this means that new residents will have a weakened attachment to traditional beliefs and values. Hence, the collective conscience will be less strong and this will generally facilitate the new organic solidarity of interdependence.

Two further points should be noted. First, cities are also on occasions centres of social pathology. Second, local or geographical loyalties will be gradually undermined with the growth of the new occupationally based division of labour; he suggests that 'geographical divisions' will generally become less and less significant in awakening feelings of attachment and loyalty.

Georg Simmel

In 'Metropolis and the city' Simmel made a number of points about living in the city (1971). First, as a result of the richness and diversity of experience, because of the multitude of stimuli, it is necessary to develop an attitude of reserve and an insensitivity to feeling. Without the development of such an attitude people would not be able to cope with such experiences caused by the high density of population. Reserve in the face of superficial contacts with the crowd is necessary for survival. The urban personality is necessarily unemotional, reserved, detached – blasé, in other words.

Second, at the same time the city assures individuals of a distinctive type of personal freedom. Simmel, like Durkheim, contrasts the modern city with the small-scale community. It is in the latter 'that the individual member has only a very slight area for the development of his [sic] own qualities and for the free activity for which he himself is responsible' (1971, p. 332). Such groups cannot, he says, give room to freedom and to the peculiarities of the inner and outer development of the individual. And this is even so in the life of the 'small town dweller'. It is only in the large metropolis that great opportunities are available for the unique development of the individual, partly because of the wider contacts available to each person in the city, and partly because cities are themselves interconnected with other cities, and so each individual is placed in exceptionally wide social contacts with people in the rest of the country and in other countries.

A third central feature of the city is that it is based on the money economy. Money is indeed both the source and the expression of the rationality and the intellectualism of the city. Both money and the intellect share a matter-of-fact attitude towards people and things. They are indifferent to genuine individuality. The typical city dweller is guided by his (sic) head and not by his heart, by calculation and intellect, not by affection and emotion. Money contributes to a levelling of feeling and attitude, to express everything in terms of a single measure.

Fourth, Simmel notes how the money economy generates a concern for precision and punctuality. This is both in the general sense that the money economy makes people more calculating about their activities and their relationships. And a concern for precision stems from the fact that the city contains so many people doing so many different things; in order for these to take place in an even moderately efficient manner there has to be some scheduling of different activities. This necessitates accurate time-keeping, precise arrangements, and a prohibition on spontaneity. Meetings with other people have to be timetabled; they are typically brief and infrequent.

Max Weber

The further crucial figure within classical sociology is that of Max Weber who makes very few references to space or geography. In some ways this is rather surprising since his younger brother Alfred Weber was one of the seminal contributors to the theory of industrial location within economic geography (1929). In general Alfred Weber argued that optimal location for a plant was uniquely determined at the site which minimized transport costs. In the two clearest examples the minimum transport cost site will be found either at the source of raw materials *or* at the place of consumption. Normally, however, it will be somewhere in between and Weber produced an interesting but flawed analysis of that scenario. He introduced the notion of an isodapane, a contour of equal transport costs, which he used to provide more or less the first analysis of the economies of agglomeration and of how they may in fact offset the tendency for firms to locate at the site which minimizes transport costs.

Max Weber by contrast was relatively critical of attempts to use spatial notions to investigate the city (see Weber 1958). He rejected analyses in terms of size and density and concentrated in the main on how the emergence of medieval cities had constituted a challenge to the surrounding feudal system. For him the city was characterized by military, economic, and political autonomy; and it was there for the first time that people came together as individual citizens. He was not concerned with the spatial organization of the city. Interestingly, Alfred Weber later came to reject an autonomous, geometric location theory and developed a rather broader cultural analysis having much in common with that of his older brother.

Early British sociology

In Britain, the early growth of sociology was particularly associated with the development of the survey. Best known here has been the tradition of the social survey, which was instituted by Charles Booth, Seebohm Rowntree, and Arthur Bowley around the turn of the century (see Kent 1985). In such studies particular towns or cities were surveyed especially with regard to the levels of identifiable poverty and deprivation in the population. It was presumed that the patterns found in particular cities could be generalized to other places since the spread of industrialization and urban growth were thought to produce homogeneity. It was only by the 1930s that it came to be realized that no particular town could be presumed to be typical of the country as a whole and social surveys became much more sensitive to locality variation. However, side by side with this development had been the growth of regional surveys, particularly as reported in the *Sociological Review*. According to one of the central figures, Patrick Geddes, not only was the objective of such surveys to study the whole social structure of a place, but it was also necessary to combine in this task the separate disciplines of geography, economics, anthropology, demography, and sociology (see Abrams 1968, Mark-Lawson 1982). Although he therefore emphasized the importance of considering the complex structure of regions, cities, and communities, relatively few such studies were completed and the regional survey did not forge the clear links between sociology and geography in the first few decades of this century that might have been possible.

1920–70: separate development

Broadly speaking, then, from World War I until about 1970, the two disciplines of sociology and geography proceeded to develop separately. The social and the spatial were viewed as distinct domains which did not have much to do with each other. Sociology and especially geography were transformed into well organized academic disciplines centred round certain mechanisms of inclusion and exclusion. Academic reputations came to be made within *each* discipline; in turn they generated new subdisciplines. The subdiscipline of *social* geography was, significantly, rather slow to develop in this period. The only subdisciplines which did cross the sociology–geography divide were those of urban and rural sociology which I shall now discuss briefly.

Urban sociology

The nature of city life was explored by many sociologists especially those working in the University of Chicago in the period between the wars. Much of this research was synthesized by Wirth in his paper 'Urbanism as a way of life' (1938). He argued that there are three causes of the differences in social patterns between rural and urban areas: size, density, and heterogeneity. With regard to the first, size, he argued, like Simmel, that the larger a settlement, the greater the variation between different areas and the more there will be segregation of different groups in different parts of the settlement. Furthermore, such increases in size reduce the chances of any two people knowing each other and this leads to a greater indifference of people towards one another. Living in large settlements involves greater social distance, a lack of spontaneous interactions, and the development of more formal agencies of control and regulation.

The effect of increased *density* (very much a spatial concept, as we saw in Durkheim) reinforces the effects of size. In particular it leads people to relate to each other on the basis of their specific rôles rather than of their personal qualities: people view each other in an instrumental fashion. Different areas develop in cities made up of those playing these different social rôles. There is a mosaic of social worlds. Increased density also leads to more formal regulation and control, especially by laws, rather than by the appropriate customs which pertain in less dense societies.

Increased *heterogeneity* means that people in cities participate in many different social circles, none of which commands their complete involvement. Individuals enjoy a different level of status within these separate circles, so much so that the urban person is unstable, disorganized, and insecure. At the same time the only way that individuals can change anything is through representation within large-scale organizations from which most people feel separated and powerless.

Writers associated with the Chicago School developed more specific analyses associated particularly with the competition for land within cities. They argued that as the population of towns grows, so there is an increased specialization of people into different economic positions, and that these different groups come to live in different sections of the town. In particular the pressure for space at the centre creates an area of high land values, and this determines the cost of land and housing in the rest of the town. These differences in land values (which means higher rents or higher prices for buying land or houses) provide the mechanisms by which different groups are distributed throughout the urban area. Burgess argued that cities are divided into concentrically organized zones (see Park & Burgess 1967). A number of geographers researched these ideas and Robson, for example, found some support for the concentric zone model (1969). In Sunderland, the northern part of the town demonstrated this pattern of concentric circles, although in the south there was a wedge-shaped pattern of development.

Another process examined by urban sociologists has been the often highly unequal access to preferred forms of housing. In a study conducted in the 1960s in the UK Rex & Moore (1969) argued that there is a shared value held by people living in cities – a desire to move to the suburbs. Different groups have different access to resources which may enable them to make such a move. These different groups are called housing classes. Such classes are those who are in a similar position in the housing market. There is of necessity competition and conflict between different housing classes for access to the generally preferred form of suburban housing.

Each of these processes generated considerable critical literature from both sociologists and geographers. A further topic in urban sociology which also generated considerable geographical interest was that of community studies. However, these really developed out of rural sociology which needs to be considered initially.

Rural sociology and the community

Corresponding to the urban way of life Redfield developed the notion of a folk society and a folk–urban continuum (1947). This generated a plethora of community studies of rural communities especially in the UK in the 1940s and 1950s (see Rees 1950, Williams 1956, Frankenberg 1957). Frankenberg (1966) summarized many such studies, maintaining that the following were the main social and spatial characteristics of life in rural communities:

(a) It is organized as a *community* with people frequently meeting together and being connected in lots of different ways; people keep meeting each other as they take on different rôles in relationship to each other.
(b) People have social networks which are close-knit; in other words, their friends know each other, as well as themselves.
(c) Most inhabitants work on the *land* or in related industries. There is a high proportion of jobs which overlap and there is a relatively simple division of labour. Most workers are relatively unspecialized farm labourers.
(d) Most people possess an *ascribed* status fixed by their family of origin. It is difficult to change such status through achievement. People are strongly constrained to behave in ways appropriate to their status. This status spreads from one situation to another, irrespective of the different activities undertaken.

(e) Economic class divisions are only one basis of social *conflict*. Various strategies develop to handle potential conflicts which cannot be avoided because people keep meeting each other. Generally, social inequalities are presumed to be justified, often in terms of tradition.

Interestingly, however, a number of community studies were undertaken in urban areas, the best known of which was that of Bethnal Green researched in the 1950s (see Young & Willmott 1957, Platt 1971). Almost three-quarters of the male labourforce were manual workers and most of the non-manual workers were shopkeepers and publicans. It was thus an overwhelmingly working-class locality. Particularly important was the rôle of relatives. The sample of 45 couples had 1691 relatives, of whom 902 lived in Bethnal Green or the neighbouring borough. Each couple had an average of 13 relatives within the borough itself. It is suggested that there was a particularly important extended family of various relatives who lived in the locality and who saw each other every day or nearly every day. The ties within the extended family were continuous, they involved *reciprocal* obligations, and they particularly involved women in the area. Of special significance were the ties linking together the grandmother with her daughter(s) and grandchildren. Paradoxically, marriage often served to bind a daughter more closely to her mother. There was a high degree of segregation between the activities of the women (more home-centred) and those of the men (more centred on work, clubs, and pubs). Major divisions of class and status between men in different occupations were not found to be significant. In another study Dennis *et al.* (1956), investigated the nature of the coal town they call Ashton. They argued that the sense of community they found stemmed from the overwhelming importance of the coalmining industry and from the shared experiences which this produced among male miners. In particular, these community relations were reinforced by the local nature of mining knowledge and experience. The colliers' skill, seniority, and even knowledge of technical terms were often not transferable from pit to pit, let alone village to village, district to district, or coalfield to coalfield. This was a factor which bound miners to their home towns and to their home collieries and was why they were generally unwilling to move elsewhere to work. Community here was thus based on the relations between male miners, a process which has meant that miners' wives have been kept in a different and subordinate position within such communities.

Finally, other studies were conducted of life in the suburbs in both British and American cities (see Willmott & Young 1960, Gans 1962, Bott 1957). The general pattern of life found there is summarized in the following points:

(a) life is centred not on the neighbourhood nor on the street but on the home;
(b) this home-centredness is reflected in the strong emphasis placed on obtaining consumer durables – on an orientation to consumption;
(c) there is relatively less emphasis on contact with relatives and relatively more on choosing, making, and keeping friends, often from a fairly wide geographical area and not just from the neighbourhood or street;
(d) where households are made up of husbands and wives there is more emphasis (compared with, say, Bethnal Green) on sharing tasks *and* sharing friends;

(e) there is fairly high participation in a variety of informal and formal organizations and in the development of friendships formed out of this voluntary participation.

New developments

There have been numerous ways in which constructive overlap has recently developed between sociology and geography. Perhaps the clearest illustration of this is to be found within the writings of the 'sociologist' Giddens and the 'geographer' Gregory and their mutual interest in post-Marxist social theory and Hägerstrand's time-geography (see Giddens (1985), Gregory Ch. 14 in this volume; and for a recent collection partly centred around realist philosophy of science, Gregory & Urry (1985)). Also, overlapping developments have occurred in the study of housing, race, gender, territoriality, and the state, all of which are discussed elsewhere in this collection. In this section I shall deal with three areas in which I consider stimulating developments to have taken place: stratification, work and industry, and urban and rural politics.

Stratification

Certainly within Britain the investigation of stratification and in particular of class has been *the* central topic of sociological investigation. Until fairly recently it was presumed by sociologists that the basic unit of investigation was the nation–state. Classes were presumed to be national, the working class was presumed to comprise all manual workers living within a given national territory (see discussion in Urry 1981a). For example, in the most extensive survey of social mobility and the class structure in the UK the focus is on male national classes and on the absolute and relative rates of mobility in and out of such classes (see Goldthorpe 1980). Goldthorpe is particularly interested in the determinants of class actions and struggle, and in this he takes as crucial the degree to which particular classes are self-recruited (that is, that sons have the same occupational status as their fathers). The contemporary (male) working class in Britain is notable here since it is now overwhelmingly self-recruited and Goldthorpe uses this fact to explain the continued solidarity of working-class partisanship, and hence its commitment to trade unionism and the Labour Party (this held true for the 1960s and 1970s, but is no longer such a strong feature in the late 1980s).

Geographically influenced literature has developed a number of criticisms of this kind of argument. First, it has been maintained that there will be all sorts of variation in subnational class structures such that no one may actually live and work in an area which possesses the contours of the national class structure (see Urry 1981a). Considerable attention has been paid as to how different localities possess distinct local stratification structures, with varying proportions of different occupations, industries, genders, and ethnic groups in places often geographically adjacent (see Cooke (1986) for a demonstration of this in the UK). It is interesting to note that Goldthorpe was himself well aware of this in his previous research on the affluent worker (Goldthorpe *et al.* 1969). Luton was chosen for this study precisely because it would provide the extreme test of the theory of embourgeoisement; if the theory did not apply in what was at the time an extremely affluent locality, then it would not be applicable in more traditional working-class localities.

Second, it has been shown that it is necessary to consider how such subnational patterns of stratification are economically structured, particularly through changes in the location of industry locally, nationally, and internationally and the effects that these have on local labourmarkets. Investigation is conducted regarding the boundaries of such markets, their size, their segmenttion, the degree to which skills and openings are matched, and so on (see on such market, Blackburn & Mann (1979), and Cooke (1983, Ch. 9) more generally). This may produce differences of interest between workers employed in locally owned and multinationally owned enterprises, or between those in private and public organizations (see Dunleavy (1980) on the latter).

Third, local social structures should be investigated for the dynamic interrelationships *between* social classes and other social forces. Classes in particular should not be viewed as means of dividing up the population and then attributing sets of attitudes to be correlated with such categories. Rather social classes should be understood as organized for action, possessing powers to effect change, and as being dynamically interrelated so forming a local social structure. It is further argued that in particular places social classes may produce distinctive sets of effects in terms of local politics, voting patterns, and levels of social provision. Analysis has been provided of 'radical regions' (see Cooke 1985), 'little Moscows' (MacIntyre 1980), and the 'neighbourhood effect' in local politics (Warde 1986).

Fourth, it is necessary to consider variations in the spatial distribution of social classes. In Italy, for example, Paci (1981) points out that the concentration of the working class in the North is higher than in virtually any other area in Europe, and this is in part relevant to an explanation of why Italy has had one of the strongest communist parties in the West. Similarly, Thrift (1986, 1987) has recently analyzed the spatial distribution of the service class in Britain and shown that it has high concentrations in parts of London and the South East which account for aspects of local culture, housing style, and costs.

Fifth, this spatial distribution of social classes may be important in generating grievances about the relatively poor allocation of certain kinds of class position to a given locality or region. Buck argues, for example, that the reason why France and Britain exhibit higher rates of regional grievance is because there is a more unequal class distribution between different regions (Buck 1979, Buck & Atkins 1978). More specifically, resentment may be generated over the lack of service-class positions as in Wales, or over the relatively few skilled working-class positions available in other parts of the country.

Sixth, debate has developed over the degree to which shared territory may be viewed as the basis for class mobilization; or, alternatively, whether territorial identifications dissipate class loyalties (see Harris 1983, Urry 1983). Clearly, the geographical literature has shown that the mobilization of social classes is always place-specific and the character of these places makes a substantial difference to those patterns of mobilization. It also seems that as the spatial structuring of societies changes so this will affect the relative importance of different territories (neighbourhoods, towns, regions) and this influences the capacity of social classes to mobilize for action (see Calhoun 1982, Urry 1986).

Finally, it has been shown that stratification in an area cannot be seriously considered without analyzing how class and gender and indeed race relations intersect. In relationship to gender this can be seen in a number of ways: that the class structure of men's jobs is quite different from that of women's jobs; that the gender of a set of places within the social division of labour is relevant to the very construction of such places and hence to the structuration of social class; that there are major forms of social inequality rooted in the unequal social relations between husbands and housewives within households; that social relations within the workplace are substantially structured by sets of patriarchal trade union and employer practices; that there are major forms of politics structured by gender particularly focused around violence, welfare, and sexuality which conventional class analysis of politics ignores; and that local politics and the local state as importantly reflect gender struggles as they do class struggles (see, for example, Institute of British Geographers, Women and Geography Study Group 1984, Murgatroyd *et al.* 1985, Crompton & Mann 1986, Walby 1986).

It should be noted that although these geographical concerns have been brought into the analysis of social class, this is not to suggest that geographers have themselves introduced such notions. In a number of cases, work on these issues has been primarily carried out by sociologists. Furthermore, the major theoretical contributions to the analysis of class have continued to be developed within sociological discourse. For example, one of the most influential theories has been that propounded by Wright whose arguments have been taken up in Massey's *Spatial divisions of labour* (1984) (to be considered below). Wright argues that it is unnecessary to assume that all positions within the social divisions of labour must fall firmly into one class or another. Instead, certain positions are to be viewed as objectively torn between classes; they are not class positions but 'contradictory class locations' (Wright 1978). He argues that there are three central processes which underlie the basic capital–labour relationship: control over the physical means of production, control over labour power, and control over investments and resource allocations. At the level of the pure capitalist mode of production there are only the class positions of capitalists and workers; the former are in control of all these processes, the latter are controlled within each. But with the development of advanced capitalism these dimensions need no longer coincide and various contradictory class locations are generated, both between the proletariat and the bourgeoisie, and between the petty bourgeoisie and the bourgeoisie/proletariat. In the former there are managers/technocrats/line supervisors, in the latter there are small employers located between the petty bourgeoisie and the bourgeoisie, and semi-autonomous employees such as researchers, teachers, lecturers, craftworkers, and so on, located between the petty bourgeoisie and the proletariat. There are also, he asserts, contradictory class locations within the political/ideological apparatuses – these involve the execution of state policies and the dissemination of ideology; hence almost one-half of the employed population in the USA are in contradictory class locations, and between one-quarter and one-third occupy such locations near the boundary of the working class. Thus he says that about two-thirds of the population constitute a basis for a socialist movement in the USA, although there are substantial conflicts of interest between the proletariat and those in various kinds of contradictory class location (see Abercrombie & Urry (1983, pp. 82–5, *et passim*) for various other recent theories of social class).

Work and industry

One particular way in which the sociology of class has been developed in the past decade or two has been through a much more detailed analysis of the nature of work. This was prompted by the publication of Braverman's *Labour and monopoly capital* (1974) which generated extremely interesting theoretical and empirical responses (see Burawoy 1979, Edwards 1979, Littler 1982, Wood 1982). Braverman argued that it is the accumulation of capital that fundamentally determines the organization of the labour process, in particular the tendency for labour to become progressively fragmented and deskilled, and for the work of conception (mental labour) to separate off from execution (manual labour) and to be embodied within functionally separate management structures. These developments occur because of the tremendous savings in the cost of labour power that capital can thereby obtain. Braverman demonstrated that through the so-called Babbage principal, capital is progressively able to obtain *precisely* those quantities and qualities of labour power that it in fact requires. The more that the labour process is fragmented, the greater the ability of capital to avoid purchasing 'unnecessarily' skilled labour to undertake deskilled work. Braverman proceeds to analyze the nature of monopoly capitalism and the consequential forms of deskilling particularly of previous craft occupations (see Cutler (1978) on Braverman's overemphasis on early capitalist labour as craft-based).

His thesis that the accumulation of capital is, through the processes of deskilling, generating a necessarily larger, less divided, and stronger working class has been challenged by a large number of theoretical and substantive analyses. *Inter alia*, it has been shown that (a) Braverman-type direct control of the labour process is only one amongst a number of managerial strategies; (b) the employment of different strategies depends at least in part upon different forms of work organization and resistance; (c) such forms of organization have important effects in dividing the workforce so that such divisions do not stem simply from the process of accumulation; (d) there are crucially important new forms of skill and control which are generated by accumulation and which are not simply deskilled (such as work involving numerically controlled machine tools); (e) there are important historical and comparative variations in the forms and degree of deskilling of the labour process through scientific management; (f) Braverman underemphasizes the internationalization of the accumulation process, which means that the deskilled labourforce is divided within spatially distinct national territories; and (g) there are absolutely crucial bases of working-class division especially around gender and ethnicity which make any class homogeneity unlikely.

These debates in turn generated considerable controversy within geography. Certain texts began to emerge which integrated the sociological study of the labour process with the geographical analysis of the spatial organization of industry. A major recent collection is that edited by Scott & Storper (1986). A centrally significant set of writings which helped bring about this partial synthesis is associated with the work of Massey, much of which is in effect summarized in her *Spatial divisions of labour* (1984; also Massey & Meegan 1982). She is keen to assert that 'geography matters' and analyzes three different spatial structures of capitalist production: (a) the locationally concentrated spatial structure with the whole production process centralized within a single geographical area and with no intra-firm hierarchies; (b) the cloning branch-plant spatial structure with relations of ownership and possession separated off in a spatially distinct managerial headquarters; and (c) the part-process spatial structure where headquarters, research and development, production of technically complex components, and final assembly, are located in spatially separated locations. Massey argues firmly against the view that there is any developmental sequence in these types, only that especially in some parts of the modern electronics and petrochemicals industries there has been a process of increased concentration upon *single* sites of plants that were previously spatially separate.

In the development of these distinct spatial structures particular emphasis is placed upon the specific form taken by the struggles between capital and labour in the different industries. In the motor vehicle industry, for example, spatial separation and variation have been as much a part of capital's strategy against labour as have changes within production. This can be seen in, for example, the reorganization of Fiat's production away from its traditional centre in Turin, where especially in the late 1960s/early 1970s there had been exceptional strike waves, to the south. This involved explicitly separating off certain parts of the production process, especially assembly work, supplies, and some subassembly and parts production. None of the higher managerial, financial, and service functions left Turin. Moreover, such shifts in industrial location have been by no means confined to the recent period. The UK footwear industry appears to have relocated out of London at the beginning of the 19th century in order to overcome rising labour costs and increasing militancy. A distinctive sexual division of labour became established in the new area of production around Northampton, whereby women were employed as subcontracted labour working at home engaged in stitching the leather uppers of the shoe on to the sole. Unlike many writers in this area Massey also analyzes the changing spatial structure in service employment. She brings out *inter alia* the important rôle of self-employment and family enterprises in this sector, the relative spatial evenness of public sector services, and the importance of the simple cloning branch-plant structure in private consumer services in the South East (see also here Daniels (1985)).

Massey then goes on to explore how these different spatial structures in different industries have the effect both of generating social inequalities between different areas and of producing distinctive local economic and social structures in various regions and localities. In the analysis of the UK coalfield areas and of Cornwall, she shows how in this process of the combination of 'layers', any particular layer, or round of investment, may produce very different effects in different areas as a result of its combination with different pre-existing structures. What we see here are national processes in combination with and embedded in particular conditions producing the uniqueness of local economic and social structures.

Massey's last argument, on the idea of rounds of investment, has generated an interesting sociological literature. Warde (1985) in particular has dubbed this a 'geological' metaphor and pointed to some major difficulties. It is necessary to elaborate further on the rules which generate different logics of location within each round of restructuring; more work needs to be done on how various classes combined together in a place to produce different political effects; and it is necessary to consider a whole variety of local effects and not only those of class. These sets of issues provide in part the context for the recent research initiative in the UK concerned with the changing urban and regional system (see Cooke 1986). In this the arguments of both geography and sociology have been brought to bear on the attempt to elucidate just how processes of economic restructuring in different industries combine together to produce locally distinct outcomes. The notion of restructuring is also problematic. Sayer has provided an extremely trenchant critique of the suggestion that capital accumulation necessarily produces a given spatial organization of production, such as the so-called new international division of labour (1985, especially pp. 14–15; and see Fröbel *et al.* (1980) on the NIDL). It is now clear that there is a wide variety of forms of restructuring that can occur within a given industry, the explanation for which requires analysis not just of the labourmarket but also of changing technologies and transformations in the product market, and that changes are more contingent than had been previously thought.

In a further study Pahl (1984) has demonstrated the importance of divisions of labour not between households but *within* the household. For example, he shows that from the early 1970s in the UK there was a marked increase in various indicators of self-provisioning by households (DIY sales for example) with extraordinarily high proportions of both men and women engaging in this form of household labour. Pahl views such work as being significantly expressive, a product more of affluence (and home-ownership) than of poverty. Overall, it is argued that the household as an economic unit is of tremendous importance, that it is necessary to examine household work strategies by which trade-offs are made between domestic self-provisioning, informal/communal and formal forms of work, and that such strategies will vary particularly by size of household and stage in the domestic cycle.

These various points are then illustrated and elaborated in the study of the Isle of Sheppey, apparently once referred to as that 'septic isle'. Two reasons for choosing Sheppey were: first, that it was thought to contain a great deal of informal work, a 'seething centre of fiddles'; and second, that by the late 1970s it had already experienced a lengthy period of deindustrialization and would hence illustrate at least some features typical of Britain as a whole in the 1980s.

Some of the main conclusions drawn with respect to the detailed examination of household work strategies are: that there is an extraordinary amount of self-provisioning; that the employment status of the female partner does not seem to affect the total amount of domestic work to be done; that there is relatively little informal work done by non-household members; that household self-provisioning and the formal purchase of services go together, rather than being substitutes for each other; that where the male is unemployed this increases the female's domestic tasks; that the main factor producing a less unequal division of domestic labour is the full-time employment of the female partner; that the more household tasks to be done, the more unequal that division and this is especially so in households with young children; that households with older members tend to have a less unequal division of domestic work; and that increasing inequalities will occur between those households where there are two or more in paid employment and there is extensive self-provisioning, and those households where both forms of labour are relatively unavailable. Overall Pahl views households as 'dynamically conservative', as relatively autonomous social units pursuing their own goals, defining their interests, and worthy of political support and encouragement.

Urban and rural politics

There is an immense range of writings concerned with especially urban politics. Most influential in the early 1970s was the perspective known as urban managerialism also particularly associated with Pahl in Britain (1975). He argued that the most fundamental process in the city concerned the distribution of scarce urban resources such as housing and transport. These necessarily have a spatial dimension which operates partly separate from the more general economic and social organization in society. However, the distribution of such resources is largely a function of the actions of those individuals who occupy strategic allocative locations in the social system. In the city there are various gatekeepers whose decisions determine degrees of access by different sections of the population to different types of urban resources. He argued that the task of urban sociology is to examine the goals and values of such 'urban managers' in order to explain the patterns of distribution. Such representative studies using this perspective include that of Harloe *et al.* (1974) who investigated a range of urban managers involved in housing provision, Ford who studied the rôle of building society managers (1975), and Elliot & McCrone (1975) who investigated the motives and values of private landlords (see also Norman 1975).

Partly in response to the supposed neo-Weberian limitations of urban managerialism, the study of the city and of urban politics was transformed by the writings of Castells. In a series of works, part sociological, part geographical, he developed a distinctive position (see Castells (1978) for a summary). Expressing this briefly, he argued that contemporary capitalist production is organized on an increasingly international or global scale. But even in this era of monopoly capitalism, there is one set of activities which is organized in and through towns and cities. This involves what he calls 'collective consumption', which consists of those services which are generally provided by the state and which are necessary for the energies and skills of workers to be sustained. This collective consumption (of transport, education, planning, health, etc.) has come to be organized by the state within towns and cities (within the urban) for two reasons. First, over the long term there has been a considerable increase in the concentration of the labourforce within urban areas. For example, in Britain throughout this century four-fifths of the population have lived in substantial urban centres. And second, there has been a long-run tendency for many of these activities to become unprofitable when provided by private industry and for them to be taken over by the state.

However, these services cannot be provided unproblematically. This is because states are rarely able to raise sufficient revenue and as a consequence there are continuing problems with the forms and level of provision, such as the amount and quality of council housing, the provision of health care, the nature of public transport, and so on. The point about all of these is that such services have become politicized because they are provided collectively. There is thus a realm of urban politics which is focused in and around these forms of collective consumption. Castells argues that urban protest always involves a number of classes organized on such issues. However, for such protest to take the form of what he terms an 'urban social movement' it must come under the leadership of the working class in that area (see Lowe 1986).

In a more recent influential work *The city and the grassroots* (1983) Castells examines a number of urban movements especially the gay and Latino movements in San Francisco and the citizens' movement in Madrid. He then derives a model from these studies to explain their relative degree of success. The movement must work on three fronts: collective consumption, community culture, and political self-management; it must define itself as an *urban* social movement; it must make use of the media, professionals, and parties; and it must be organizationally independent of such parties (see discussion in Pickvance (1985)). One problem about this schema is that the notion of the urban is presumed to be self-evident (see Dunleavy 1982). In fact Pickvance suggests that there are three important characteristics of the urban: collective consumption (as in Castells); local-level political processes (see Saunders 1986); and spatial proximity (see Scott 1980). Existing formulations are unsatisfactory because they only focus upon one of these. Pickvance further argues that Castells's model ignores some centrally significant contextual factors: rapid urbanization, state responses to the demands of the movements, political context, the rôle of the middle class in resource mobilization and its objective work and residential situation, and the general economic and social conditions which affect the general disposition to political activism (Pickvance 1985, pp. 39–44; Castells 1985, pp. 55–6).

The other studies of urban social movements should be noted. First, Saunders shows the importance of middle-class urban protest in the southern and eastern areas of Croydon, what he terms the 'Deep South' (1980). The overall effect of this suburban social movement has been to preserve low levels of rates in Croydon and hence low levels of collective consumption, to keep large areas of green belt land in the south of the borough, and to maintain the low density of housing in the area. Thus when new housing schemes have been approved in South Croydon only three to five houses per acre have been allowed. In other parts of Croydon housing has been built with up to five times that density. Saunders maintains that the southern residents have benefited greatly from the successful maintenance of these policies in the local council. At the same time there have been many losers, especially the low-income groups who live in the north and west of the borough. These groups lose out due to the relatively poor levels of collective consumption in the borough, the higher density of housing in their area, and the concentration of industry and office development in their areas of the town.

Second, Dunleavy has investigated the development of high-rise flats in Britain in the 1960s (1981). As is well known, they represented a massive decline in the standards of public housing provision. They were overwhelmingly unsuitable for families with young children and for old people, their two largest groups of inhbitants. Many of them have deteriorated so rapidly that they can no longer be let or have even been knocked down. However, in general most of the people who have been affected by these developments have not protested about the policy of building high-rise flats. This is true both of the people whose older houses were knocked down in order to build tower blocks, and of the people who lived in the new blocks. Dunleavy then considers one particular area where there was in fact protest, in Newham in London following the partial collapse of one tower block, Ronan Point. A movement sprang up to prevent further rehousing in such blocks. However, the dominant Labour group refused to discuss this Beckton protest, the organizers of which then turned to the Conservatives for help. A central government inquiry report found that such blocks did indeed contain unacceptable weaknesses under certain conditions, but Newham Council did nothing to stop work on the contract for the nine blocks in question. Indeed, the companies concerned both received a contract for more work. Throughout the Labour Council refused to meet the Beckton protestors who were forced to engage in various forms of demonstration and publicity-seeking protest in order to get a hearing of any sort. These tactics alienated their Conservative allies and the protesters maintained that under no circumstance would they move into the flats being offered. However, within a few months this protest collapsed – partly because in despair some people left the area, some were rehoused in other parts of the borough, while two members of the protest committee accepted rehousing – at which point the media coverage ceased. In short, the protest showed, first, that even a strong Labour council would not take the protests of its own supporters at all seriously, but second, that, if given the chance, people would have opted for a quite different system of housing than that which they were offered, in this case high-rise tower blocks. By contrast, lack of protest on such issues is normal, demonstrating that most people most of the time are powerless in the face of such situations, constantly being reduced by the seemingly inexorable exercise of routine power by the public housing system (Dunleavy (1981); see also Mollenkopf (1983) on the paradoxical political effects of American protest and community activism during the 1960s).

In the study of rural life a major contribution in Britain has been made by the sociological investigations of Newby (see e.g. Newby 1977, 1979). He argues that there is one aspect of the rural which does distinguish such communities and that it is the ownership of property. It is the organization of relationships of property that shapes the nature of the rural social structure. In the 19th century most land in Britain was owned by very few landowners and was worked by tenant farmers who in turn employed a large number of agricultural workers. Over the course of this century there has been an extensive increase in owner-occupation so that about 67 per cent of farm land is now farmed by its owners, there has been a substantial increase in the ownership of agricultural land by large institutions, especially those within the City of London, and the average farm size has considerably increased. As a result there has been an irreversible rationalization of agriculture. Instead of it being a way of life centred in a 'green and pleasant land' (see Newby 1979), it has become a business where farms are organized to produce profits. Agriculture has become industrialized particularly with the growth of factory-farming. One aspect of these developments has been the increased mechanization of farm work. This has had quite dramatic effects in reducing the workforce employed on farms, at the same time that the size of most farms has increased considerably in terms of acreage farmed. Because of this reduction, mechanization has *increased* the variety of skills that must be mastered – it has reduced rather than increased the division of labour.

This reduction in the average numbers of farm workers has also weakened the bases of agricultural trade unionism, reduced the bureaucratization of farms, and lessened the social distance between farmers and their workers. Indeed, the connections between farmers and their worker(s) have also been strengthened by the arrival of people from urban areas. The result has been to produce an encapsulated rural community particularly organized around the farm and farming and defined in opposition to the urban middle-class newcomers. Farmers and farm labourers have come to form a community within a community and this has reinforced a kind of nostalgia for the past when rural societies were thought to be organized as relatively undivided communities. Rural politics has come to be partly organized in terms of this opposition between the urban newcomers and existing farmers and labourers. Many of these themes have then been examined in a variety of different rural areas (see Bradley & Lowe (1984), and Pahl (1965)).

Conclusion

There is not the slightest doubt that the last decade or so has seen a growing overlap between the concerns of sociology and of at least parts of human geography. This is true both of the mainly empirical research considered in this chapter, as well as of the more theoretical material. In conclusion it is worth noting three areas in which future further constructive overlap is likely to occur.

First, there will clearly be extensive debate in the next few years about just what sort of society is evolving. It would seem that modern societies are no longer simply to be viewed as industrial and that changes are occurring of a broadly post-industrial sort. However, many of the specific formulations of this notion (see Touraine 1974) have been extensively criticized and attention is increasingly being devoted to the analysis of the nature of service industry and of service occupations (see Gershuny 1978, Gershuny & Miles 1983, Daniels 1985). This requires investigation of the forms of spatial reorganization in different service sectors (see Urry 1987); of the social relations characteristic of working within such service industries (see e.g. *Service Industries Journal*); and of the changing preferences of consumers for receiving a given service in a variety of different forms: individual/collective, public/private, marketed/non-marketed, socialized/privatized (see Saunders 1986, Ch. 8). More generally, it has been argued that modern industrial societies are increasingly 'disorganized'. The existing social and spatial fix of some major countries, based upon dominant manufacturing industry, politics structured by social class, the growing size of firms, the dominant rôle of large cities, and the central organizing rôle of national states, has begun to be reversed – a trend that will have major effects on social and political life in the late 20th century (see Offe 1985, Lash & Urry 1987).

Second, and connected to the last point, there is an increasing awareness that patterns of social life are the product of highly complex processes. They cannot simply be read off from either social or spatial structures on the one hand, or from the character of individual human action on the other. Essential to the analysis of such outcomes is the necessity of investigating a wide variety of forms of individual resistance and social struggles which in turn have to be related to the multiple forms of both social and spatial inequality systematically generated in modern societies. Also essential is an analysis of the complex modes by which such struggles are *transmuted* into socio-spatial outcomes, outcomes which are often unintended or even in contradiction to the apparent objectives of the individual or group in question. The social world is thus of immense complexity and opacity and still reveals little of its contradictory workings to the investigator.

Third, there is an increasing awareness of the irreducibly global character of contemporary social experience. This can be seen in a variety of aspects: the awesome dependence of human existence upon the relatively unpredictable decisions of the leaders of the two superpowers; the extraordinary importance of electronically transmitted information which enables geographically distant entities to be organizationally unified; the at least fragile growth of state organizations which transcend the individual nation–state; and the growth of means of mass communication which can simultaneously link 20–30 per cent of the world's population in a shared cultural experience. All these developments make the conventional study of sociology confined to a single society increasingly irrelevant, while no properly human geography could even get off the ground without taking such social–spatial transformations fully into account.

References

Abercrombie, N. & J. Urry 1983. *Capital, labour and the middle classes*. London: Allen & Unwin.

Abrams, P. 1968. *The origins of British society, 1834–1914*. Chicago: University of Chicago Press.

Blackburn, B. & M. Mann 1979. *The working class in the labour market*. London: Macmillan.

Bott, E. 1957. *Family and social network*. London: Tavistock Publications.

Bradley, T. & P. Lowe (eds) 1984. *Locality and rurality: economy and society in rural regions*. Norwich: GeoBooks.

Braverman, H. 1974. *Labor and monopoly capital*. New York: Monthly Review Press.

Buck, T. 1979. Regional class differences: an international study of capitalism. *International Journal of Urban and Regional Research* **3**, 516–26.

Buck, T. & M. Atkins 1978. Social class and spatial problems. *Town Planning Review* **49**, 209–21.

Burawoy, M. 1979. *Manufacturing consent*. Chicago: University of Chicago Press.

Calhoun, C. 1982. *The question of class struggle*. Oxford: Basil Blackwell.

Castells, M. 1978. *City, class and power*. London: Macmillan.

Castells, M. 1983. *The city and the grassroots*. London: Edward Arnold.

Castells, M. 1985. Commentary on C. G. Pickvance's 'The rise and fall of urban movements'. *Environment and Planning D, Society and Space* **3**, 55–61.

Cawson, A. & P. Saunders 1983. Corporatism, competitive politics and class struggle. In *Capital and the state*, R. King (ed.), 8–27. London: Routledge & Kegan Paul.

Cooke, P. 1983. *Theories of planning and spatial development*. London: Hutchinson.

Cooke, P. 1985. Class practices as regional markers: a contribution to labour geography. In *Social relations and spatial structures*, D. Gregory & J. Urry (eds), 213–41. London: Macmillan.

Cooke, P. 1986. Global restructuring. Local response. London: ESRC.

Crompton, R. & M. Mann (eds) 1986. *Gender and stratification*. Cambridge: Polity Press.

Cutler, A. 1978. The romance of 'Labour'. *Economy and Society* **7**, 74–95.

Daniels, P. 1985. *Service industries*. London: Methuen.

Dennis, N., F. Henriques & F. Slaughter 1956. *Coal is our life*. London: Eyre & Spottiswoode.

Dunleavy, P. 1980. The political implications of sectoral cleavages and the growth of state employment, Part 1. *Political Studies* **28**, 370–83.

Dunleavy, P. 1981. *The politics of mass housing, 1945–1975*. Oxford: Clarendon Press.

Dunleavy, P. 1982. The scope of urban studies in social science. Course D202, Units 3/4. Milton Keynes: Open University Press.

Durkheim, E. 1984. *The division of labour in society*. Cambridge: Polity Press.

Edwards, R. C. 1979. *Contested terrain*. London: Heinemann.

Elliot, B. & D. McCrone 1975. Landlords as urban managers: a dissenting opinion. In *Proceedings of the Conference on Urban Change and Conflict*, M. Harloe (ed.). London: Centre for Environmental Studies.

Ford, J. 1975. The role of the building society manager in the urban stratification system. *Urban Studies* **12**, 295–302.

Frankenberg, R. 1957. *Village on the border*. London: Cohen & West.

Frankenberg, R. 1966. *Communities in Britain*. Harmondsworth: Penguin.

Fröbel, F., J. Heinrichs & O. Kreye 1980. *The new international division of labour*. Cambridge: Cambridge University Press.

Gans, H. 1962. *The urban villagers*. New York: Free Press.

Gershuny, J. 1978. *After industrial society?* London: Macmillan.

Gershuny, J. & I. Miles 1983. *The new service economy*. London: Frances Pinter.

Giddens, A. 1985. Time, space and regionalisation. In *Social relations and spatial structures*, D. Gregory & J. Urry (eds), 265–95. London: Macmillan.

Goldthorpe, J. 1969. *The affluent worker in the class structure*. Cambridge: Cambridge University Press.

Goldthorpe, J. 1980. *Social mobility and class structure in modern Britain*. Oxford: Clarendon Press.

Gregory, D. & J. Urry (eds) 1985. *Social relations and spatial structures*. London: Macmillan.

Harloe, M., R. Issacharoff & R. Minns 1974. *The organization of housing*. London: Heinemann.

Harris, R. 1983. Space and class: a critique of Urry. *International Journal of Urban and Regional Research* **7**, 115–21.

Harvey, D. 1982. *The limits to capital*. Oxford: Basil Blackwell.

Institute of British Geographers, Women and Geography Study Group 1984. *Geography and gender: an introduction to feminist geography*. London: Hutchinson.

Kent, R. 1985. The emergence of the sociological survey, 1887–1939. In *Essays on the history of British sociological research*, M. Bulmer (ed.), 52–69. Cambridge: Cambridge University Press.

Lash, S. & J. Urry 1987. *The end of organized capitalism*. Cambridge: Polity Press.

Littler, C. 1982. *The development of the labour process in capitalist societies*. London: Heinemann.

Lowe, S. 1986. *Urban social movements: the city after Castells*. London: Macmillan.

MacIntyre, S. 1980. *Little Moscows*. London: Croom Helm.

Mark-Lawson, J. 1982. Social surveys 1880–1939: some problems in the construction of historical evidence. Unpublished PhD dissertation. University of Lancaster.

Marx, K. 1973. *Grundrisse*. Harmondsworth: Penguin.

Marx, K. & F. Engels 1888. *The Manifesto of the Communist Party*. Moscow: Foreign Languages Press.

Massey, D. 1984. *Spatial divisions of labour*. London: Macmillan.

Massey, D. & R. Meegan 1982. *The anatomy of job loss*. London: Methuen.

Mollenkopf, J. H. 1983. *The contested city*. Princeton, NJ: Princeton University Press.

Murgatroyd, L., M. Savage, D. Shapiro, J. Urry, S. Walby & A. Warde 1985. *Localities, class, gender*. London: Pion.

Newby, H. 1977. *The deferential worker*. London: Allen Lane.
Newby, H. 1979. *Green and pleasant land?* Harmondsworth: Penguin.
Norman, P. 1975. Managerialism: a review of recent work. In *Proceedings of the Conference on Urban Change and Conflict*, M. Harloe (ed.). London: Centre for Environmental Studies.
Offe, C. 1985. *Disorganized capitalism*. Cambridge: Polity Press.
Paci, M. 1981. Class structure in Italian society. In *Contemporary Italian society*, D. Pinto (ed.), 206–22. Cambridge: Cambridge University Press.
Pahl, R. 1965. *Urbs in rure*. London School of Economics, Geographical Papers 2. London: LSE.
Pahl, R. 1975. *Whose city?* Harmondsworth: Penguin.
Pahl, R. 1984. *Divisions of labour*. Oxford: Basil Blackwell.
Park, R. & E. Burgess 1967. *The city*. Chicago: University of Chicago Press.
Pickvance, C. 1985. The rise and fall of urban movements and the role of comparative analysis. *Environment and Planning D, Society and Space* **3**, 31–53.
Platt, J. 1971. *Social research in Bethnal Green*. London: Macmillan.
Quaini, M. 1982. *Geography and Marxism*. Oxford: Basil Blackwell.
Redfield, R. 1947. The folk society. *American Journal of Sociology* **52**, 293–308.
Rees, A. D. 1950. *Life in a Welsh countryside*. Cardiff: University of Wales Press.
Rex, J. & R. Moore 1969. *Race, community and conflict*. Oxford: Oxford University Press.
Robson, B. 1969. *Urban analysis*. Cambridge: Cambridge University Press.
Saunders, P. 1980. *Urban politics*. Harmondsworth: Penguin.
Saunders, P. 1986. *Social theory and the urban question*, 2nd edn. London: Hutchinson.
Sayer, R. A. 1985. Industry and space: a sympathetic critique of radical research. *Environment and Planning D, Society and Space* **3**, 3–30.
Scott, A. 1980. *The urban land nexus and the state*. London: Pion.
Scott, A. & M. Storper (eds) 1986. *Production, work, territory*. London: Allen & Unwin.
Simmel, G. 1971. *On individuality and social forms*. Chicago: Chicago University Press.
Smith, N. 1984. *Uneven development. Nature, capital and the production of space*. Oxford: Basil Blackwell.
Soja, E. 1985. The spatiality of social life: towards a transformative retheorisation. In *Social relations and spatial structures*, D. Gregory & J. Urry (eds), 90–127. London: Macmillan.
Storper, M. & R. Walker 1983. The theory of labour and the theory of location. *International Journal of Urban and Regional Research* **7**, 1–43.
Thrift, N. J. 1986. Localities in an international economy. ESRC Workshop, UWIST, Cardiff, September.
Thrift, N. J. 1987. The geography of late twentieth-century class formation. In *Class and space*, N. J. Thrift & P. Williams (eds), 207–53. London: Routledge & Kegan Paul.
Touraine, A. 1974. *The post industrial society*. London: Wildwood House.
Urry, J. 1981a. Localities, regions and social class. *International Journal of Urban and Regional Research* **5**, 455–74.
Urry, J. 1981b. Sociology as a parasite: some vices and virtues. In *Practice and progress*, P. Abrams (ed.), 25–38. London: Allen & Unwin.
Urry, J. 1983. Some notes on realism and the analysis of space. *International Journal of Urban and Regional Research* **7**, 22–7.
Urry, J. 1986. Class, space and disorganised capitalism. In *Politics, geography and social stratification*, K. Hoggart & E. Kofman (eds), 16–23. London: Croom Helm.
Urry, J. 1987. Some social and spatial aspects of services. *Environment and Planning D, Society and Space* **5**, 5–26.
Walby, S. 1986. *Patriarchy at work*. Cambridge: Polity Press.
Warde, A. 1985. Spatial change, politics and the division of labour. In *Social relations and spatial structures*, D. Gregory & J. Urry (eds), 190–212. London: Macmillan.

Warde, A. 1986. Space, class and voting in Britain. In *Politics, geography and social stratification*, K. Hoggart & E. Kofman (eds), 33–61. London: Croom Helm.
Weber, A. 1929. Theory of the location of industries. In *The writings of Alfred Weber*, C. J. Friedrich (ed.). Chicago: University of Chicago Press.
Weber, M. 1958. *The city*. Chicago: Free Press.
Williams, M. W. 1956. *The sociology of an English village: Gosforth*. London: Routledge & Kegan Paul.
Willmott, P. & M. Young 1960. *Family and class in a London suburb*. London: Routledge & Kegan Paul.
Wirth, L. 1938. Urbanism as a way of life. *American Journal of Sociology* **33**, 57–71.
Wood, S. (ed.) 1982. *The degradation of work*. London: Hutchinson.
Wright, E. O. 1978. *Class, crisis and the state*. London: New Left Books.
Young, M. & P. Willmott 1957. *Family and kinship in East London*. London: Routledge & Kegan Paul.

13 *Modern geography, Western Marxism, and the restructuring of critical social theory*

Edward Soja

After a prolonged period of virtual isolation and separate development, modern geography and Western Marxism crossed paths during the 1960s to begin what is now promising to become a mutually transformative encounter. At first, the connection between these two essentially 20th-century disciplines and discourses was peripheral and built primarily on a one-way flow of ideas. A distinctively *Marxist geography* took shape from an infusion of Western Marxist theory and method and formed part of a new critical human geography arising in the 1970s in response to the increasingly establishmentarian positivism of mainstream geographical analysis. The Marxist critique jostled some of the foundations of modern geography, but it remained inward-looking, unsettled in its critical stance, and largely unnoticed outside the disciplinary discourse.

Since 1980, however, the scope of the encounter between modern geography and Western Marxism has expanded and the flow of ideas has begun to move in both directions. A broader and deeper critical debate on the appropriate theorization of space – or what can be more concretely described as the *spatiality of social life* (Soja 1985a, 1985b; see also the entry on 'Spatiality' in Johnston *et al.* (1986)) – has been reaching into and challenging many of the long-established traditions of Western Marxism, while simultaneously training the conceptual and institutional frameworks of modern geography. Out of this expanding contemporary debate has come a compelling call for a significant reformulation of social theory based on a radical change in the ways we look at, conceptualize, and interpret not only space itself but the whole range of relationships between space, time, and social being; between human geography, history, and society.

In his recent writings, Harvey has given this call one of its most explicit and focused definitions:

> The historical geography of capitalism has to be the object of our theorizing, historico-geographical materialism the method of inquiry (1985, p. 144).

This objective is easier to state than accomplish, as Harvey is quick to note, but the statement is none the less of major importance. For the first time a powerful argument is developing around the need for an explicitly historical *and geographical* materialism, a far-reaching rethinking of basic Marxist principles that would allow the 'making of geography' to combine with the 'making of history' as the interactive presuppositions of social consciousness, social structure, and social action. In this argument, historical geography becomes much more than just an evolving outcome or product of human agency; or merely a conditioning framework of constraints and limitations. It is also the essential *medium* through which social relations are materialized, made concrete, constituted. From this provocative premise linking space, time, and being comes the claim that the historical geography of capitalism – appropriately defined to capture the interactive spatiality and historicity of social life – must become the 'object of our theorizing'.

Had the call for an historico-geographical materialism remained confined to the developing debate within Marxist geography, it would still deserve major attention here as an important wellspring of new models and alternative critical approaches to contemporary geographical analysis. But something else has been happening since 1980 which interferes with and complicates such a direct and focused discussion. There has been an unprecedented generalization of the debate on the appropriate theorization of space and time, geography and history, in social theory as well as in broader realms of critical discourse in art, architecture, literature, film, and popular culture. Today, the debate has become too important to be left only to those identified by the conventional labels of 'geographer' and 'Marxist'. Other critical observers of the contemporary scene must be drawn into the discussion.

In addition, it is becoming increasingly clear that the insertion of space into historical materialism and the wider frameworks of critical theory involves much more than a simple incremental adaptation, another new variable or model to be creatively assimilated into the old narrative forms. Critical theory and Western Marxism have been so muted with regard to space for so long that the inclusion of a theoretically meaningful spatial dimension may not be possible without shattering many well established interpretive assumptions and approaches, especially those associated with a deeply engrained primacy of historical versus geographical mode of explanation and critique. Similarly, modern geography has been so introverted and cocooned with respect to the construction of critical social theory and so confined in its definition of historical geography that it may be incapable of adjusting to the contemporary reassertion of space without a radical deconstruction and reconstitution. This double bind gives additional force to the claim that the encounter between Western Marxism and modern geography is promising to become mutally transformative.

At one level, this review chapter will present a straightforward history of the evolving encounter between modern geography and Western Marxism – a description of the formative tracks of Marxist geography and the sources of its contemporary turn towards a distinctively historico-geographical materialism. To do so will require reaching back to the turbulent period before and after the turn of the 19th century, the last *fin de siècle*, when the specialized foundations of modern geography and an equally modernized Western Marxism were firmly established and wrapped around a conceptualization of social theory which left little room for spatial thinking and geographical modes of explaining social phenomena. How this stranglehold of theoretical modernism was eventually broken through the contemporary reassertion of space in social theory and an emerging critique of *historicism* (defined here as the primacy of historical versus geographical explanation, the privileging of temporal over spatial forms of critical interpretation and social causality) will bring the discussion up to the present, at the edge of a new *fin de siècle*.[1]

There is another argument, however, which will cut across this story-line and recompose the historical narrative to focus on the particularities of the contemporary moment and the pressures for significant transformation being felt by both modern geography and Western Marxism. Here too the notions of *deconstruction* and *reconstitution* come to mind, especially as recently decribed by the English literary critic, Eagleton (1986, p. 80):

> To deconstruct . . . is to reinscribe and resituate meanings, events and objects within broader movements and structures; it is, so to speak, to reverse the imposing tapestry in order to expose in all its unglamorously dishevelled tangle the threads constituting the well-heeled image it presents to the world.

The discussion which follows will also attempt to 'reverse the imposing tapestry' presented in standard intellectual histories of modern geography and Western Marxism, and to reinscribe and resituate the debate on the theorization of space and time, geography and history, within broader movements and structures. Here too there will be a need to see the connections between the contemporary context and earlier periods of disruptive crises and transformative *restructuring* – in both the material conditions which define the historical geography of capitalism and in the theoretical frameworks constructed to explain and interpret these changing material conditions over time and space.

In the perspectives provided by recent analyses of the latest round of crisis and restructuring in the historical geography of capitalism – the empirical focus for most of the chapters in this book – we can also see the initial stirrings of a new and different interpretation of the contemporary moment. It is an interpretation that moves us from deconstruction alone to a potential reconstitution, to the possibility that what we are seeing and experiencing today is the birth of a *postmodern* critical social theory that effectively incorporates the best of modern geography and Western Marxism but transcends their confining disciplinary identities and traditions.[2] Keep this possibility in mind as you read through the following narrative, for it shapes the text from beginning to end.

Situating the encounter of modern geography and Western Marxism

Sequences of modernity, modernization, and modernism

In *All that is solid melts into air: the experience of modernity*, Berman (1982) explores the multiple reconfigurations of social life that have punctuated the historical geography of capitalism over the past 400 years. At the heart of his interpretation is an especially revealing comparison of the nature and experience of modernity around the end of the 19th century and again today, as we approach the next *fin de siècle*. Berman broadly defines modernity as 'a mode of vital experience', a collective sharing of a particularized sense of 'the self and others', of 'life's possibilities and perils'. In this definition, special place is given to the ways we think about and experience time and space, history and geography, events and localities, the immediate period and region in which we live. Modernity is thus comprised of both context and conjuncture, the *specificity* of being alive, in the world, at a particular time and place, a vital sense of what is *contemporary*. As such, it becomes a useful general term to capture the specific and changing meaning of the three most basic and formative dimensions of human existence: space, time, and being; the spatial, temporal, and social orders of human life.[3]

In the last *fin de siècle*, often stretched to include the years between 1880 and 1920 (or, to choose different turning points, between the defeat of the Paris Commune and the success of the Russian Revolution), the world changed dramatically. Industrial capitalism survived its predicted demise through a radical social and spatial restructuring which both intensified (deepened, as in the rise of corporate monopolies) and extensified (widened, as in the expansion of imperialism) its encompassing production relations and divisions of labour. Accompanying the rise of this 'different' political economy of capitalism was also an altered or restructured 'culture of time and space' (see Kern 1983), filled with ambitious visions and projects for changing the future. Each took shape from the shattered (deconstructed?) remains of an older order as the very nature and experience of modernity was significantly reconstituted.

Berman and others see in the contemporary period the onset of a similar transformative, if not necessarily revolutionary, process. As occurred roughly a century ago, there is now a complex dialectic developing between urgent socio-economic *modernization*, sparked by system-wide crises; and responsive cultural *modernism* aimed at making sense of, and gaining control over, the changes taking place in the material world. Modernization and modernism interact under these conditions of intensified crisis and restructuring in a tense and unsettled relation in which everything seems to be 'pregnant with its contrary', in which all that was once assumed to be solid 'melts into air', a description Berman borrows from Marx and represents as an essential feature of the vital experience of modernity in transition.

Modernization can be directly linked to the many different 'objective' processes of structural change associated with the ability of capitalism to develop and survive, to reproduce successfully its fundamental social relations of production and distinctive divisions of labour. These restructuring processes are continuous, but become especially critical and accelerated during periods of deep and systemic global crisis, such as the Age of Revolution between 1830 and 1848, the so-called Long Depression before the turn of the 19th century, the Great Depression between the two World Wars, and, it now seems abundantly clear, the period since the late 1960s.[4] Berman lists the key source shaping these periodically intensified modernizations of capitalism, each contributing toward the restructuring of modernity itself:

> the industrialization of production, which transforms scientific knowledge into technology, creates new environments and destroys old ones, speeds up the whole tempo of life, generates new forms of corporate power and class struggle; immense demographic upheavals, severing millions of people from their ancestral habitats, hurtling them halfway across the world into new lives; rapid and often cataclysmic urban growth; systems of mass communi-

> cations, dynamic in their development, enveloping and binding together the most diverse people and societies; increasingly powerful national states, bureaucratically structured and operated, constantly striving to expand their powers; mass social movements of people and peoples, challenging their political and economic rulers, striving to gain control over their lives; finally, bearing and driving all these people and institutions along, an ever-expanding, drastically fluctuating capitalist world market (Berman 1982, p. 16).

Here then is the awesome catalogue of forces which define the 'creative destruction' so closely associated with the survival of capitalism – and which Harvey has made so geographical in strategy and outcome in his notion of the search for a 'spatial fix' (Harvey 1982). It is also an effective description of the contemporary period of crisis-induced restructuring in the political economy and historical geography of capitalist development.

Modernism as oppposed to modernization has been less assiduously explored in Marxist geography. In general terms, modernism refers to the cultural, ideological, reflective and, I would add, theory-forming reactions to modernization and restructuring. It covers an immense variety of more subjective visions, values, and action programmes in art, literature, science, philosophy, and political practice which are unleashed by the disintegration of an inherited order and the projected possibilities of a restructured modernity. Modernism is thus the explicitly evaluative, culture-shaping and situated consciousness of modernity and is itself roughly able to be split into periods in conjunction with the historical rhythms of intensified capitalist crisis, restructuring, and modernization.

The last *fin de siècle* was a particularly fertile spawning ground for powerful new modernisms which began as avant-garde modern movements in almost every field of discourse and creativity. Many of these diverse modern movements not only survived a significant mid-century restructuring (through the Great Depression and World War II) but became entrenched and hegemonic in the postwar period, albeit faced with strengthening countercurrents challenging this hegemony. Looking back, Marxism–Leninism can be seen as one of these *fin de siècle* modern movements, a reinvigorating and avant-garde restructuring of historical materialism in theory and practice, a modernized Marxism. Its rigidification under Lenin's successors allowed it to survive a long series of crises, but also split the movement geographically into a narrowed traditional core and a more theoretically (if not practically) innovative periphery. The latter is what is characteristically defined as Western Marxism, removed enough from an increasingly orthodox Marxism–Leninism to be distinctive, but too close to represent an entirely autonomous modern movement of its own.

Arising in part as a reaction to the restructuring of Marxism was the consolidation of the 'Western', 'modern', or from a Marxist standpoint 'bourgeois' social sciences, including (but only just) what I have called modern geography (with stress on the latter's shared disciplinary origins in late 19th-century modernisms). The social sciences also developed an internal division through the 20th century between an increasingly orthodox (and hegemonic) core tradition based on an instrumental and largely positivist appropriation of natural science methods in social analysis and theorization; and a less rigidly compartmentalized and critical variant centred, ironically like Marxism, in historical modes of explanation and an emphasis on the power of human consciousness and social will.[5]

The broad movements and structures defined by this complex interplay of modernism and modernization provide an appropriately comprehensive framework within which to situate and trace the encounter between Western Marxism and modern geography. The roots of each of these two 20th-century modern movements are more clearly identified and positioned historically within the intellectual, political, and institutional struggles which rose in the late 19th century as competitive reinterpretations of how best to theorize, explain, and induce progressive changes in the modern social order. We can now begin to fill in this still rather skeletal framework by attaching the subsequent development of critical social theory (and, eventually, Marxist geography) to the particular conceptualizations of space–time–being, geography–history–society, which rose in those formative years of the last *fin de siècle*.

The subordination of space in social theory, 1880–1920

Kern, one of a growing number of specialists in what has come to be called the 'history of consciousness', describes the *fin de siècle* as a period which generated a distinctively different culture and consciousness of space and time from that which preceded it:

> From around 1880 to the outbreak of World War I a series of sweeping changes in technology and culture created distinctive new modes of thinking about and experiencing time and space. Technological innovations including the telephone, wireless telegraph, x-ray, cinema, bicycle, automobile, and airplane established the material foundation for this reorientation; independent cultural developments such as the stream-of-consciousness novel, psychoanalysis, Cubism, and the theory of relativity shaped consciousness directly. The result was a transformation of the dimensions of life and thought (Kern 1983, pp. 1–2).

What is not so clear from the writings of the historians of consciousness was the concurrent consolidation among theorists of society and social being of an interpretive prioritization of time over space, history over geography. Actually lived experience may not have induced such a prioritization, but at each level of philosophical and theoretical discourse, from ontology (how we define the nature and essence of being in the abstract) and epistemology (the study of how we know or accumulate reliable knowledge about the world) to the explanation of empirical events and the interpretation of specific social practices, the historical 'imagination' seemed to be annihilating the geographical.

This growing subordination of the spatiality of social life as a generative source of social theorizing and the associated tendency toward an *historicism of theoretical consciousness* was so complete by the end of this period that it was to remain virtually unseen and unquestioned for almost half a century afterwards. Historicism in various forms became the epistemological centrepiece for most of the newly formed social sciences (especially in their more liberal or critical variants) and in the modernized Marxism which also was consolidated and codified at the same time. In the wake of these two opposing modern movements, modern geography as it took shape was squeezed out of the competitive battleground of theory construction and the attempt to make sense of the dramatic changes affecting then contemporary society and social life. A few residual voices were heard, but the once much more central rôle of geographical analysis and explanation was reduced to little more than describing the stage-setting where the real social actors were 'making history'.

Social theorization thus came to be dominated by a narrowed and streamlined historical materialism, stripped of its more geographically sensitive revisionisms (e.g. the utopian and anarchist socialisms of Fourier, Proudhon, Kropotkin, Bakunin); and a set of compartmentalized social sciences, each becoming increasingly positivist, instrumental (in the sense of serving to 'improve' capitalism rather than to overthrow it), and, with a few exceptions, less attentive to the formative spatiality of social life. In at least one of the disciplinary orthodoxies which consolidated around the turn of the century, neoclassical economics, the subordination of space was so great that its most influential theoreticians proudly produced visions of a depoliticized economy that existed as if it were packed on to the head of a pin, in a fantasy world with virtually no spatial dimensions. Real history was also made to stand still in neoclassical economics, although time in the abstract was attended to as part of a generalized notion of causal process and sequential change – a form of temporalism rather than historicism rooted in natural science models of antecedent cause and subsequent effect.

For the most part, critical social theory (liberal and Marxist) came to be centred around interpretations of the temporal dynamics of modernization and modernism. Modernization was conceptualized in Marxian political economy first of all in the revolutionary transition from feudalism to capitalism. But there was a problematic transition also unfolding in the *fin de siècle*, a profound restructuring of the capitalist world which had to be understood theoretically and politically (considered much the same thing in Marxist epistemology). The interpretation of this transition came to be dominated by a Leninist vanguard responding to the rise of monopoly capital, corporate power, and the imperialist state. There was great sensitivity to geographical issues in the writings of Lenin, Luxemburg, and others, especially to the complexity and possible necessity of geographically uneven development (*cum* imperialism) in the survival of capitalism. None the less, *fin de siècle* Marxism remained solidly encased in historicism. The motor behind uneven development was still quintessentially historical, the making of history through the unfettering struggle of social classes. The geography of this process was seen primarily as an external constraint or an almost incidental outcome. History was the emotive 'variable container'. Geography, as Marx put it, was an 'unnecessary complication'.

In the social sciences, modernization was conceptualized around a somewhat similar historical rhythm initiated with the origins of capitalism and the Industrial Revolution, and moving through an epochal transition at the end of the 19th century. This latter transition was typically expressed as a grand and telling passage from tradition to modernity. For the major theoreticians, modernity had indeed arrived, for better and worse, and it too demanded to be understood as the dominant theoretical and political referrant. What Marxism–Leninism saw as the rise of imperialism the social sciences began to interpret as the time-lagged diffusion of development (*qua* capitalist modernity) to the undeveloped, traditional, not yet fully modernized parts of the world. Key theoreticians such as Weber and especially Durkheim paid attention to human geography and the geographically uneven development of society. But ultimately, geography remained an adjunct, a reflective mirror, an external backdrop to an essentially historical drama driven forward by individual and/or collective will and consciousness.

For both dominant streams, modernism represented a response to the question: what *now* is to be done to take control and direct (or redirect) the restructured modernization of capitalist society and its expansive diffusion to the rest of the world? Answers to this question in the *fin de siècle* would hold sway through the 20th century as hegemonic programmes for social progress, one rooted in revolutionary socialist strategy (reinforced in due course by events in Russia) and the other attached to the possibilities of planned liberal reform primarily under the aegis of the capitalist state (also reinforced by the successful reforms of the 'progressive era'). Both of these progressive modern movements were to be shaken by crisis and doubt in their separate spheres of influence. After World War II, however, the Chinese billion recharged the revolutionary socialist programme while the booming welfare states invigorated the liberal reform movement to the point that some proponents claimed complete conquest of crisis, the end of the need to restructure capitalism again.

The key argument I wish to establish in this broad picture of modernization and modernism is not only that spatiality was subordinated in social theory, but that the *instrumentality of space* was increasingly lost from view in political and practical discourse. The politics and ideology embedded in the social construction of human geographies and the centrally important rôle which the manipulation of these geographies played in the survival of capitalism seemed to become increasingly invisible and mystified left, right, and centre. Looking back precisely to this period, Lefebvre captured as well as anyone the essence of what was missing from explicit theoretical and political attention. His words are important for they also carry with them a critical social philosophy of space that would eventually shape the encounter between modern geography and Western Marxism and help pave the way to the reassertion of space in social theory:

> Capitalism has found itself able to attenuate (if not resolve) its contradictions for a century, and consequently, in the hundred years since the writing of *Capital*, it has succeeded in achieving 'growth.' We cannot calculate at what price, but we know the means: *by occupying space, by producing a space* (Lefebvre 1976, p. 21).

> If space has an air of neutrality and indifference with regard to its contents and thus seems to be 'purely' formal, the epitome of rational abstraction, it is precisely because it has already been the focus of past processes whose traces are not always evident on the landscape. Space has been shaped and moulded from historical and natural elements, but this has been a political process. Space is political and ideological. It is a product literally filled with ideology (Lefebvre 1976, p. 31).

The modernity which was shaped in the turn of the century was built around a 'spatial fix', the restructuring of an encompassing, encaging, exploitative, and illusive geographical landscape which increasingly impinged upon the routines of daily life and channelled those diverse processes which Berman so insightfully identified: the industrialization of production, demographic upheaval, rapid urban growth, the increasing power of the bureaucratically structured state, the challenge of mass social movements, the expansion of the capitalist world market.

This was not a sudden development, nor should it be viewed as conspiratorial, completely successful, or entirely unseen by those experiencing it. Many of the avant-garde movements of the *fin de siècle*, in poetry and painting, novel writing and literary criticism, architecture, and what then represented progressive urban and regional planning, perceptively sensed the instrumentality of space and the changing geography of capitalism. So too did the Marxist analysis of imperialism. But within the consolidating realms of social science and scientific socialism an obstructive veil was being pulled over the politics and ideology of space and, by the middle of the 20th century, the veil had become almost totally opaque.

Why this happened is not easy to answer. How it happened is just now being discovered and explored in any detail. Part of the story is probably related to the explicit theoretical rejection of environmentalism and all physical–external explanations of social processes and the formation of human consciousness, with human geography being thrown out with the bath water of environmental determinism. Another part of the story almost surely has to do with the modernist political strategies which were developing at this time. Those seeking the demise of capitalism, for example, tended to see in spatial consciousness and identity – in localisms, regionalism, nationalism – a dangerous fetter on the rise of a united world proletariat or a deviant and utopian 'fetishism' of territorial society, equally necessary to combat especially at the level of the capitalist state. Those seeking reformist solutions saw in similar phenomena an uncontrollable inefficiency, over-eager rebelliousness (when in opposition to national patriotism), and a possible threat to the expectantly benevolent power of the state and instrumental social 'science'. There is so much more which needs to be explored in this *fin de siècle* theoretical subordination and depoliticization of space. No definitive explanation of its origins yet exists for it has only become clearly visible in the past few years.

Interlude: the involution of modern geography

By the 1920s, the isolation of modern geography from the production of social theory was well advanced. For most of the next 40 years, geographical thinking turned inwards and seemed to erase even the memories of earlier engagements with the mainstreams of social theorization. Only the ghost of Kant effectively remained alive from that distanced past and its privileged apparition was used to lead the academic discipline of geography (although not all of its adherent geographers) into further isolation. After all, who better to wrap geography in a warm cocoon of intellectual legitimacy than the greatest philosopher in centuries who also professed to be a geographer of sorts?

Geography settled into a position within the modern academic division of labour which distinguished it (and rationalized its separation) from both the specialized and 'substantive' disciplines in the social sciences (where social theory was assumed to originate) and from history, its allegedly co-equal partner in filling up 'the entire circumference of our perception', as Kant put it. Geography and history were ways of thinking, schemata which co-ordinated and integrated all sensed phenomena. But by this time, putting phenomena in a temporal sequence (Kant's *nacheinander*) had become much more significant and revealing to social theorists of every stripe than putting them beside each other in space (Kant's *nebeneinander*). History and historians had taken on a critical interpretive rôle in modern social theory: an integrative and cross-disciplinary responsibility for the study of *development and change*, whether expressed in the biography of individuals, the explanation of particular events, or the modernization of whole societies. The historian as social critic and observer, history as a privileged interpretive perspective, became familar and accepted in academic and popular circles. In contrast, geography and geographers were left with little more than the detailed description of outcomes, what came to be described as the areal differentiation of phenomena.

The exceptional theoretical acquiescence of mid-century human geography has been frequently noted. Here and there, geographers individually contributed to theoretical debates in the social sciences and scientific socialism, drawing mainly on the continuing strength of physical geography and the occasional appeal of historians to limited environmental explanations of historical events. But the discipline as a whole turned inwards, abstaining from the great theoretical debates on the explanation of social-historical processes. It was almost as if a high wall had been raised between time, being, and society on one side, space on the other, keeping human geography and the social production of space apart from history, its making, and its makers.

Modern geography was reduced primarily to the accumulation, classification, and theoretically innocent representation of factual material describing the areal differentiation of the Earth's surface – what in more contemporary terms might be called the *outcomes* of geographically uneven development. Accurate packages of such information continued to be of use to the state, in the East and West, for military intelligence, economic planning, and imperial administration. This attachment to the state made political geography the most active source of what little spatial theorizing there was (Mackinder 1904, 1919), at least until the aberrant episode of German *Geopolitik* made non-fascist geographers think twice about venturing too far into the realms of political theory, state policy formation, and international relations.

Modern human geography took three prevailing forms. One emphasized the old environmental 'man–land' tradition and sought associations between physical and human geographies on the visible landscape. Another concentrated on the locational patterns of phenomena topically organized to reflect the compartmentalization of social science (e.g. economics, sociology, political science, etc., but *not* political economy). A third form aimed at synthesizing everything in sight through a comprehensive and typically encyclopaedic regionalization of phenomena, an approach considered by most geographers to be the distinctive essence of the discipline. What characterized all three of these analytical forms was the explanation of geographies *by* geographies, geographical analysis turned into itself, describing outcomes of processes whose deeper theorization was left to others.

While this involution occurred, the main currents of Western Marxism and liberal social science increasingly lost touch with geography and the geographical imagination, especially in the English-speaking world. A few residual pockets of insightful geographical analysis survived – in the evolutionary urban ecology of the Chicago School; in the urban and regional planning doctrines which consolidated in the interwar years (Friedmann & Weaver 1979, Weaver 1984); in the regional historiography and attention to environmental detail of the French Annales School and a few North American and British historians (see 'Annales School' and 'Frontier thesis' in Johnston *et al.* 1986); in the writings of Lenin, Luxemburg, Trotsky, and others on geographically uneven development (Smith 1984); in the work of Gramsci on the regional question, local social movements, and the state (Soja & Hadjimichalis 1979). But even in these pockets historical interpretations predominated over the geographical. In any case, each one was to dwindle in impact and importance through the Great Depression and the war years, so that by 1950 their specifically geographical insights were only dimly perceptible. At this point, with postwar recovery and economic expansion in full flow throughout the advanced capitalist (and socialist) world, the despatialization of social theory was at its peak.

Prelude: French connections within Western Marxism

There is no good historical geography of the despatialization of social theory up to the 1960s. The sounds of silence are difficult to pick up. There is available, however, in the works of Anderson (1976, 1980, 1983), an excellent critical survey of Western Marxism which almost inadvertently chronicles the loss of spatial consciousness in mainstream Marxism after the Russian Revolution and simultaneously sets the stage, again without necessarily intending to do so, for understanding how and why the pertinent spatiality of social life began to be rediscovered in the 1960s. Anderson's work can be used to 'reinscribe and resituate' the origins of the encounter between Western Marxism and modern geography.

From 1918 to 1968, Anderson argues, a new post-classical Marxist theory crystallized and redirected historical materialist interpretations of what I have called modernity, modernization, and modernism. This retheorization was unevenly developed geographically, finding its primary homelands in France, Italy, and Germany, 'societies where the labour movement was strong enough to pose a genuine revolutionary threat to capital' (Anderson 1983, p. 15). In Britain and the United States no such revolutionary challenge was apparent, while in the East a rigid Stalinist economism left little room for redirection and reinterpretation. For Anderson, the 'founding fathers' of this countercurrent were Lukacs, Korsch, and Gramsci, while following in their wake were such figures as Sartre and Althusser in France; Adorno, Benjamin, Marcuse and others associated with the Frankfurt School in Germany; and Della Volpe and Colletti in Italy.

This Latinate and Frankfurt-based movement shifted the institutional and intellectual terrain of Marxist theory, rooting it more than ever before in university departments and research centres, and in a resurgent interest in philosophical discourse, questions of method, a critique of bourgeois culture, and such subjects as art, aesthetics, and ideology (which were lodged in the classically neglected superstructural realms of capitalism). More traditional infrastructural themes having to do with the inner workings of the labour process, struggles at the workplace over the social relations of production, and the laws of motion of capitalist development tended to be given relatively less attention, along with more conventionally political topics as the movement of the world economy, the structure of the capitalist state, and the meaning and function of national identity.

To Anderson, Marxism seemed to be moving backwards, from economics through politics to philosophy, reversing the consummate path taken by Marx. But by the 1970s, this 'grand Western Marxist tradition' had 'run its course', in Anderson's words. It was replaced by 'another kind of Marxist culture, primarily oriented towards just those questions of an economic, social or political order that had been lacking from its predecessor' (Anderson 1983, p. 20). It also took on a different geography, becoming centred in the English-speaking world rather than in Germanic or Latin Europe. As a result, 'the traditionally most backward zones of the capitalist world, in Marxist culture, have suddenly become in many ways the most advanced' (Anderson 1983, p. 24).

There is something important hidden, however, in this historical picture of the development of Western Marxism: a primarily French debate on the theorization of space that would establish the foundation for the reconnection of Marxism and geography and eventually spread to shape the rising tide of anglophonic Marxist theory and analysis. It arose precisely from the Latinate interest in philosophical discourse and questions of method, for it sought to reconstruct the very foundations of historical materialism and the legacy of Marx, to open up Marxism to a materialist interpretation of spatiality to match its materialist interpretation of history. Sartre's 'search for a method' in his increasingly Marxist existentialism and Althusser's anti-historicist rereading of Marx were key steps in this direction, although their explicit assertion of space was weak and submerged under starkly opposed but equally 'totalizing' philosophical projects. One figure, more than any other, built critically upon the openings provided by both existential phenomenology and structuralism to take the debate on the theorization of space directly into the heart of Western Marxism. This was Henri Lefebvre.[6]

After virtually ignoring Lefebvre in his earlier works, Anderson, seemingly out of nowhere, discovers him *In the tracks of historical materialism* (1983) but remains perplexed about what to do with his discovery:

> No intellectual change is ever universal. At least one exception, of signal honour, stands out against the general shift of positions in these years. The oldest living survivor of the Western Marxist tradition I discussed, Henri Lefebvre, neither bent nor turned in his eighth decade, continued to produce imperturbable and original work on subjects typically ignored by much of the Left [Lefebvre, 1967 and 1974 are noted, both still untranslated into English]. The price of such constancy, however, was relative isolation (Anderson 1983, p. 30).

Little more is mentioned of Lefebvre by Anderson, who thereby misses the prefigurative source of Marxist geography, the first steps toward a genuinely historical and geographical materialism, and the archetype for the recent parade of prominent social theorists who have suddenly discovered the centrality of space, reconnected it to time and being, and recognized the necessity to reconstruct (sympathetically, for the most part) the very foundation of Marxist theory and method (Giddens 1979, 1981, 1984, Poulantzas 1978, Jameson 1984).

As Anderson notes, Lefebvre's 'imperturbable and original work' moved constantly against the grain of Western Marxism and isolated his influence. Lefebvre's early formulations concentrated on the 'mystification of consciousness' (Lefebvre & Guterman 1936) and what he called 'everyday life in the modern world', (Lefebvre 1946, 1961, 1968a). This combination was a clear reference to what he saw as the characteristic features of the capitalist modernity that consolidated around the *fin de siècle* and was forcefully sustained by a 'bureaucratic society of controlled consumption' centred in the modernized capitalist state. When Lefebvre is recognized at all by most Western Marxists, he is remembered for these relatively less deconstructive early formulations. But Lefebvre did not stop there.

In the 1960s, Lefebvre began reconstructing his ideas on capitalist modernity, focusing first on the urbanization of consciousness, the struggle for control over *le droit à la ville* (the right to the city), and the degree to which the transformation of capitalism would require an 'urban revolution' (Lefebvre 1968b, 1971). No Marxist scholar had ever before given such theoretical and political emphasis to the urbanization process and, through it, to the specifically spatial organization of capitalist societies. As Lefebvre kept trying to explain, his use of the terms urban and urbanism stretched well beyond the immediate confines of cities. It was a summative metaphor for what he would later and much more directly call *the social production of space* (1974). It was this process, and its rôle in the *reproduction* of social relations, which he argued lay at the strategic heart of the survival and growth of capitalism in the 20th century. The instrumentality of space, of the social construction and configuration of human geographies, was thereby brought back into view as an urgent focus for theoretical and political attention.

Lefebvre valiantly tried to argue that this was not an original discovery, that the same argument could be found in the combined writings of Hegel, Marx, and Lenin. But whatever its antecedents, Lefebvre's assertive theorization of the production of space represented a shattering challenge to the orthodoxies of Marxist thought and above all to its hegemonic historicism. This more than anything else tended to narrow and isolate Lefebvre's influence, for it seemed thoroughly indigestible to mainstream Western Marxism (particularly its most historicist anglophonic tradition). After the end of the Paris uprisings in 1968, which many saw as a political test of Lefebvre's ideas, even some of his closest followers began to abandon him and leap on to what were then the two great bandwagons of French Marxism, the recalcitrant structuralism of Althusser and the politicized existentialism of Sartre. Significantly, it was this post-1968 Lefebvre-you-have-failed-us mood that coloured the translation of Lefebvre's work into anglophonic Marxism. The major example was the influential Althusserian treatise, *The urban question*, written by Castells (1977), a former student of Lefebvre in Paris.[7]

All this blunted and distorted Lefebvre's ideas, which he continued to develop 'neither bent nor turned in his eighth decade'. There is so much to capture and summarize in the more than a dozen major books he has published since 1970, but one passage, from one of the very few of his works translated into English, states his ultimate argument most emphatically and directly:

> The dialectic is back on the agenda. But it is no longer Marx's dialectic, just as Marx's was no longer Hegel's. . . . The dialectic today no longer clings to historicity and historical time, or to a temporal mechanism such as 'thesis–antithesis–synthesis'. . . . To recognise space, to recognise what 'takes place' there and what it is used for, is to resume the dialectic; analysis will reveal the contradictions of space (Lefebvre 1976, pp. 14, 17).

From the 1960s to the present, the encounter between modern geography and Western Marxism has revolved around – and finally landed upon – this configured dialectic, this insistent demand to change the ways in which we think about the relations between space, time, and being; geography, history, and society; the production of space, the making of history, and the constitution of social relations.

Marxist geography and the restructuring of social theory

Modernity began to come apart in the 1960s, initiating significant shifts and turns in the nature and interpretation of modernization and modernism. The long postwar boom ended in explosions of the marginalized – blacks and other ethnic minorities, women, students, the unemployed, and rebellious territorial communities in urban neighbourhoods, economically depressed regions, and superexploited Third World states, all expressing discontent with their subordinate position within the modern order (Lefebvre 1976). Almost simultaneously, the world economy edged toward what Mandel (1978) called the 'second slump', the second global crisis since the *fin de siècle* origins of the modern era. Looking back, there seems no doubt that during this time the world entered another period of comprehensive *restructuring*, filled with perplexing conflicts between the old and the new, between 'inherited' and 'projected' orders. This contemporary restructuring process, which continues still, forms the material backdrop to a concurrent reconfiguration of social theory and the associated development of Marxist geography through three overlapping phases.

Adding Marx to modern geography: the first critique

The origins of Marxist geography hinged upon a reconnection of spatial form (e.g. the apparent outcomes of geographically uneven development) to social processes, to the organizational structures and practices which constitute human societies. This reconnection began to be asserted in principle during the late 1950s, when the so-called 'quantitative-theoretical revolution' arose from within modern geography's introverted and essentially atheoretical cocoon. This increasingly technical and mathematically constituted version of geographical description, however, differed only superficially from Kantian tradition and was eventually reabsorbed without a major internal transformation. This 'advanced' form of modern geography (perhaps late modern would be the appropriate term) grounded explanation primarily in social physics, statistical ecologies, and narrow appeals to the ubiquitous friction of distance. But after all was said and done, outcomes continued to explain outcomes in an infinite regression of geographies upon geographies, one set of mappable variables 'explaining' another. The adopted positivist cloak merely relegitimated geography's fixation on empirical appearances and involuted description.

But a crack had opened as geographers collectively became more aware of their isolation and began to search for new attachments outside their old cocoon. What they found 'outside' in the 1960s was a material world increasingly torn by conflict, crisis, and almost inexplicable change; an academic environment which was more politicized and acceptably 'radical' than it had been for generations; and a theoretical discourse which had turned significantly against positivism and was exploring critical alternatives drawn from the 'Grand Houses' of European social theory. In this context of incipient restructuring, parts of geography also began to radicalize in the late 1960s, led by contributors to the new journal *Antipode* and inspired by a series of leftward turns taken by some of the most prominent anglophonic geographers of the time. Harvey's dramatic shift in direction from the positivist ecumenicism of *Explanation in geography* (1969) to the avowedly Marxist *Social justice and the city* (1973) was particularly pathbreaking and influential, especially to the generation of young geographers so recently taught to pay close attention to Harvey's work.

Although more heterogenous at first, radical geography rapidly moved toward a dedicated 'Marxification' of geographical analysis, the insertion of Marxist principles and concepts into the late modern disciplinary cocoon. Historical materialism became the preferred route to connect spatial form with social process, human geography with class analysis, the description of geographical outcomes with the explanatory power of Marxian political economy. One by one, the familiar objects of geographical description were subjected to a Marxist analysis and interpretation: the patterns of urban land rent and land use, the variegated forms of the built environment, the location of industry and transport routes, the evolution and ecology of urbanization, the functional hierarchy of cities, the mosaic of uneven regional development, the diffusion of innovations, the formation of cognitive or 'mental' maps, the inequalities in the wealth of nations, the formation and transformation of geographical landscapes.

What was being inserted and applied had several important features. At the core was a radical political economy based essentially in Marx's *Capital*. But carried along with it were three more contemporary (and often complexly intertwined) schools of interpretation. Within the rising tide of anglophonic Marxism described by Anderson there was both a powerful (largely British) historicism averse to speculative theorizing and attached to pragmatic empirical analysis; as well as a brash and venturesome (mainly North American) 'neo-Marxism' springing from a presumed necessity to update Marxian principles and drawing from more unconventional sources of insight. Virtually incomprehensible to the former but provocatively inspirational to the latter was a third stream of ideas from an ebbing but still influential French Marxist tradition dominated by Althusserian structuralism (but also including a few Lefebvrian echoes).

The structuralist 'reading' was particularly attractive to Marxist geography. As an apparently rigorous epistemological rationalization for digging under the surface appearance of phenomena (spatial outcomes) to discover explanatory roots in the structured and structuring social relations of production, the structuralist perspective fits very well into the formative project of marxifying geographical analysis. It was also instructively anti-positivist; its anti-humanistic stance was an effective tool against the growing influence of behavioural and phenomenological geographies (which often presented themselves as anti-Marxist); and it opened up the superstructural realm which seemed to contain so much of what geographers looked at. Above all else, structuralism's programmatic anti-historicism (accompanied by its compelling spatial metaphors) allowed more room for geography – indeed beckoned it – to enter into mainstream theoretical debates, to make synchronic versus diachronic analysis essentially spatial. Most of the alternative perspectives seemed almost anti-spatial in contrast, arising as they did from a long tradition of intellectual apartheid between geographical and socio-political analysis. Although it was often disguised and rarely explicit, a structuralist epistemology was infused through the early development of Marxist geography.

Two scales of analysis and theorization dominated the initial conjunction of Marxian political economy and critical human geography. Marxist geographers contributed significantly to the formation of a specifically *urban* political economy, building heavily upon Harvey's increasingly formalized marxification of urban geography and Castells's monumental adaptation of Althusser, Lefebvre, and Touraine, a French sociologist and theorist of 'social movements'. In an almost entirely separate sphere, the analysis of more global patterns of geographically uneven development (and underdevelopment) produced a new and increasingly spatial political economy of the international division of labour and the capitalist 'world system' of cores and peripheries. This new *international* political economy was based primarily on neo-Marxist critiques of conventional theories of development and modernization raised by specialists working not in the advanced capitalist cities but in the contexts of the peripheralized Third World. Its most influential exponents (e.g. Frank, Wallerstein, and Amin) included very few Marxist geographers and, although neo-Marxist development theory incorporated a significant spatial emphasis, there was little explicit debate on the theorization of space – in marked contrast to urban political economy and sociology, where such debate abounded from the very beginning.

The new political economies of urbanization and international development attracted many adherents, but it was not long before the peculiar mixture of perspectives created serious epistemological problems, especially with regard to the theorization of space and spatiality. Marxist geography teetered uncomfortably between a pragmatic and anti-speculative historicism (which traditionally rejected explicitly geographical explanation and was equally uncomfortable with what many saw as an excessive new emphasis on consumption and exchange versus production); and a neo-Marxist structuralism (which all too easily seemed to breed determinisms, annihilate the politically conscious subject, and expel the theoretical primacy of historical explanation). On the one hand, it pursued the relatively 'safe' project of inserting Western Marxism into modern geography; but on the other, it was beginning to look in the other direction, toward the need to spatialize historical materialism, to bring critical human geography into the interpretive core of Western Marxism. This was another matter altogether. Amidst great confusion, a second phase in the development of Marxist geography had begun.

The provocative inversion: adding a geography to Western Marxism

The marxification of geographical analysis continued through the 1970s and a similar, often intertwined, application of Marxist principles, concepts, and methods also took place within the fields of urban, regional, and international development theory and planning.[8] But by the end of the decade a fractious debate had arisen over the difference that space makes to the materialist interpretation of history, the critique of capitalist development, and the politics of socialist reconstruction. Arguments spun back and forth between those who sought a more flexible and dialectical relation between space and society, geographical and historical materialisms (Soja & Hadjimichalis 1979, Soja 1980, Peet 1981); and those who saw in this effort a theoretical degeneracy, a dithering radical eclecticism, a politically dangerous and divisive spatial fetishism irreconcilable with class analysis and historical materialism itself (Anderson 1980, Eliot Hurst 1980, Smith 1979, 1980, 1981). To some sympathetic outside observers, Marxist geography seemed to be destroying itself from within, leading one to argue 'why geography cannot be marxist' (Eyles 1981) and another to lament what he saw as a premature abandonment of spatial explanation in radical geographical analysis (Gregory 1981).

A growing movement developed in Marxist geography as well as within urban and regional studies which appeared to be concluding that space and spatiality could fit into Marxism only as a reflective expression, a product of the more fundamental social relations of production and aspatial (but preeminently historical) 'laws of motion' of capital (Walker 1978, Massey 1978, Markusen 1978). The geography of capitalism was a worthy subject, geographically uneven development was an interesting outcome of the history of capitalism, but neither was a *necessary* ingredient in Marxist theory nor an inherent part of the core principles of historical materialist analysis. This rekindling of Marxism orthodoxy was reinforced by a broader-based 'critique of the critique' in Western Marxism, denouncing the theoretical inadequacies and overinterpretations of the Althusserian devil and his neo-Marxist attendants. A symbolic banner for this attack was provided by the title of Thompson's passionate anti-structuralist reassertion of the primacy of history, *The poverty of theory* (1978; see Brenner (1977) for an incisive and direct critique of neo-Marxist development theory from a similar perspective).

Despite the power of this reassertive historicism, the project of spatializing Marxist theory remained alive. Its main motivating expression was the increasingly forceful argument that the organization of space – the forming of spatial structures, the unevenly developed geographical landscape – was not only a social product but also rebounded to shape social relations, social practices, social struggle. In *Ideology, science, and human geography*, perhaps the decade's most insightful and comprehensive reinterpretation of geographical explanation, Gregory put the argument this way:

> The analysis of spatial structure is not derivative and secondary to the analysis of social structure . . . rather, each requires the other. Spatial structure is not,

therefore, merely the arena within which class conflicts express themselves but also the domain within which – and in part through which – class relations are constituted (Gregory 1978, p. 120).

There was thus a complex and problem-filled dialectic between the production of human geographies and the constitution of social relations which needed to be recognized and opened up for theoretical and political interpretation. To do so meant that the construction of human geographies, the produced spatiality of social life, needed to be seen as simultaneously contingent *and* conditioning, as both an outcome *and* a medium of the making of history, as part of an historical *and* geographical materialism (versus an historical materialism applied to geographical questions).

This provocative inversion of social causality, this call for a recombination of history and geography, of space, time, and being as *intercontingent* relationships was extraordinarily difficult to accept within both modern geography and Western Marxism. The former habitually avoided making the spatial contingency of society and behaviour too explicit (except via the physical force of the friction of distance), for fear of promoting another form of environmental determinism. The latter, especially at a time of reassertive orthodoxy, tended to see again only another attempt to impose an external constraint upon the freedoms of class consciousness and social struggle in the making of history. Just as important, the 'provocative inversion' had not yet been backed by a more formal theoretical elaboration or by a body of cogent empirical analysis. Between *Social justice and the city* in 1973 and *Limits to capital* in 1982, there were very few major books produced by Marxist geographers and almost nothing written in English that systematically and explicitly advanced the call for an historico-geographical materialism.[9] It is no wonder then that well into the 1980s Marxist geography continued to dance a tedious gavotte around the appropriate theorization of space, spinning and turning to avoid what was eventually to become insistently obvious: that Marxism itself had to be critically restructured to incorporate a salient and central spatial dimension. The inherited orthodoxies of historical materialism left almost as little room for space as the rigid cocoons of bourgeois social science. Making geography Marxist was thus not enough. Another round of critical rethinking was necessary to *spatialize Marxism*, to recombine the making of history with the making of geography, a 'critique of the critique of the critique' which would usher in a new phase in the encounter between modern geography and Western Marxism.

Deconstruction and reconstitution: passages to postmodernity?

Bringing up to date the development of historico-geographical materialism and the attendant retheorization of the intercontingency of space, time, and being is a necessarily eclectic exercise. It now seems clear that geographical modes of analysis, explanation, and social criticism have become more centrally attached to contemporary political and theoretical debates than at any other time in this century. But the attachment derives from many different sources, takes a variety of forms, and resists easy synthesis. It is also still very tentative and limited in its impact, for the spatialization of social theory has only just begun and its initial impact has been highly disruptive. Harvey acutely observed the disintegrative effects of spatialization, especially with respect to the discipline of sociology and the defensive sociologism of Saunders' otherwise insightful survey, *Social theory and the urban question* (Saunders 1981):

> the insertion of concepts of space and space relations, of place, locale, and milieu, into any of the various supposedly powerful but spaceless social theoretical formulations has the awkward habit of paralyzing that theory's central propositions. . . . Whenever social theorists actively interrogate the meaning of geographical and spatial categories, either they are forced to so many ad hoc adjustments that their theory splinters into incoherence or they are forced to rework very basic propositions. Small wonder, then, that Saunders (1981, p. 278), in a recent attempt to save the supposed subdiscipline of urban sociology from such an ugly fate, offers the extraordinary proposition, for which no justification is or ever could be found, that 'the problems of space . . . must be severed from concern with specific social processes' (Harvey 1985c, p. xiii).

As Harvey also notes, the same can be said for Marxist theory more generally and, by extension and with some irony, for Marxist geography as well. In the contemporary restructuring of social theory, Marxist geography has been losing its identity and distinctiveness as the spatializing project expands well beyond the conjunction of Marxian political economy and critical human geography. And in many cases, the newcomers are even bolder in their assertions of the *contemporary* significance of spatiality than are Marxist geographers. Here, for example, is a passage written by the English cultural critic, Berger, published in 1974:

> We hear a lot about the crisis of the modern novel. What this involves, fundamentally, is a change in the mode of narration. It is scarcely any longer possible to tell a straight story sequentially unfolding in time. And this is because we are too aware of what is continually traversing the story-line laterally. That is to say, instead of being aware of it as an infinitely small part of a straight line, we are aware of it as an infinitely small part of an infinite number of lines, as the centre of a star of lines. . . .
>
> Such awareness is the result of our constantly having to take into account the *simultaneity and extension of events and possibilities*. . . .
>
> There are many reasons why this should be so: the range of modern means of communication: the scale of modern power: the degree of personal political responsibility that must be accepted for events all over the world: the fact that the world has become indivisible: the unevenness of economic development within that world; the scale of the exploitation. All these play a part. *Prophesy now involves a geographical rather than historical projection; it is space, not time, that hides consequences from us*. . . . Any contemporary narrative which ignores the urgency of this dimension is incomplete and acquires the oversimplified character of a fable (Berger 1974, p. 40; emphasis added).

The parallels with Lefebvre and Berman are interesting. Similar echoes can be found in the more recent writings of the leading American Marxist literary critic, Jameson, in an important article on 'Postmodernism, or the cultural logic of late capitalism':

> postmodern (or multinational) space is not merely a cultural ideology or fantasy, but has genuine historical (and socio-economic) reality as a third great original expansion of capitalism around the globe (after the earlier expansions of the national market and the older imperialist system, which each had their own cultural specificity and generated new types of space appropriate to their dynamics). . . . We cannot [therefore] return to aesthetic practices elaborated on the basis of historical situations and dilemmas which are no longer ours . . . *the conception of space that has been developed here suggests that a model of political culture appropriate to our own situation will necessarily have to raise spatial issues as its fundamental organizing concern* (Jameson 1984, pp. 88–9; emphasis added).

John Urry, a sociologist whose earlier work on theoretical realism (Keat & Urry 1981) provided an insightful pathway to the current spatialization of social theory, joins the growing chorus. He begins by criticizing the very influential work of another 'newcomer' to the spatializing project, Giddens (1979, 1981; see also Soja 1983):[10]

> [Giddens] deals relatively little with the theorization of space; in particular [he] tends to neglect the problems of explaining the causes and consequences of recent transformations in the spatial structuring of late capitalism. Moreover, . . . this omission is particularly serious since *it is space rather than time which is the distinctively significant dimension of contemporary capitalism*, both in terms of its most salient processes and in terms of a more general social consciousness. As the historian of the *longue durée*, Braudel, argues 'All the social sciences must make room for an increasingly geographical conception of mankind' (Urry 1985, p. 21; emphasis added).

Two more observations, written in the late 1970s by two of the most influential European Marxist theorists of late capitalism, are indicative of the growing importance attached to space not only in sociology and cultural criticism but within the contemporary mainstream of Marxian political economy. In each case, the theoretical challenge presented has tended to be neglected by most Marxists unable to contend with its unsettling implications. The first is from Mandel, who more than any other contemporary figure has shaped our understanding of the periodization and *regionalization* of capitalist development:

> The unequal development between regions and nations is the very essence of capitalism, on the same level as the exploitations of labour by capital (Mandel 1976, p. 43).

This is no casual aside, for Mandel's work has systematically and effectively elaborated upon the crucial rôle of *geographically uneven development* in the historical survival of capitalism (Mandel 1974, 1980). Jameson's work on the cultural logic of postmodernism draws directly from Mandel and much of the current research on economic restructuring, regional political economy, and changing spatial divisions of labour adapts some variant of Mandel's periodization–regionalization (Massey 1984, Smith 1984, Soja *et al.* 1983, Soja 1985b).

The quest for a materialist historical geography also became central in the last major book published by Poulantzas before his death. In *State, power, socialism* (1978), Poulantzas presents a political theorization of the origins and contemporary nature of the capitalist state which rests upon the interacting spatial and temporal matrices of social life. In his depiction of the spatial matrix, he carries forward Lefebvre's key arguments about the distinctive but contradictory features of capitalist spatiality: its tendency toward both fragmentation and homogenization, differentiation and equalization, the essential problematic of geographically uneven development that becomes a primary area for the activity of the capitalist state (see also Hadjimichalis 1986, Soja 1985a, 1985b, Smith 1984):

> Separation and division in order to unify; parcelling out in order to structure; atomization in order to encompass; segmentation in order to totalize; closure in order to homogenize; and individualization in order to obliterate differences and otherness. The roots of totalitarianism are inscribed in the spatial matrix concretized by the modern nation–state – a matrix that is already present in its relations of production and in the capitalist division of labour (Poulantzas 1978, p. 107).

Poulantzas goes on to argue that these spatial and temporal matrices, concretized as historical geographies, are both *presuppositions* and *embodiments* (a 'primal material framework') of the most fundamental properties of capitalist society. They may be examined as 'outcomes' but they have a much deeper meaning:

> For its part, Marxist research has up to now . . . considered that transformations of space and time essentially concern ways of thinking: it assigns a marginal role to such changes on the grounds that they belong to the ideological–cultural domain – to the manner in which societies *represent* space and time. In reality, however, transformations of the spatio-temporal matrices refer to the materiality of the social division of labour, of the structure of the State, and of the practices and techniques of capitalist economic, political and ideological power; they are the *real substratum* of mythical, religious, philosophical or 'experiential' representations of space and time (Poulantzas 1978, p. 98).

This collection of observations from outside a narrowly bounded Marxist geography is indicative both of the extent to which the reassertion of space in social theory has captured a much wider range of contributors over the past decade; and of the growing convergence among very diverse points of view round the *power-filled instrumentality of space in the contemporary restructuring of social life*, of capitalist modernity. With this rising chorus of outsiders pushing forward a deconstruction and reconstitution of historical materialism, even the most adamant critics of earlier attempts to spatialize Marxism have turned around in the 1980s. In *Spatial divisions of labour*, for example, Massey describes her path through the 'provocative inversion':

> My basic aim had been to link the geography of industry and employment to the wider, and underlying structures of society . . . . The initial intention, in other words, was to start from the characteristics of economy and society, and proceed to explain their geography. But the more I got involved in the subject, the more it seemed that the process was not just one way. It is also the case – I would argue – that understanding geographical organisation is fundamental to understanding an economy and society. The geography of society makes a difference in the way it works (Massey 1984, p. x).

Smith similarly strays – ever so fleetingly – off the path of Marxist formalization with the following observation from *Uneven development* (with its Lefebvre-like subtitle, 'Nature, capital and the production of space'):

> Occupying the common ground between the geographical and political traditions, a theory of uneven development provides the major key to determining what characterizes the specific geography of capitalism. . . . But one cannot probe too far into the logic of uneven development without realizing that something more profound is at stake. It is not just a question of what capitalism does to geography but rather of what geography can do for capitalism. . . . From the marxist point of view, therefore, it is not just a question of extending the depth and jurisdiction of marxist theory, but of pioneering a whole new facet of explanation concerning the survival of capitalism in the twentieth century (Smith 1984, p. xi).

The most dramatic turnabout again appears to come from Harvey, who incongruously combined in *The limits to capital* (1982) the crowning achievement of the formalistic marxification of geography and the opening salvos for the necessary deconstruction and reconstitution of this very achievement. Looking back, he is understandably concerned that most readers seemed to be somewhat confused over the message being presented in the deliberate double-meaning of the term 'limits':

> curiously, most reviewers passed by (mainly, I suspect, out of pure disciplinary prejudice) what I thought to be the most singular contribution of that work – the integration of the production of space and spatial configurations as an active element within the core of Marxian theorizing. That was the key theoretical innovation that allowed me to shift from thinking about history to historical geography and so to open the way to theorizing about the urban process as an active moment in the historical geography of class struggle and capital accumulation (Harvey 1985c, p. xii).

He also becomes more explicit in his recent work about the historicism so deeply embedded in Western Marxism and the difficulties this presents for the development of historico-geographical materialism:

> Marx, Marshall, Weber, and Durkheim all have this in common: they prioritize time and history over space and geography and, where they treat of

> the latter at all, tend to view them unproblematically as the stable context or site for historical action. . . . The way in which the space relations and the geographical configurations are produced in the first place passes, for the most part, unremarked, ignored. . . . Marx frequently admits of the significance of space and place in his writings . . . [but] geographical variation [i.e. uneven development] is excluded as an 'unnecessary complication'. His political vision and his theory are, I conclude, undermined by his failure to build a systematic and distinctively geographical and spatial dimension into his thought (Harvey 1985a, pp. 141–3)

Yet there is a lingering ambivalence (or perhaps caution?) in Harvey's work and that of other Marxist geographers who seem to share the same project of constructing an appropriately historical and geographical materialism. The critical spatial silence in Marxist theory is recognized, as is the exaggerated primacy of historical explanation (what I have defined as historicism). There is also almost unanimous agreement that broader cultural variables need to be incorporated into a spatialized Marxian political economy. Harvey goes as far as to describe himself as an urban *flaneur*, a 'restless analyst' of the 'unpredictable collision of experiences and imagination' in his recent analysis of the 'urbanization of consciousness' (1985c). But there is a residual conservativeness as well, an unwillingness to engage in too deep a deconstruction of historical materialism and the grand narrative forms of Marxist analysis, lest their fundamental power be drained away in a faddish new eclectism.

Harvey defines the half-way house of what might be called late modern Marxist geography by making the grand narrative of Marxist theory off limits to deconstruction and reconstitution:

> Geographical space is always the realm of the concrete and the particular. Is it possible to construct a theory of the concrete and the particular in the context of the universal and abstract determinations of Marx's theory of capitalist accumulation? This is the fundamental question to be resolved (Harvey 1985a, p. 144).

This assertive formulation of the project of historico-geographical materialism tends to deflect the need critically to rethink Marx's 'universal and abstract determinations' and also concentrates the 'difference that space makes' (Sayer 1985) solely in the realm of the concrete, the particular, the empirical. The overarching theorization is assumed to be in order with the adjectival injection of 'geographical' into historical materialism. The contemporary challenge thus becomes one of demonstration rather than deconstruction, a flexing of strengthened theoretical muscles to show the power of a newly spatialized Marxism in contending with the concrete and the particular both in the past (e.g. 19th-century Paris (Harvey 1985c)) and in the present (e.g. the rise of 'flexible accumulation' and the 'passage to postmodernity' (Harvey 1987)).

Other contemporary descendants of Marxist geography, often pushed to disclaiming the Marxist label, set no such limits to the potential theoretical repercussions arising from the new empirics and its associated search for a mesolevel analysis which mediates between universal and abstract determinations and the concrete and particular (Scott & Storper 1986). The late modern stance, with its protected Marxian narrative, is rejected in favour of a freer, more flexible and eclectic exploration of the empirical dynamics of restructuring, in particular the dramatic changes which have been taking place over the past 20 years in the technology, corporate organization, global distribution, and labourmarket impact of capitalist industrial production. This empirical research arena is shared with the late modernists, but the approaches taken are more diverse, less constrained by disciplinary orthodoxies, and frequently couched in a rhetoric which is deeply critical of inherited forms of Marxist geographical analysis.

However this current clash of perspectives works itself out, it has produced a new zest for detailed empirical research centring on the dynamics of contemporary restructuring processes, the most recent round of modernization in the changing political economy of capitalism. Simultaneously, it has revived interest in industrial production and the labour process, in marked contrast to the focus on issues of consumption and distribution which dominated postwar urban, regional, and international development research. And it has also contributed to a reinvigorated *regional geography* first through the development of a regional political economy (which was missing in the earlier phases of Marxist geography) and more recently through the use of regional synthesis to tie together the multiple levels of restructuring: of the built environment and the spatiality of everyday life; of localities and labour labourmarkets; of urban form and metropolitan spatial structures; of the mosaic of uneven regional development and the rôle of the national state; of international finance and the global organization of capital and industrial production; of explicitly spatial divisions of labour at every geographical scale, from the household to the world economy. Infused throughout this empirical research has been a growing attention to issues and emphases which had been stubbornly neglected in the past, especially the social constitution and spatio-temporal structuring of *gender* relations and what, in conjunction with the social production of space, can be called the social production of *nature*. No detailed references need be given here, for they appear in the specified contexts of the other chapters in this book.

But what of the larger theoretical challenge that I have thematically traced through the encounter between modern geography and Western Marxism? Is the reassertion of space in critical social theory being sidetracked by the pronounced empirical turn? Or may this be the most fruitful pathway to rigorous theoretical restructuring? Which are the most important organizing themes for the retheorization of space, time, and being? Historico-geographical materialism as defined by Harvey? The space–time structuration theory of Giddens? The periodizing and regionalizing 'long waves' of Mandel? The changing 'cultural logic' of modernity and postmodernity described by Berman and insightfully translated into the present by Jameson? The new spatial dialectic outlined by Lefebvre? The meso-level analytics of industrial restructuring, changing spatial divisions of labour, and geographically uneven development? The forceful connection between space, time, and *power* that figures so prominently in Poulantzas's analysis of politics and the state? The related 'spatial analytics' of power devised by Foucault and only now being perceptively dug up by geographers (see Gregory (1989) and this volume)?

What a rich tableau of possibilities for the approaching *fin de siècle*! If there is more to be said in conclusion, it must be to urge an openness and flexibility of creative vision, a resistance to paradigmatic closure and rigidly categorical thinking, a capacity to *combine* theoretical perspectives rather than prune and prioritize them. Flexible specialization is the pervasive theme of the contemporary period, whether it be in the technology, and new organizational forms of a post-Fordist political economy, the cultural logic of postmodernism in art and ideology, or the reassertion of space in a post-historicist critical social theory.

In each of these *three restructurings*, new modes of thinking about and experiencing space and time are at the centre of the emerging changes and are demanding new forms of understanding and interpretation that depart significantly from the hegemonic modernisms which have prevailed throughout most of the 20th century. This is why we must move beyond the disciplinary rigidity of a modernist Marxist geography still clinging to theoretical and political practices which have been elaborated around conditions and dilemmas which may no longer exist – and to a grand narrative unable to contend with the possibility that today it is space and geography, much more than time and history, that hide things from us.

But there is another clash of perspectives developing *within* what might be called the postmodern project, one which revolves around the political implications of postmodernization and postmodernism. A powerful neoconservative variant of flexible specialization and postmodern reconstruction is currently on the ascendant in contemporary political culture and empirical social science. It is using the deconstruction of modernism to draw an obfuscating veil over the instrumentality of the three restructurings, making any opposition appear to be extremism, the very hope of resistance to seem absurd. Often armed with whimsy and pastiche, this reactionary anti-modernism tends to smother all progressive political projects in artful games with space and time, geographies and histories, in an effort to celebrate the contemporary as the best of all possible worlds. The end of modernism is proclaimed in this 'rush to the post' as if the creation of a radical and resistant postmodern political culture were impossible, as if the problems of poverty and exploitation had disappeared in this new era of globalism and high technology.

An alternative radical political culture must therefore be developed to lift this gratuitous veil and reveal the instrumental power of space in each of the three restructurings: in the political economy of post-Fordism, in the ideology and aesthetics of cultural postmodernism, and in the postmodernization of critical social theory. This will require moving beyond rigorous empirical descriptions which imply scientific understanding but often hide political meaning; beyond a simplistic anti-modernism which abandons its progressive projects entirely; beyond an anti-Marxism which rejects all the insights of historical materialism in the wake of a 'post-Marxist' rhetoric. Above all, it will need to move beyond a merely intellectual reassertion of space in social theory to the creation of a politicized spatial consciousness and a resistant spatial praxis. The most important new models to arise from the provocative encounter between modern geography and Western Marxism are thus still to be produced.

Notes

1 Williams (1983) defines historicism in three senses: (a) neutral – a method of study using facts from the past to trace precedents of current events; (b) deliberate – an emphasis on variable historical conditions and contexts as a privileged framework for interpreting all specific events; and (c) hostile – an attack on all interpretation or prediction which is based on notions of historical necessity or general laws of historical development. My definition does not reject the power and importance of historical explanation as a method, but identifies historicism with an overconcentration on this method to the specific exclusion of geographical modes of explanation, i.e. an interpretive prioritization of time and history over space and geography. A complete reversal of this prioritization is also reductionist, leading to a spatialism or geographical determinism. The key point is to *combine* the powers of historical and geographical explanation without assigning an *a priori* primacy of one over the other. The relevance of this argument to the development of historico-geographical materialism will become clearer in the subsequent discussion.

2 The term postmodern, as it is used here, does not imply a complete break with modern geography and Western Marxism but rather a critical reworking of these modernist disciplines and perspectives – a major *theoretical restructuring* directly associated with other contemporary restructuring processes. In recent years, a controversial debate has developed primarily in cultural theory and criticism around *postmodernism* as a distinctly different critical and aesthetic perspective attuned to the particularities of the contemporary world. For those who wish to delve more deeply into these debates – especially now that they have begun to impinge upon the evolving encounter between modern geography and Western Marxism – I recommend Jameson (1984); the essays by Jameson, Habermas, and others in Foster (1983) and *New German Critique* (1984); Lyotard (1986); Eagleton (1986); and Hutcheon (1987). All these references deal specifically with the politics of postmodernism and the potential for both radical and reactionary responses to the cultural restructuring of contemporary capitalist societies. The emergence of a postmodern critical social theory, it should be added, is likely to evoke similar potentialities for both radical and reactionary postmodern politics.

3 It can be argued that space, time, and being are the abstract dimensions which together comprise all facets of human existence, just as space, time, and matter are the essential qualities of the physical world. More concretely specified, each of these abstract existential dimensions is *socially constructed* to shape empirical reality and simultaneously to be shaped by it. Thus the spatial order of human existence arises from the production of space or *the construction of human geographies*; the temporal order is concretized in what Marx described as *the making of history*; and the social order (being-in-the-world) revolves round the *constitution of society*, the production and reproduction of social relations and practices. Social theory in its broadest sense becomes the specification and interpretation of the relationship between these concretizing processes. Reducing social theorization to the making of history is the basis for historicism (see n. 1); reducing it to the making of human geographies is spatial determinism; taking any one of the many aspects of society as the sole source of social theory and explanation creates a host of other 'isms': economism, psychologism, idealism, technologism, etc. Again, the key point of this argument is the need to combine theoretical perspectives to make sense of the empirical world.

4 These prolonged phases of crisis and accelerated restructuring (followed in the past by extended periods of rapid economic expansion in capitalist societies), have become increasingly influential historical demarcations in contemorary social theory, political economy, and critical human geography. Precisely how and why they occur is still controversial and there exist many different conceptualizations of

the historical and geographical 'rhythm' they describe. The most pertinent of these conceptualizations for the present discussion builds primarily upon Mandel's interpretation of 'long waves' of capitalist development (Mandel 1975, 1980). Also compatible with this approach in my view are Massey's (1984) analyses of successive 'layers of investment' and changing 'spatial divisions of labour'; the sequencing of 'regimes of accumulation' and 'modes of regulation' described by the French Regulationist School (Aglietta 1979, Lipietz 1986); and Harvey's notion of the search for a 'spatial fix' as part of a series of reconfigurations in the geographical landscape of capital (Harvey 1982). The specifically *cultural* restructuring arguments of Berman and others add another important dimension to this *periodization* and *regionalization* of capitalist development.

5 See Hughes (1958), still one of the best descriptions of the formation of this countertradition around the turn of the century; and Mills (1959), the most explicit attachment of the sociological imagination to the interpretive meaning of history and biography, to critical historiography.

6 Sartre and Althusser are the archetypal representatives of two streams of critical thought which through the 20th century tended to intensify a discordant split within Western Marxism over the relative importance of agency versus structure and modernism versus modernization. The *hermeneutics* stream, combining existentialism, phenomenology, and other forms of interpretive sociology, was centrally concerned with the subjectivity, intentionality, and consciousness of knowledgeable human agents not only 'making history' but also shaping the political culture of everyday life and modern capitalist society. In contrast, the *structuralist* emphasis concentrated primarily upon the more objective conditions and forces operating to shape the underlying logic of capitalist development and modernization. The potential complementarity and ability to combine of these two streams was ignored until recently as each attacked the reductionist tendencies of the other. Also lost in the scuffles were the possibilities that each was offering for the reassertion of space in Marxist theory. Through a critical appropriation of the strengths of structuralism and existential phenomenology (via the most stinging attacks on their reductionist tendencies), Lefebvre combined the interpretation of agency and structure, modernism and modernization, subjectivity and objectivity, around the social production of space for the first time. With little reference to Lefebvre's work, Gregory (1978) and Giddens (1979, 1981) entered into remarkably similar projects some time later, the first with his discussion of 'structural', 'reflexive', and 'committed' explanation in geography and 'critical science'; the second in his ontological assertion of 'space–time structuration' and a 'contemporary critique' of historical materialism (see also Soja 1980, 1983, 1985a).

7 It is important to note again that the original French version of this book, *La Question urbaine*, was published in 1972, before Lefebvre's more clearly definitive works (Lefebvre 1972, 1973, 1974). Harvey's first discovery of Lefebvre (1973) leaned heavily on Castells (Soja 1980, Martins 1983); and even today, the most extensive treatment of Lefebvre's work written in English (Saunders 1981) fails to list his pivotal work, *La Production d l'espace* (1974) and his more recent writings (Lefebvre 1976–8, 1980) in its bibliography.

8 Since the late 1960s, it has become increasingly difficult to separate critical planning theory and debates on the nature of urban and regional planning practice from the unfolding encounter between modern geography and Western Marxism and the debates on the theorization of space. Indeed, the anglophonic tradition of planning education throughout the 20th century has been one of the most important places for the preservation of practical geographical analysis and critical spatial theory – as much if not more than the institutionalized discipline of geography itself. Today, there are probably more radical or Marxist geographers involved in the education of

planners and the practice of planning, in proportion to their numbers, than from any other comparable specialization within modern geography.

9 There were several important editorial collections which presented a radical spatial perspective (Peet 1977, Dear & Scott 1981, Carney *et al.* 1980) but remarkably few original and synthesizing book-length works written by individual, unambiguously Marxist *geographers*. The closest approximations are Olsson (1980), Scott (1980), Brookfield (1975), and Gregory (1978).

10 Giddens's work deserves much more attention than has been given to it in this chapter – and perhaps somewhat less attention than many contemporary spatial theoreticians have been giving to it as the foundation for the reincorporation of space in social theory. Giddens's major achievement has been the insertion of space into the typically spaceless *ontology* of being, consciousness, and human action, an important step toward developing an appropriately spatialized epistemology, theory-formation, and empirical analysis. The space–time structuration theory which comes out of this ontological reconfiguration moves further toward these objectives, but it is not yet a completed project and there are alternatives to be considered, especially with regard to making political and theoretical sense of *contemporary* spatiality and societal restructuring, as Urry notes.

References

Aglietta, M. 1979. *A theory of capitalist regulation*. London: New Left Books.

Anderson, J. 1980. Towards a materialist conception of geography. *Geoforum* **11**, 171–8.

Anderson, P. 1976. *Considerations on Western Marxism*. London: Verso.

Anderson, P. 1980. *Arguments within English Marxism*. London: Verso.

Anderson, P. 1983. *In the tracks of historical materialism*. London: Verso.

Berger, J. 1974. *The look of things*. New York: Viking Press.

Berman, M. 1982. *All that is solid melts in to air*. New York: Simon & Schuster.

Brenner, R. 1977. The origins of capitalist development: a critique of neo-Smithian Marxism. *New Left Review* **104**, 25–92.

Brookfield, H. 1975. *Interdependent development*. London: Methuen.

Carney, J., R. Hudson & J. Lewis (eds) 1980. *Regions in crisis*. London: Croom Helm.

Castells, M. 1977. *The urban question*. London: Edward Arnold.

Dear, M. & A. Scott (eds) (1981). *Urbanization and urban planning in capitalist society*. London and New York: Methuen.

Eagleton, T. 1986. *Against the grain: essays 1975–1985*. London: Verso.

Eliot Hurst, M. 1980. Geography, social science and society: towards a redefinition. *Australian Geographical Studies* **18**, 3–21.

Eyles, J. 1981. Why geography cannot be Marxist: towards an understanding of lived experience. *Environment and Planning A* **12**, 1371–88.

Foster, H. (ed.) 1983. *The anti-aesthetic*. Port Townsend, WA: Bay Press: (also published in 1985 as *Postmodern culture*. London: Pluto Press.)

Friedmann, J. & C. Weaver 1979. *Territory and function*. Berkeley and Los Angeles: University of California Press; London: Edward Arnold.

Giddens, A. 1979. *Central problems in social theory*. London and Basingstoke: Macmillan.

Giddens, A. 1981. *A contemporary critique of historical materialism*. London and Basingstoke: Macmillan.

Giddens, A. 1984. *The constitution of society: outline of the theory of structuration*. Cambridge: Polity Press.

Gregory, D. 1978. *Ideology, science and human geography*. London: Hutchinson.

Gregory, D. 1981. Human agency and human geography. *Transactions of the Institute of British Geographers* **6**, 1–18.

Gregory, D. 1984. Space, time and politics in social theory: an interview with Anthony Giddens. *Environment and Planning D, Society and Space* **2**, 123–32.
Gregory, D. 1989. *The geographical imagination*. London: Hutchinson.
Gregory, D. & J. Urry (eds) 1985. *Social relations and spatial structures*. London: Macmillan.
Hadjimichalis, C. 1986. *Uneven development and regionalism*. London: Croom Helm.
Harvey, D. 1969. *Explanation in geography*. New York: St Martin's Press.
Harvey, D. 1973. *Social justice and the city*. Baltimore: Johns Hopkins University Press.
Harvey, D. 1982. *The limits to capital*. Oxford: Basil Blackwell.
Harvey, D. 1984. On the history and present condition of geography: an historical materialist manifesto. *Professional Geographer* **36**, 1–11.
Harvey, D. 1985a. The geopolitics of capitalism. In *Social relations and spatial structures*, D. Gregory & J. Urry (eds), 126–63. London: Macmillan.
Harvey, D. 1985b. *The urbanization of capital*. Baltimore: Johns Hopkins University Press.
Harvey, D. 1985c. *Consciousness and the urban experience*. Baltimore: Johns Hopkins University Press.
Harvey, D. 1987. Flexible accumulation through urbanization: reflections on 'post-modernism'. in the American city. *Antipode* **19**, 260–86.
Hughes, H. S. 1958. *Consciousness and society: the reconstruction of European social thought 1890–1930*. New York: Vintage Books.
Hutcheon, L. 1987. Beginning to theorize postmodernism. *Textual Practice* **1**, 10–31.
Jameson, F. 1984. Postmodernism, or the cultural logic of late capitalism. *New Left Review* **146**, 53–92.
Johnston, R. J., D. Gregory & D. M. Smith 1986. *The dictionary of human geography*. Oxford: Basil Blackwell.
Keat, R. & J. Urry 1981. *Social theory as science*. London: Routledge & Kegan Paul.
Kern, S. 1983. *The culture of time and space 1880–1918*. Cambridge, Mass.: Harvard University Press.
Lefebvre, H. 1946. *Critique de la vie quotidienne*, reissued 1958. Paris: L'Arche.
Lefebvre, H. 1961. *Fondements d'une sociologie de quotidienne*. Paris: L'Arche.
Lefebvre, H. 1968a. *La Vie quotidienne dans le monde moderne*. Paris: Gallimard.
Lefebvre, H. 1968b. *Le Droit à la ville*. Paris: Anthropos.
Lefebvre, H. 1971. *La Revolution urbaine*. Paris: Gallimard.
Lefebvre, H. 1972. *La Pensée Marxiste et la ville*. Paris: Casterman.
Lefebvre, H. 1973. *La Survie du capitalisme*. Paris: Anthropos.
Lefebvre, H. 1974. *La Production de l'espace*. Paris: Anthropos.
Lefebvre, H. 1976, *The survival of capitalism*. London: Allison & Busby.
Lefebvre, H. 1976–8. *De l'Etat*, 4 vols. Paris: Union Générale D'Editions.
Lefebvre, H. 1980. *Une Pensée devenue monde: faut-il abandonner Marx?* Paris: Fayard.
Lefebvre, H. & N. Guterman 1936. *La Conscience mystifiée*. Paris: Gallimard.
Lipietz, A. 1986. New tendencies in the international division of labor: regimes of accumulation and modes of regulation. In *Production, work, territory*, A. Scott & M. Storper (eds), 16–40. Boston: Allen & Unwin.
Lyotard, J.-F. 1986. *The postmodern condition: a report on knowledge*. Manchester: Manchester University Press.
Mackinder, H. 1904. The geographical pivot of history. *Geographical Journal* **23**, 421–37.
Mackinder, H. 1919. *Democratic ideals and reality: a study in the politics of reconstruction*. London: Constable.
Mandel, E. 1975. *Late capitalism*. London: Verso.
Mandel, E. 1976. Capitalism and regional disparities. *Southwest Economy and Society* **1**, 41–7.
Mandel, E. 1978. *The second slump*. London: Verso.

Mandel, E. 1980. *Long waves in capitalist development.* Cambridge: Cambridge University Press.
Markusen, A. 1978. Regionalism and the capitalist state: the case of the United States. *Kapitalistate* **7**, 339–62.
Martins, H. 1983. The theory of social space in the work of Henri Lefebvre. In *Urban political economy and social theory*, R. Forrest, J. Henderson & P. Williams (eds), 160–85. Aldershot: Gower.
Massey, D. 1978. Regionalism: some current issues. *Capital and Class* **6**, 106–25.
Massey, D. 1984. *Spatial divisions of labour: social structures and the geography of production.* London and Basingstoke: Macmillan.
Mills, C. W. 1959. *The sociological imagination.* New York: Oxford University Press.
New German Critique 1984. Modernity and Postmodernity **33**, 1–269.
Olsson, G. 1980. *Birds in egg/eggs in bird.* New York: Methuen.
Peet, R. (ed.) 1977. *Radical geography: alternative viewpoints on contemporary social issues.* Chicago: Maaroufa Press.
Peet, R. 1981. Spatial dialectics and marxist geography. *Progress in Human Geography* **5**, 105–10.
Poulantzas, N. 1978. *State, Power, Socialism.* London: Verso.
Saunders, P. 1981. *Social theory and the urban question.* London: Hutchinson.
Sayer, A. 1985. The difference that space makes. In *Social relations and spatial structures*, D. Gregory & J. Urry (eds), 49–66. London: Macmillan.
Scott, A. 1980. *The urban land nexus and the state.* London: Pion.
Scott, A. & M. Storper (eds) 1986. *Production, work, territory.* Boston: Allen & Unwin.
Smith, N. 1979. Geography, science, and post-positivist modes of explanation. *Progress in Human Geography* **3**, 356–83.
Smith, N. 1980. Symptomatic silence in Althusser: the concept of nature and the unity of science. *Science and Society* **44**, 58–81.
Smith, N. 1981. Degeneracy in theory and practice: spatial interactionism and radical eclecticism. *Progress in Human Geography* **5**, 111–18.
Smith, N. 1984. *Uneven development.* Oxford: Basil Blackwell.
Soja, E. 1980. The socio-spatial dialectic. *Annals of the Association of American Geographers* **70**, 207–25.
Soja, E. 1983. Redoubling the helix: space–time and the cultural social theory of Anthony Giddens. (Review essay.) *Environment and Planning A* **15**, 1267–72.
Soja, E. 1985a. The spatiality of social life: towards a transformative retheorization. In *Social relations and spatial structures*, D. Gregory & J. Urry (eds), 90–127. London: Macmillan.
Soja, E. 1985b. Regions in context: spatiality, periodicity, and the historical geography of the regional question. *Environment and Planning D, Society and Space* **3**, 175–90.
Soja, E. & C. Hadjimichalis 1979. Between geographical materialism and spatial fetishism: some observations on the development of marxist spatial analysis. *Antipode* **11**, 3–11.
Soja, E., R. Morales & G. Wolff 1983. Urban restructuring: an analysis of social and spatial change in Los Angeles. *Economic Geography* **59**, 195–230.
Thompson, E. P. 1978. *The poverty of theory and other essays.* London: Merlin.
Urry, J. 1985. Social relations, space and time. In *Social relations and spatial structures*, D. Gregory & J. Urry (eds), 20–48. London: Macmillan.
Walker, R. 1978. Two sources of uneven development under advanced capitalism: spatial differentiation and capital mobility. *Review of Radical Political Economy* **10**, 28–37.
Weaver, C. 1984. *Regional development and the local community: planning, politics and social context.* Chichester: Wiley.
Williams, R. 1983. *Keywords: a vocabulary of culture and society.* London: Fontana.

14 *The crisis of modernity? Human geography and critical social theory*

Derek Gregory

'You've got to keep abreast of science in the pamphlets and published papers, and always be on the alert, to point out improvements . . .'.

Gustave Flaubert, *Madame Bovary*

'I don't much care for coincidences. There's something spooky about them: you sense momentarily what it must be like to live in an ordered, God-run universe, with Himself looking over your shoulder and helpfully dropping coarse hints about a cosmic plan. I prefer to feel that things are chaotic, free-wheeling, permanently as well as temporarily crazy – to feel the certainty of human ignorance, brutality and folly.'

Julian Barnes, *Flaubert's parrot*

Introduction

Modernism and postmodernism have become the buzz-words of the 1980s: so much so that their constant repetition threatens to dull the sensibilities. But important issues are at stake, and it is not enough to dismiss the postmodern challenge as symptomatic of either careerism or capitalism as so many critics seem content to do. To many of them, postmodernism is simply the product of disciplinary politicking, in which ascendant academics are once again jostling for new stalls in the intellectual marketplace; to others, it is little more than the cultural logic of late capitalism, the designer label on the baggage of flexible accumulation. But if these twin objections are superimposed a different reading becomes possible and postmodernism can be seen to mark an unprecedented crisis of intellectual activity within the contemporary crisis of modernity.[1] The debate then has the liveliest of implications for the project of a critical human geography (and critical theory more generally) because it raises acute questions about the very possibility of critique itself. More directly: is there anything which entitles us to redeem the Enlightenment project and reaffirm our faith in rationality? to think of human history as an epic of progress? to be optimistic about the advance of science and the improvement of the human condition? Or are the resources of modernity so exhausted that we feel only incredulity toward those who seek to breathe new life into them? Ought we, instead, to take nihilism seriously and unmask the 'will to power' which supposedly lies behind all systems of reason?[2]

One of the few geographers to have been bothered by questions of this sort was Sauer. Reflecting on the social sciences in the middle of this century, he asked:

> Do we think that we dominate time, as an upward spiral that we have under control, our increasing knowledge confidently shaping its development? Or is this faith that we are shaping progress by material skills and building an ever expanding system in truth the great 'phantasm' of our day, the 'brave new world'? Have we set up an economy of waste, which we call the miracle of American production? Can we disregard our deficit spending of natural resources because we shall continue the triumph of mind over matter? Are other times and other places of importance only in so far as they can be related to our egocentric and ephemeral positions? Are we the cleverest people of all time or the blindest because we think neither of whence we came nor whither we are going? (Sauer 1963, p. 387).

The questions were serious ones, but they remained largely unanswered. To be sure, Sauer was deeply conservative and his pessimism was, in part, a threnody for a lost world to be recaptured through a reactionary cultural geography. But concerns like these are not intrinsically conservative, any more than post-modernism is intrinsically neoconservative, and they can be given a much more radical edge by connecting them to Marxist and post-Marxist critiques of modernity. I raise them at the outset and in this particular form because most of what follows is about the recovery of what has come to be called the *theoretical* imperative: but no viable critical theory can suppress the *practical* imperative either, and the debate over modernism and postmodernism throws these moral and ethical questions into sharp relief.[3]

That said, much of the discussion of modernism and postmodernism is, well, *difficult*, and the temptation to dismiss the whole business as a bagatelle is, I suspect, very real. Accordingly, I have tried to make my own account as accessible as I can by beginning not with any summary definitions but by constructing and deconstructing one possible history of geography. Although this is not an unduly abstract description, it is, of necessity, one whose shapes are stark and unrelieved by shading. This means that it runs the risk of implying that one has to choose between the two motifs: that one must either defend the metanarratives of modernism or celebrate the diverse 'local knowledges' of postmodernism. But I see no need to put my head in that particular cleft stick. In the second and subsequent sections, therefore, I soften the original outlines by filling in some of the more prominent figures. My purpose is not to play one off against the others, because I have no desire to substitute one absolutism for another. Instead, I seek to clarify the *tension* between modernism and post-modernism.[4] To the extent that I am successful, however, I will also – and of necessity – have to show that human geography is not a spectator at some intellectual carnival; on the contrary, it is vitally implicated in the transformation of a much wider intellectual landscape.

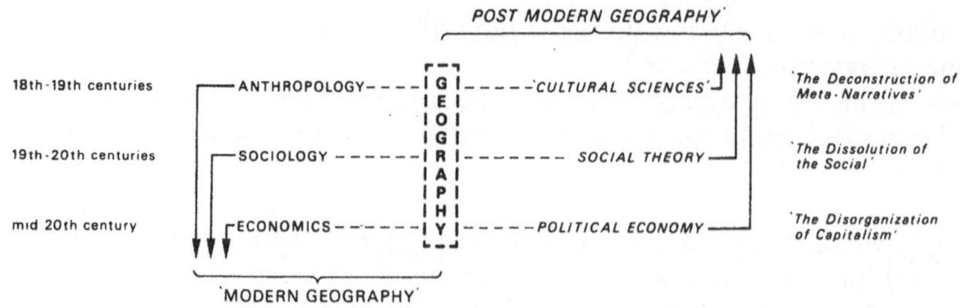

Figure 14.1 The making and unmaking of modern human geography.

Maps of an intellectual landscape

I want to locate my argument within the intellectual landscape whose contours are sketched in Figure 14.1. I should say at once that this rough map – like all maps – is inevitably partial and, no doubt, partisan, but I use it because it draws attention to what I take to be some of the salient features in the making (and unmaking) of modern human geography.[5] In so far as it seeks to do so by describing the relations between human geography and the other human sciences, then one other disclaimer is necessary: I would not want this simple diagram to be read as an endorsement of the academic division of labour. Boundaries between disciplines are obviously not purely arbitrary constructions, since it is always possible to provide some sort of explanation of how they have come to be drawn this way rather than that; but neither do they mark necessary separations, whose lines somehow chart natural breaks of slope in the intellectual landscape. Disciplines are, at bottom, social constructions, and although my sketch map recognizes the power of these divisions within the bourgeois academy (as it must do), it should not be taken to legitimize their individual claims to the terrain which it represents.

In using human geography to say something about modernism and post-modernism, however, the map is partial in quite another sense: these movements have their own historical geographies.[6] A more rigorous account would obviously have to pay more attention to these differences than I am able to do here, and to that extent I am aware that what follows is a thoroughly *modern* exercise in intellectual cartography.

Towards modern human geography

As the left-hand side of Figure 14.1 shows, I cut into the history of geography in the second half of the 18th century. I do this not because earlier developments are unimportant or uninteresting (quite the reverse), but because I think the origins of a distinctively *modern* geography are to be found there. In this I follow Stoddart, who dates the transformation to 1796, the year in which Cook first entered the Pacific. Cook's voyages were undertaken in the company of a group of illustrators, collectors, and scientists whose work displayed three features of decisive significance for the formation of geography as what Stoddart calls an 'objective science': a tripartite concern for realism in description, for systematic classification in collection, and for the comparative method in explanation. Stoddart's (1986, p. 35) central point is that it was 'the extension of [these] scientific methods of observation, classification and comparison to peoples and societies that made our own subject possible'. But the point needs to be sharpened. This was no simple 'extension' because, as Foucault and others have shown, it had the most radical of consequences for the constitution of the human sciences and their conception of the human subject; and its logic was powered by more than the ship of reason, since these nominally scientific advances were at the same time the spearheads of colonialism and conquest: of subjection. If my emphases are different from Stoddart's, however, at least we agree that modern geography can be represented as a 'European science' and, just as important, that its trajectory cannot be separated from what Wolf (1982) once called 'the people without history'. It follows that some of the most seminal cross-fertilizations during this period were those between anthropology and geography, loosely defined, and that what later came to be known as anthropogeography has to be acknowledged as a tradition of basic importance to the modern discipline.

I do not propose to trace this thread through the various designs of succeeding generations, even though it is, of course, indelibly present in the writings of Humboldt and Ritter (and, for that matter, Ratzel and Sauer). Instead, I want to disentangle the spinning of a second thread towards the end of the 19th century: between sociology and geography. In this case the definitions can be more precise, since the interactions between the two disciplines were, in part, a way of clarifying their different destinations. To Durkheim, for example, sociology has to subsume the study of the spatial structure of society – its *morphologie sociale* – and any account of the constitution of social life would necessarily incorporate many of the elementary theorems of Vidal de la Blache's *geographie humaine*. In reply, the Vidalian School insisted on the independent integrity of a human geography predicated not only on these horizontal relations but also on the vertical relations between society and nature (Berdoulay 1978). Across the Atlantic, Park and the Chicago School were drawn in to the border dispute. Park was also impressed by the salience of spatial structure for the constitution of social life, but he regarded geography as a 'concrete science' whose interest in spatial relations was necessarily confined to the individual and the irregular: to what, in a neo-Kantian vocabulary, would be called the idiographic. Like history (and, interestingly enough, anthropology), Park believed that geography made no concession to 'the mania for classification' because it was not properly concerned with the elucidation of general processes or universal principles. That project was reserved for the 'abstract sciences' like sociology and human ecology, which, though they were built upon the foundations of history and geography, were nevertheless essentially nomothetic in their orientation. If sociology and geography subsequently developed more or less in isolation, therefore, their supposedly distinctive characteristics were shaped in important ways by the *fin-de-siècle* encounter between them (Entrikin 1980).

By the mid-20th century a third strand had been spun alongside (even on top of) the other two: between economics and geography. Before Word War II this was, with some isolated exceptions, resolutely empirical in its cast, but the emergence of the so-called 'new geography' in the postwar decades gave it a tighter theoretical twist. If this allowed geography to become an abstract, nomothetic science, however, one which could even claim the models (though significantly not the ethnographies) of the Chicago School as its own, it was a geography which was still sharply set off from the social. Most of the formulations that were developed during this period were drawn from classical location theory and, even when they were reworked, they remained partial or general models of a space economy. Few of the contributions to the original *Models in geography* addressed cultural or social questions, for example, even in passing, and the geometric order which they supposed to be immanent in the human landscape was derived, for the most part, from the economy: from transcodings of an abstract calculus of profit and loss into an equally abstract 'friction of distance' (Chorley & Haggett 1967). To be sure, several commentators complained that human geography could not be reduced to regional science in this way, but the hegemony of neoclassical economics was sufficiently powerful to ensure that the formal exploration of spatial structure would not be caught up in tangled webs of social relations. Human geography was spatial science.

These paragraphs are little more than caricatures, but three features of modern geography ought to be clear enough:

(a) All three strands were bound in to a *naturalism*. One might expect this to have assumed special significance in a discipline like ours, where human geography is yoked to physical geography, but none of the other disciplines I have mentioned escaped the shadow of the natural sciences either. It would not be difficult to identify the impress of evolutionary biology on anthropology; sociology was no stranger to biological analogies, of course, and this was often conjoined to an interest in social physics; and the calculus of neoclassical economics was deliberately modelled on statistical mechanics. It is scarcely surprising, in consequence, that identical links should have been forged in human geography; from Vidal de la Blache's seemingly simple organicism to Hägerstrand's vision of time-geography as a 'situational ecology'; from essentially Newtonian models of spatial interaction to technically more sophisticated models of dissipative structures within abstract landscapes. By the end of the 1960s, as Stoddart had insisted (and, as a matter of fact, continues to insist), much of geography was shaped by the mainstream of the natural sciences

(b) All three strands were woven around a conception of science as *totalization*. It is one of the characteristics of modern science, so I shall claim, that explanation is supposed to reside in the disclosure of a systematic order whose internal logic imposes a fundamental coherence on the chaos of our immediate impressions. In the modern human sciences, this typically involved constructing a 'centre' around which social life revolves, rather like a child's mobile. In some versions concepts of social system or social structure were the central pivot, in other versions concepts of human being or human subjectivity took centre stage. Both of these can be read in human geography too, but the progressive formalization of spatial science gave a central place to geometry: to abstract orders in mathematical spaces. Whatever the arena, however, explanations were (in a special sense) simple. This may seem outrageous to those who worked in the mills of locational analysis, oiling the wheels of technique and technology, but it is surely not inaccurate. Spatial science was part of an economy of explanation which put a high premium on the *simplicity* of its explanations.

(c) When these three strands are spliced together the emergence of spatial science becomes not a sudden break, paradigmatic or otherwise, but instead a continuation and a *culmination* of modern geography.[7]

Towards postmodern human geography?

Let me now borrow Eagleton's phrase from Soja (see above, p. 320) and 'reverse the imposing tapestry'. To deconstruct the history of modern geography in this way is not an exercise in pure scholasticism, because its 200-year trajectory has *in fact* been reversed – and in under two decades. This is something of an exaggeration, but as the right-hand side of Figure 14.1 shows, it is at least plausible to unravel my three threads in reverse order to pick out an altogether different design.

Human geography was probably the last of the human sciences to take Marxism seriously, in the English-speaking world at any rate, but in the early 1970s one of the dominant voices in the critique of spatial science was that of political economy. In his *Theoretical geography* Bunge (1962) had predicted that the 'science of space' (geography) would find the 'logic of space' (geometry) a sharp tool; but political economy now seemed to offer a still sharper set of instruments. Its cutting-edge was the 'logic of capital', scrawling its signs on the surface of the landscape. These studies of the uneven development of capitalism were limited by the uneven development of Marxism itself, however, and Harvey's (1982) careful delineation of *The limits to capital* marks not only the bounding contours of capitalist development but also an important lacuna in Marx's master-work. No matter how ingenious the exegesis, or how assiduously the footnoes and manuscript jottings are pasted back into the main text, Marx is still left saying little about space. Marxism is not closed around Marx's own writings, of course, but Western Marxism in general maintained the same strategic silence about the spatiality of capitalism.[8] And yet many commentators would now argue that the production of space is focal to the development of capitalism in the late 20th century: that its new regime of 'flexible accumulation' depends upon the production of a tense and turbulent landscape whose dynamics are so volatile, whose space economies are so disjointed, that one is entitled – perhaps even compelled – to speak of the emergence of 'disorganised capitalism' (Harvey 1987, Lash & Urry 1987). These twin descriptions have highly specific meanings, and they are by no means concordant with one another, but they are fragments of a still larger mosaic; of Castells's 'new relationship' between space and society, formed by the hypermobility of capital and information cascading through a vast world system, where 'space is dissolved into flows', 'cities become shadows' and places are emptied of their local meanings; of Jameson's 'post-modern hyperspace', part of the cultural logic of late capitalism, whose putative 'abolition of distance' renders us all but incapable of comprehending – of *mapping* – the decentred communication networks whose global webs enmesh our daily lives (Castells 1983, p. 314, Jameson 1984).

There are important differences between these authors, but they share more than an emphasis on the spatiality of capitalism. Common to all of them is a recognition of the politics of resistance and, ultimately, of transcendence. Yet the political project of historical materialism, in both its classical and its modern forms, has been skewered by what Anderson calls variously a 'permanent oscillation' and a 'potential disjuncture' between two principles of explanation: the structural logic of the mode of production and the conscious and collective agency of human subjects (Anderson 1983, p. 34). Marx knew that people make history, of course, but he also recognized that they do not do so just as they please nor in circumstances of their own choosing. Towards the end of the 1970s one group of authors, working out the implications of this (unobjectionable) formulation, moved beyond Marx's original theses – though his writings remained indispensable to many of their arguments – to fashion what Giddens termed a 'post-Marxist' theory of structuration.

In its most developed versions, this (still programmatic) social theory turns on the claim that people make not only history but *geography*: that time–space relations are not incidental to the constitution of societies. The interpenetrations of agency and structure can only be grasped, so Giddens now argues, by explicating the ways in which time and space are 'bound into' the conduct of social life.[9] A number of commentators have drawn attention to the radical consequences of claims of this sort for human geography; but they are just as revolutionary in their implications for conventional social theory. For they make sociology's master-concept – society – immensely unstable. Instead of presupposing that societies are totalities with clear-cut boundaries, it becomes necessary to show how social relations are stretched across varying spans of time and space. This is evidently not the radical denial of structure which some critics claim to find in Giddens's writings, and although the problematic of time–space distanciation speaks directly to the theoretical 'dissolution of the social' which Lash and Urry have identified in social theory more generally, it is not, I think, symptomatic of the radicalization of agency which they regard as immanent in the reconfiguration of contemporary social theory (Storper 1985, Lash & Urry, 1986). As Giddens has constantly accentuated, the whole thrust of structuration theory has been to dismantle the dualism between agency and structure and replace it with a duality. In so far as this requires a sustained attack on the dualism between society and space, however, the dissolution of the social is more properly seen, in my view, as the simultaneous dissolution of the spatial. The theoretical separations between the two cannot be sustained. If, as Mann (1986, p. 4) urges, 'societies are much *messier* than our theories of them', this is, in large measure, a direct consequence of their spatiality. All the time that human geography remained committed to a spatial science that put a premium on geometric order – on pattern – it was, of necessity, prevented from making any serious contribution to this extraordinary project. But many writers are now much more sceptical about the ways in which the conventional vocabularies of human geography and social theory act, as Thrift (1985a) puts it, 'to complete the incomplete, to structure the partially structured [and] to order the only partly ordered.'

Arguments of this sort challenge the traditional ways of posing the 'problem of geographical description'. Since the Enlightenment this has usually been cast in Kantian terms and, indeed, the dualism between the social and the spatial is itself a Kantian one. The formalization of spatial science made the distinction between the two sharper still. It was supposed to correspond to Carnap's division between substance languages (in which individuals are identified by their position within a matrix of substantive properties) and space–time languages (in which individuals are located within a Cartesian co-ordinate system). Crossing the divide was difficult, even dangerous.[10] But the linguistic turn in the human sciences traces a much wider arc. It describes a hermeneutic circle and often intersects with a pragmatism that directs attention towards the construction of meaning as an irredeemably practical activity that takes place in a multiplicity of different settings. When the linguistic turn is conceived in these terms, the problem of geographical description becomes more than a matter of abstract logic: it becomes a question of reading and decoding, of literally *making sense* of different places and different people. Interpretative anthropology is no stranger to concerns of this kind, of course, and in cultural geography many writers would now regard Geertz's arguments for 'thick description' as, if not a commonplace, then at any rate one of the orthodoxies of interpretative enquiry. Attempts to understand what Geertz now calls 'local knowledge' – to reinstate the ethnographic tradition that modern geography all but erased – require just that sensitivity to setting which characterizes Geertz's own writings (Geertz 1983, 1984).[11]

But the linguistic turn revolves around more than the recovery of meaning: it is also about the re-presentation of meaning. As such, it speaks directly to the so-called crisis of representation in the human sciences. In particular, it requires us to attend not only to the theoretical categories which inform our accounts but also to the textual strategies through which they are conveyed. For what do ethnographers do? asks Geertz. 'They write' (Geertz 1973, Clifford & Marcus 1986, Marcus & Fisher 1986). And if, as the previous paragraphs suggest, we are now obliged to describe the disjointedness of late capitalism or the loose-knit texture of everyday life, then we evidently need to be on our guard against genres which make covert assumptions about closure and coherence. One of the strengths of deconstruction, and of postmodern literary theory more generally, is precisely its ability to alert us to this duplicity of the text. Conventional narrative forms are especially suspect. Jameson (1980) claims that their 'comprehensible order' conceals a fundamentally contradictory social world, and from a parallel perspective White suggests that the value we typically ascribe to narration 'arises out of a desire to have real events display the coherence, integrity, fullness and closure of an image of life that is and can only be imaginary'. On these readings, it is scarcely surprising that ideologies should so often assume a narrative form (Jameson 1980, White 1987, Thompson 1984a). But objections of this kind are not confined to the avant-garde of literary criticism. They can be registered with equal force against ostensibly 'factual' accounts which assume a narrative form. These too are 'fictions', in the literal sense of 'something made'. Where literary theory is at its most subversive – and where it reaches most deeply into the heart of the human sciences – is in disclosing *other* ways in which those accounts might be made. They have their own problems, of course; and postmodern ethnography is still in its infancy, while experiments in geographical description have barely begun.[12] But the significance of the current connection between human geography and the cultural sciences (a flag of convenience) flows directly from the licence it issues for experimentation with different modes of representation which might more adequately render the play of difference.

I have taken more time over these sketches, though they are obviously still very rough and ready, and I can now draw out three features from their collective portfolio. Although these correspond to (and contrast with) my previous characterization of modern geography, I want to emphasize that they do not stand in any simple opposition to it. In part, clearly, this is because none of the moments which I have identified constitutes, in and of itself, a postmodern geography. Taken together they may well mark out a path towards postmodernism, but they are far from signalling (let alone celebrating) its arrival. But it is also, and more importantly, because postmodernism needs to be seen not as a negation of everything that went before but rather as a critical commentary upon it. I make this point again, and press it home in the paragraphs below, because several recent commentaries seem to miss it altogether. In what follows, therefore, I will pay particular attention to those writers who have sought to defend a thoroughly modern Marxism against the incursions of poststructuralism, deconstruction and postmodernism.

First, all three moments recognize the *specificity* of the human sciences – the fact that they seek to understand a world which is meaningful to the people who live within it – and, in different ways, endorse the reflexivity that is intrinsic to what Lawson (1985) calls 'the post-modern predicament'.[13] At the same time, however, they do not completely repudiate the canons of naturalism. Some of the most interesting work in the triangle between geography, political economy, and social theory claims to depend upon a particular version of the philosophy of realism, for example, and when Bhaskar (1979), one of its principal architects, talks about 'the possibility of naturalism' he is rejecting those attempts which seek, without qualification, to model the human sciences on the natural sciences and simultaneously refusing to rule out the admission of some of the protocols of the natural sciences into the human sciences. This is far from straighforward, even on Bhaskar's own terms, but it poses a special problem for anyone who wants to read work of this kind as somehow symptomatic of postmodernism. One of the basic aims of Bhaskar's project – and, presumably, of empirical enquiries based upon it – is the recovery of ontological depth; and yet one of the obsessions of postmodernism is 'depthlessness' or, as Jameson (1984) puts it, the replacement of depth models by the play of 'multiple surfaces', each one shimmering off another. This accounts for much of modern Marxism's hostility towards postmodernism, of course, but it remains the case that none of the three moments which I have singled out, Marxist or otherwise, is shackled to the baggage-trains of a traditional naturalism.

Second, Marxism is further distanced from postmodernism by its commitment to the concept of totality, and even the ostensibly post-Marxist theory of structuration retains the concept in a modified form (Jay 1984). If most of the writers I have mentioned continue to insist on the necessity for some basic 'principles of intelligibility', however, this does not immediately implicate them in the search for a single systematic order which I attributed to modern science. Realism may well require the disclosure of various structures involved in the production and reproduction of social life, but by its very nature it brings into focus the differentiated and stratified character of the world. For this reason, enquiries informed by philosophies of this kind rarely offer reductive explanations. The ebb and flow of human history is not reduced to the marionette movements of a single structural principle – whatever its location – and the differences which make up human geographies are not explained by some central generating mechanism. All three moments emphasize heterogeneity: they draw attention to a plurality of structures and a variety of outcomes. In doing so, they map out a terrain of considerable *complexity*.

Third, the trajectory which I have set out is clearly a *critique* rather than a continuation of spatial science, so much so, in fact, that some of the earliest contributions (and particularly those influenced by an unreconstructed political economy) registered a withdrawal from any form of spatial analysis. This has now changed, I think fortunately, not least because the recognition (and representation) of heterogeneity – inside Marxism and out – requires a sensitivity to the spatiality of social life. I am aware, however, that in describing these various changes I have, in effect, offered my own master-narrative: a sequence of events in which, as Peter Shaffer once put it, 'moments seem to clip together like magnets'. If this makes my own account as vulnerable to deconstruction as any other, so be it.[14]

Postmodernism, polyphony, and social theory

So much for context. I now want to consider the relations between human geography and social theory in more detail. These are among the most prominent features on many maps of the contemporary intellectual landscape, and Dear (1988) has recently argued that human geography is in crisis precisely because it has progressively disengaged from the mainstream of philosophy, the social sciences, and the humanities. I am not so sure, not least because what Dear sees as a weakness in human geography – its plurality and fragmentation, its lack of a clearly defined core – is simultaneously supposed to be a source of strength for social theory. Whatever one makes of this, Dear is certainly not alone in thinking that the most effective response to the postmodern challenge is to realign human geography with social theory. And in making this claim he is adamant that there can be no Grand Theory for human geography (or any other discipline): rather, postmodernism entails a respect for the 'creative tension' between *different* theories. That may well be so (though it is hardly necessary to invoke postmodernism to make the point), but it is far from easy to sustain a dialogue between competing theoretical traditions – to put the hermeneutics of intellectual enquiry into practice, and many of the most interesting and innovative attempts to do so – to transcend the antinomies of traditional social theory – continue to revolve around the construction of Grand Theory.

Much depends on what one means by the term, of course, since one of the most striking characteristics of the so-called return of Grand Theory is an emphasis upon the *local* and the *contingent* (Skinner 1985). This is one of the most insistent motifs in the debate over modernism and postmodernism too, and it clearly intersects with many of the most basic concerns of human geography. In the remainder of this chapter, therefore, I want to contrast the writings of two authors who are especially sensitive to the polyphony which characterizes contemporary social theory: Habermas and Giddens. A number of parallels can be drawn between them, but I will pay particular attention to their differing views about the importance of concepts of spatiality and spatial structure. My main argument can be simply stated: Habermas is silent about space, whereas Giddens makes time–space relations central to social theory. I do not want to leave matters there, however, because I think one can still learn a great deal from Habermas. So I also propose to indicate some of the ways in which his project can be reformulated to take time–space relations into account and to suggest some of the ways in which Giddens's project needs to be reformulated as well. It should be obvious that in doing so I hope that the project of a critical human geography is strengthened too.

Pragmatism and place

Rationality, relativism, and the social construction of meaning

In one sense, Habermas's contribution to the formation of a critical human geography has been largely philosophical. The critique of spatial science was launched from a philosophical plane and depended, in large measure, on the deployment of various post-positivist philosophies of science. Many of them – I am thinking of phenomenology, realism, and structuralism in particular – opened a much wider space for 'space' than positivism, but they were first enlisted to assault the canons of spatial science and to put a new floor under geographical enquiry more generally. Certainly, Habermas's early writings were mobilized by several commentators to attack the exclusively technical conception of science embodied in conventional spatial analysis and to argue for a human geography committed, like Habermas's critical theory, to the project of human emancipation. But in another sense I suspect that one of the most enduring lessons to be learned from Habermas will turn out to be the need to downgrade our expectations of purely philosophical examinations. A critique of epistemology, Habermas once declared, is only possible as a social theory. No special privileges can be accorded to philosophy, post-positivist or otherwise: it has no peculiar power to legislate between other discourses but must instead enter into an 'interpretative' relationship with the human sciences (Habermas 1987c).

That said, Habermas's purpose is undoubtedly to reclaim, in a more modest form than the Enlightenment's original version, a cardinal commitment to a form of critique grounded in a universalistic account of reason. Other writers who, like Habermas, also want to think 'after Philosophy with a capital P', as Rorty (1979, 1982) puts it, would be much more sceptical about amibitions of this kind. Rorty himself has endorsed a pragmatism that would cut the ground from beneath all 'foundational' discourses. This offers more than the spectacle of a sociology of knowledge (important though that is), because one of the *raisons d'être* of the human sciences is surely to comprehend the 'otherness' of other cultures. There are few tasks more urgent in a multicultural society and an interdependent world, and yet one of modern geography's greatest betrayals was its devaluation of the specificities of places and of people. It is in exactly this sense and for precisely this reason that the renewal of cultural geography is now so obviously dependent on Geertz's (1973) 'thick description': on the ethnographic attempt to render the layers of meaning in which social actions are embedded less opaque, less refractory to our own (no less particular) sensibilities, yet without destroying altogether the strangeness which draws us to them in the first place (Cosgrove & Jackson 1987).

But if we allow – as Geertz (1989) insists we do – that this is not an argument for an unqualified relativism, then its hermeneutic impulse – what Rorty calls the desire to 'keep the conversation going' – is not so very far from Habermas as it might seem. For it was in the pragmatists' vision of a participatory democracy underwritten by an intrinsically fallible process of self-correcting enquiry that Habermas discovered the kernel of what was later to become his master-concept of 'communicative action'. 'Like Rorty,' Habermas remarked recently, 'I have for a long time identified myself with the radical democratic mentality which is . . . articulated in American pragmatism' (1985, p. 198).

Pragmatism emphasizes the social construction of meaning through the practical activity of human beings. This makes it pregnant with methodological implications for, as Bernstein suggests

> The shift of orientation from the foundation paradigm to that of inquiry as a continuous self-corrective process requires us to re-think almost every fundamental issue in philosophy . . . . [The human being] as inquirer, as a participant in a community of inquiry, is no longer viewed as 'spectator' but rather as an active participant and experimenter. [The human being] as agent comes into the foreground here because human agency is the key for understanding all aspects of human life, including human inquiry and knowledge (1972, p. 177).

Human agency is a far from simple concept and I will return to its complexities shortly. But what is of more moment for my present purposes is the close association between pragmatism (when conceived in these terms) and symbolic interactionism.

Symbolic interactionism and the geography of the life-world

Habermas's debt to symbolic interactionism is well-known. He regards Mead's writings as being of decisive importance for the paradigmatic restructuring of social theory around communicative action, and his own discussion of the deformation of the modern life-world depends on a critical appropriation of some of Mead's basic theses (Habermas 1987c).[15] But I attach so much significance to symbolic interactionism for a different set of reasons. In human geography its perspectives have opened up progressively wider avenues of enquiry which radiate from a central concern with the social construction of *place*. A number of different threads have been tangled together in the course of these investigations, and by no means all of them register the advances which Habermas attributes to Mead, but taken together they demarcate a strategic silence in Habermas's writings.

Most accounts of the geography of the life-world would, I think, represent place as what Ley calls (in some part following Mead) 'a negotiated reality, a social construction by a purposeful set of actors'. And this is a two-way street, 'for places in turn develop and reinforce the identity of the social group that claims them'. In Ley's view, therefore, reflection on the meaning of place in the production and reproduction of social life reveals a profound relationship between landscape and identity. For example:

> The separation of subject from object is a characteristic of the anomie of the post-industrial metropolis. Dehumanized urban settings comprise little more

> than space, little more than geometry; they excite no commitment, they are not 'for' a collectivity. They are the landscape realization of a mode of analysis which has emphasized functional materialism; such analysis translated into a planning paradigm creates a landscape of objective meaning only, a form without a subject (Ley 1986, p. 15).

This phrasing owes as much to constitutive phenomenology as it does to symbolic interactionism, and it must be said that other commentators are unhappy with these descriptions. Relph in particular insists that

> They may be dehumanising, but it is quite wrong to describe modern landscapes as 'dehumanised', or devoid of human content. On the contrary, they are entirely the products of rational human intelligence doing its best to control uncertain environments . . . . There is almost nothing in them that has not been conceived and planned so that it will serve those human needs which can be assessed in terms of efficiency or improved material conditions. But there is almost nothing in them that can happen spontaneously, autonomously or accidentally, or which expresses human emotions and feeling. If this absence diminishes the quality of our lives it does so sadly, quietly, unobtrusively, rather than with some overt and brutal denial (Relph 1981, p. 104).

The differences are substantial, but anyone familiar with Habermas's later writings will recognize that these separate claims can be convened within his plenary theory of communicative action. And a manoeuvre like this is more than a mapping from one vocabulary to another. It indicates that what Habermas now calls the 'colonization' of the life-world by the selective rationalization of the social system *is embedded in the production and reproduction of the 'placeless' landscapes of modernity*. These are not superficial but *structural* features of the late 20th-century world. Yet Habermas says virtually nothing about them.

Place, power, and the time-geography of everyday life

This must be pressed still further. Most interpretative sociologies have been tacitly aware of the significance of the settings in which social action takes place, but this has rarely been elucidated in any systematic fashion. One obvious exception is Goffman's remarkable series of studies of social interaction, which pays considerable attention to the time–space 'zoning' of encounters, but even Goffman is unable (or unwilling) to provide an account of the ways in which the strands of social interaction thread out into wider skeins of social reproduction. I take this to be one of the signal achievements of Hägerstrand's time-geography (Giddens 1987a, Pred 1981).

Hägerstrand conceives of time and space as resources on which individuals have to draw in the day-to-day conduct of social life. It is still true, I think, that Hägerstrand is more interested in the realization of projects than in their constitution, and to that extent the physicality of his 'web model' (Fig. 14.2) distances it from any interpretative tradition (Gregory 1985). Even so, his more recent writings seem to be characterized by a much greater ethnographic sensibility and, for the first time, one begins to glimpse the symbolic landscapes

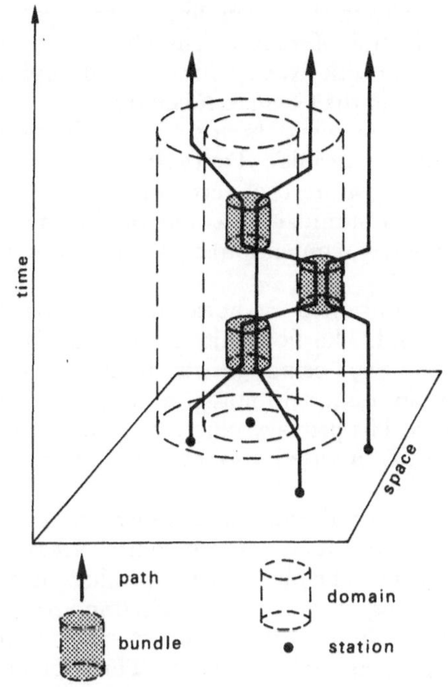

Figure 14.2 Hägerstrand's web model.

through which his individuals are supposed to trace their paths in time and space (Hägerstrand 1982). Any rigorous explication of those landscapes will, I believe, involve more than a simple extension of the existing problematic of time-geography. Several openings are possible, but I want to indicate just two of them.

One could, I suppose, deepen the phenomenological roots of Hägerstrand's project. Drawing on Heidegger and Husserl, for example, Pickles has made out a plausible case for a conception of human spatiality which, although he makes no mention of it, is not altogether dissimilar to Hägerstrand's. He argues that our most immediate experiences of the world are not cognitive abstractions of separate objects – the standard assumption of spatial science – but rather 'constellations of relations and meaning' which we encounter in the course of our everyday activities. Heidegger called those constellations, generically, 'equipment', and suggested that the latter's most important characteristic was that it was 'ready-to-hand'. In Pickles's view, such a conception is essentially non-parametric: distant from the formalized metrics of the physical world, it literally 'brings home' what Hägerstrand would call the *contextuality* of human being. To be sure, Hägerstrand would not, I suspect, want to make the distinctions between the human and the physical sciences that are so central to Pickles's thesis. But he would have no difficulty in accepting that human spatiality is related 'to several concurrent and non-concurrent equipmental contexts', and that in consequence it has the character of a 'situating' enterprise in which we 'make room for' and 'give space to' congeries of equipment. He has himself drawn attention to the intermingling of 'presences and absences' in the production and reproduction of social life, and his own conception of space – in Swedish, *rum* – resonates with exactly that sense of 'inward-directed finitude' that Pickles (1985) hears in the German *Raum* (room).

I want to pursue a somewhat different route, however, and to connect Hägerstrand's project to Foucault's. Foucault's writings are indebted to Heidegger too, of course, but instead of reconstructing this discursive triangle I will confine myself to a summary account of the more obvious points of connection (and, ultimately, contention) between Foucault and Hägerstrand:[16]

(a) Both writers are concerned, at bottom, with what Foucault calls 'bio-power'. Hägerstrand, like Foucault, predicates his analysis on the corporeality of the human being, on its 'bodily existence', and seeks 'to incorporate certain essential biological and ecological concepts' into human geography. His primary objective is the way in which bodies are (in Foucault's terms) 'inserted into' the routinized templates of everyday life.

(b) This is not an unregulated process, however, and Hägerstrand identifies a series of 'authority constraints' which control access to an often unstable hierarchy of domains. The outcome is a product of what Foucault would call the play of power, in which, so Hägerstrand strongly implies, there is no centre: the threads of power are 'subject-specific', fibrous and capillary and constitute 'pockets of local order'. This elementary web model is paralleled in Foucault's own account of the 'net-like organization' through which power is exercised. 'Not only do individuals circulate between its threads,' says Foucault, but 'they are always in the position of simultaneously undergoing and exercising this power.' And he too insists that power is deployed through a hierarchy of different domains ('disciplinary spaces'), 'from the great strategies of geopolitics to the little tactics of the habitat'.

(c) It ought to be axiomatic that Hägerstrand's authority constraints stipulate specific modes of conduct within these domains; yet Hägerstrand says little if anything about them. The contrast with Foucault could not be greater, for what concerns him is the multiplicity of ways in which bodies are, so to speak, constituted *as* individuals *through* these systems of disciplinary space. This is far removed from Ley's presumptive reciprocity between people and places; indeed, its claims are foreign to humanistic geography and to humanism more generally. Certainly, Foucault agrees that 'the individual is not to be conceived as a sort of elementary nucleus, a primitive atom . . . on which power comes to fasten or against which it happens to strike' – one of the standard humanist objections to Hägerstrand's physicalist vocabulary – but this is precisely because 'it is already one of the prime effects of power that certain bodies, certain gestures, certain discourses, certain desires, come to be identified and constituted as individuals.'

Habermas is not indifferent to claims of this sort, even if his hostility towards Foucault's project means that he develops them in a radically different direction and in a radically different vocabulary. As Giddens points out, Habermas's theory of communicative action 'does not posit a self-sufficient subject, confronting an object world, but instead begins from the notion of a symbolically structured life-world *in which human reflexivity is constituted* (Giddens 1987b, p. 236).[17] The impulse is much the same, I think, though the emphasis on the life-world is evidently symptomatic of symbolic interactionism and sharply separates the theory of communicative action from any poststructuralism.

This is not to say that Habermas's project can be directly assimilated to symbolic interactionism. He conceives of his work as a reconstruction of historical materialism and this plainly requires him to address more than the life-world. In his early writings he objected to the primacy of Marx's 'paradigm of production' and insisted on the co-equal importance of a 'paradigm of communication'; and if, as some commentators complain, he subsequently tipped the scales in favour of communication, he has nevertheless continued to insist on the importance of analyzing the structural templates of late capitalism. His theory of communicative action concerns not only social integration, therefore, but also system integration. Like many other writers, however, Habermas's discussion of these matters is conducted in the notations of what a previous generation called the 'sociological imagination': that is to say, it is concerned, *grosso modo*, with the intersections between biography and history. And yet, as I now want to show, an inclusive account of the constitution of social life is equally dependent upon the geographical imagination.

Social integration, system integration, and the spatiality of social life

Habermas and the reconstruction of historical materialism

One of the principal foci of Habermas's reconstruction of historical materialism is the conjunction between 'societies and the acting subjects integrated into them'. I cannot discuss the details of the scheme here, but I want to draw attention to three of its most important features which Habermas first set out in detail in the early 1970s.[18]

(a) Habermas makes a basic distinction between **social integration** and **system integration**. This conceptual couple was registered in his previous writings in innumerable ways, most obviously in the debates with Gadamer (over hermeneutics) and with Luhmann (over systems theory), and its two halves derive from radically different theoretical traditions:

> We speak of social integration in relation to the systems of institutions in which speaking and acting subjects are socially related. Social systems are seen here as *life-worlds* that are symbolically structured. We speak of system integration with a view to the specific steering performances of a self-regulated *system*. Social systems are considered here from the point of view of their capacity to maintain their boundaries and their continued existence by mastering the complexity of an inconstant environment. Both paradigms, life-world and system, are important. The problem is to demonstrate their interconnection (Habermas 1976, p. 4; author's emphasis).

Table 14.1 Society, organization principle and crisis in Habermas.

Type of society	Organization principle ⟶	Characteristic form of crisis
PRIMITIVE	Kinship relations	Ecological-demographic crisis
TRADITIONAL	Class domination through the state	Political crisis
EARLY CAPITALISM	Class domination through capital-labour relation	Economic crisis
↓ LATE CAPITALISM	Class domination through state mediation of (corporate) capital-labour relation	Socio-cultural crisis

And as far as Habermas is concerned, the ultimate nature of their interconnection is *hierarchical*, in that 'the life-world persists before its transformation into an alienated system and persists even after that has occurred' (Jay 1984, p. 485).

(b) Habermas retains the asymmetric relation between system and life-world when he connects the conceptual couple to a theory of crisis. He reverts to the original, medical sense of the term to emphasize that a crisis is both 'objective' and 'subjective', both 'caused' and 'experienced'. He accepts that crises issue from unresolved 'steering problems' – the dislocation of modes of system integration which are, in various vocabularies, meat and drink to most Marxist and post-Marxist theories – but then, consistent with the priorities established by his philosophical anthropology, he argues that a crisis will only be *realized* (in the fullest sense of the word) to the extent that human subjects 'feel their social identity threatened'. This is not a purely private and personal matter. It depends, so Habermas says, on whether existing modes of social integration are capable of providing resources adequate to the task of resolving the system's steering problems. Their ability to do so is circumscribed by their institutional cores, which Habermas conceptualizes as a series of 'organization principles' which, as Table 14.1 shows, can be connected through to characteristic forms of crisis (Habermas 1976, 1979).

(c) If, as Table 14.1 also shows, this provides Habermas with the bare bones of an evolutionary schema, its 'developmental logic' (as he calls it) depends as much on the constitution of human beings as knowledgeable and capable human subjects as it does on the transformation of the structures of the life world:

> Social systems, by drawing on the learning capacities of social subjects, can form new structures in order to solve steering problems that threaten their continued existence. To this extent the evolutionary learning process of society is dependent on the competences of the individuals that belong to them. The latter in turn acquire their competences not as

> isolated monads but by growing into the symbolic structures of their life-worlds (Habermas 1979, p. 154).

These learning processes take place, as one might expect, in the twin dimensions of technically useful knowledge and moral–practical consciousness, but, as one might also expect, Habermas insists on the primacy of moral–practical consciousness:

> The introduction of new forms of social integration – for example, the replacement of the kinship system with the state – requires knowledge of a moral–practical sort and not technically useful knowledge that can be implemented in rules of instrumental and strategic action. It requires not an expansion of our control over external nature but knowledge that can be embodied in structures of interaction – in a word, an extension of the autonomy of society in relation to our own, internal nature (Habermas 1979, p. 146).

Put like that, I imagine that Habermas's faith in the Enlightenment project – in the project of modernity – must seem as plain as a pikestaff. But one needs to be very careful about the conceptual status of this scheme (especially when presented in such an abbreviated form). Habermas does not mistake this 'argumentation sketch' for an historical account, and neither should we. In any event, he has subsequently revised and refined many of its details, as I will show shortly. But his basic account of social integration and system integration should be clear enough to allow a preliminary comparison with Giddens's version of the same conceptual couple.

Giddens and the deconstruction of historical materialism

Giddens concedes that he has learned much from Habermas's writings, but he deliberately describes structuration theory as a 'deconstruction' of historical materialism – not altogether accurately, I think – in order to distance himself from Habermas. This does not mean that he rejects all of Marx's formulations, but it does mean that he is much more sceptical about the possibility of maintaining any formal continuity with the overall framework in which they are set. I want to concentrate on one aspect of this. Like Habermas, Giddens draws a distinction between social integration and system integration. But he formulates it in a strikingly different fashion in order to disclose what he calls the 'binding' of time and space into social life.

The architecture of Giddens's scheme is summarized in Figure 14.3.[19] Giddens argues that the continuity of everyday life depends, in large measure, on routinized interactions between people who are *co-present* in time and space. This is what he means by **social integration**, but it is also what society meant before the 18th century: simply the company of others. Although Giddens is critical of several of Hägerstrand's theorems, and rightly so, he nevertheless suggests that time-geography provides an exemplary notation through which the characteristic shapes of **time–space routinization** can be captured.

But one of the basic objectives of structuration theory is to show how 'the limitations of individual "presence" are transcended by the stretching of social relations across time and space'. In other words, social practices reach beyond the here and now to include interactions with others who are *absent* in time or space. This is what Giddens means by **system integration**, but it is also what

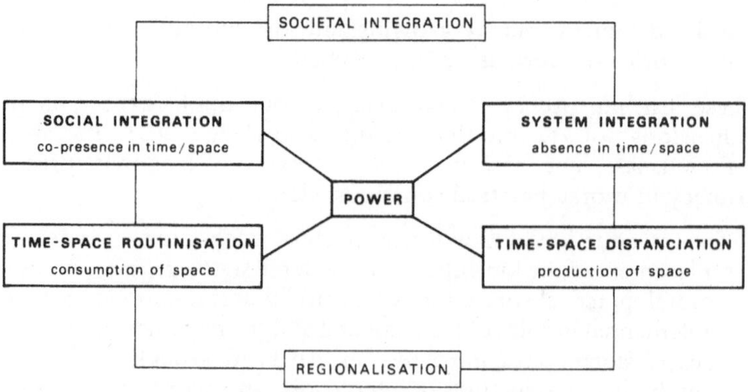

Figure 14.3 Time–space relations and structuration theory.

society came to mean after the 18th century: the larger world stretching away from the human body and the human being. Giddens calls this process of 'stretching' **time–space distanciation**, and uses it to refer to the ways in which social systems are 'embedded' in time and space.

A similar distinction can be found in Schutz & Luckmann's (1974) account of what they call the 'spatial arrangement of the everyday life-world'. They distinguish between the *primary zone* of operation – 'the province of nonmediated action . . . the primary world within reach' – and the *secondary zone* of operation, 'the world within potential reach' which 'finds its limits in the prevailing technical conditions of a society'. But they are concerned with the structures of the life-world and develop their argument in the vocabulary of constitutive phenomenology, whereas Giddens objects to the persistent failure of approaches of this sort to recognize the centrality of power to the production and reproduction of social life. Power, so he claims, is a chronic feature of social life, and its exercise is logically implicated in the transformative capacities of human agency.[20]

Power and space are closely linked in Giddens's societal typology. He argues that the exercise of power involves drawing upon structures of domination whose resources sustain dominion either over the social world (authoritative resources: roughly speaking, the political) or over the material world (allocative resources: roughly speaking, the economic). These take different forms in different societies, but more important is the fact that they intersect in different ways in different types of societies. For it is through these intersections – through the differential mobilization of authoritative and allocative resources – that different social systems stretch across different spans of time and space in different ways (Fig. 14.4).

Giddens sorts some of the most powerful media of time–space distanciation into two sets of 'episodes' (though there are undoubtedly others). The first set includes the emergence of writing, which was almost everywhere closely identified with the origin of the state and the formation of class-divided societies, and which opened up spheres of interaction far beyond the spoken circles of oral cultures and traditional tribal societies; and the diffusion of printing and other communications technologies, which dramatically enhanced

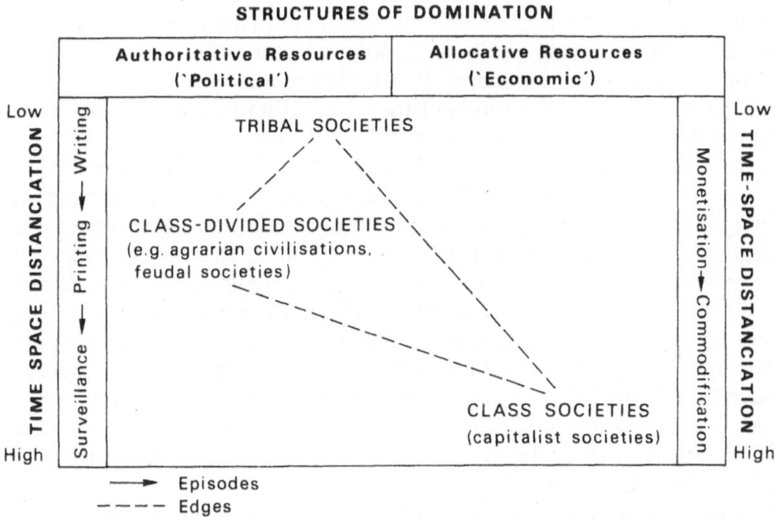

Figure 14.4 Time–space distanciation and the mobilization of resources.

the surveillance capabilities of the state and allowed it to penetrate much more deeply into the day-to-day lives of its subject populations. These various changes correspond to the mobilization of authoritative resources.

The second set includes the emergence of money – Simmel repeatedly drew attention to 'the power of money to bridge distances' – and the subsequent universalization of the money economy, which Giddens (like Marx) regards as the cutting-edge of the so-called commodification of everyday life in the class societies of modern capitalism. These various changes correspond to the mobilization of allocative resources.

In the late 20th century these two sets intersect in innumerable ways and Giddens uses them to illuminate the project of modernity: to show how it is that our world is so different from anything that went before. His focus is much the same as Habermas's, therefore, but there are nevertheless substantial differences between them.[21]

Crisis, capitalism, and modernity

The concept of crisis is of basic importance to historical materialism, but Habermas and Giddens deploy it in different ways.

Habermas uses it to account for *both* the transition from one type of society to another *and* the constitution of capitalist society in particular. The evolutionary logic of his scheme depends on what Giddens would call an 'internalist' model of social change. Societies are assumed to be clearly defined entities, set off one from another, and social change is brought about by the resolution of crises which are in some sense 'internal' to them. Giddens, by contrast, intends the concept of time–space distanciation to draw attention to the fact that the extension and closure of societies across time and space is deeply problematic.

In human geography questions of this sort have usually been conceived in exclusively technical terms to do with calibrating the 'friction of distance' or measuring 'edge effects' on maps. But Giddens makes them focal to social theory by insisting that questions of time–space distanciation replace, to some substantial degree, traditional conceptions of 'totalities'. In doing so, he displaces the concept of crisis from the central position it occupies in Habermas's evolutionary typology. Although Giddens (1984) argues that the structural principles of human societies 'operate in contradiction', he is adamant that 'no single and sovereign mechanism' can be specified that will account for the major transitions between societal types: in his view, capitalism is not the climax of an evolutionary succession at all. If the episodes which Giddens describes are supposed to constitute a non-evolutionary sucession of societies, however, they are given a strongly evolutionary cast. They connote a progressive widening of systems of interaction which, as several commentators have recognized, is congruent with some of the most basic criteria for evolutionary theory in general and of historical materialism in particular (Wright 1983, Thompson 1984, Callinicos 1985).

Habermas covers some of the same ground, and in his discussion of the invention of writing, the advent of printing, and the diffusion of electronic media he accentuates the ways in which these developments have produced a 'progressive emancipation of communication from specific contexts' (the original meaning of distanciation). But he is less interested in the spatiality of these progressions than in the 'condensation' of communicative action which they promote (Habermas 1987a, p. 184). This is causally connected, so Habermas suggests, to the modern reduction of the life-world to a mere satellite of the system: and it is to the pathological consequences of this that his theory of communicative action is principally addressed.

Communicative action and the colonization of the life-world

In the 1970s, as part of his reconstruction of historical materialism, Habermas (1975) had sketched out a simple model of the crisis tendencies of late capitalism and, in particular, of the significance of legitimation crisis for modern class societies (Fig. 14.5a). But he subsequently admitted that this model was at best obscure, because he was unable to connect the paradigms of 'life-world' and 'system'. Since then he has provided a detailed account of the pathologies of modernity which considerably clarifies his earlier remarks. As Figure 14.5b shows, he continues to distinguish between economic and political–administrative subsystems and, indeed, he still recognizes the chronic recurrence of both economic crises and rationality crises. But he now emphasizes that these distinctions are paralleled by (and connected to) the formation of 'private' and 'public' spheres within the life-world. The interdependencies between them are secured through a network of exchange relations, dominated by monetarization and bureaucratization, and which, in Giddens's terms, would roughly correspond to the mediating functions of allocative and authoritative resources (Habermas 1987a, pp. 318–31).

Habermas's debt to Parsons and Luhmann is obvious, but specifying the model in this way allows him to make the distinction that had previously eluded him

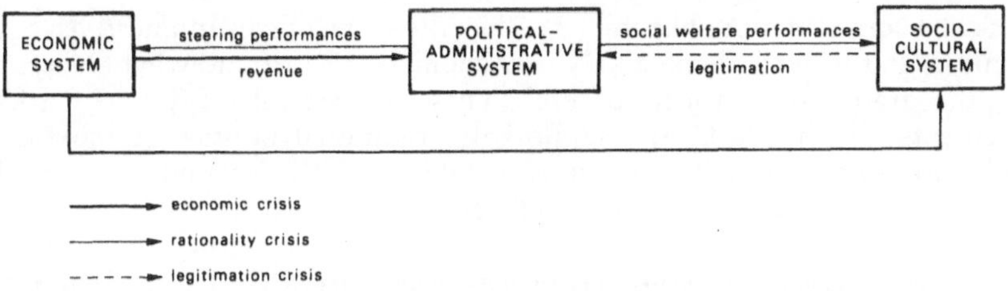

Figure 14.5a The crisis tendencies of late capitalism (after Habermas).

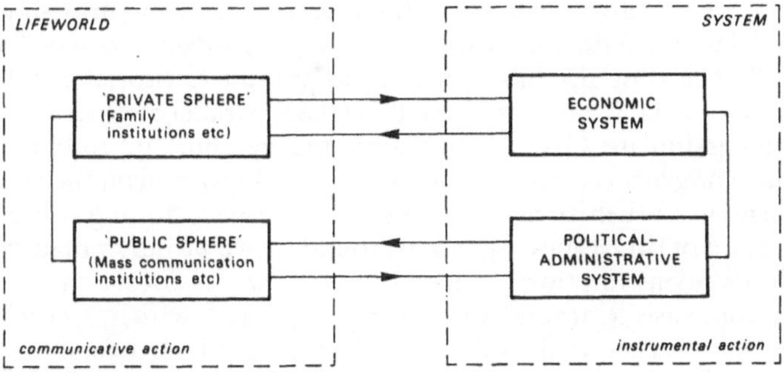

Figure 14.5b Life-world and system (after Habermas).

> between the deficits that inflexible structures of the life-world can give rise to in maintaining the economic and political systems on the one hand, and manifestations of deficiencies in the reproduction of the life-world itself on the other. Empirically the two are connected . . . but it makes sense to separate analytically the *withdrawal of motivation* affecting the occupational system and the *withdrawal of legitimation* affecting the system of domination, on the one side, from the *colonisation of the life-world* that is manifested primarily in phenomena of loss of meaning, anomie and personality disorders, on the other side (Habermas 1982, p. 280; emphasis in original).

These distinctions are unsatisfactory in a number of ways, I think, not least as a result of their implicit but unexamined gender subtext (Fraser 1985), but they are also more than analytic devices. Several critics have noted that Habermas's original model of legitimation crisis has had only a limited purchase on events since the 1970s and is quite unable to account for social struggles peculiar to the 1980s.[22] The 'colonisation of the life-world' now allows Habermas to extend the empirical scope of his thesis and to address these complaints with an unprecedented directness.

Habermas claims that the colonization of the life-world occurs as soon as monetarization and bureaucratization reach 'beyond their mediating roles and penetrate those spheres of the life-world which are responsible for cultural transmission, socialization and the formation of personal identity'. These extensions are made through the contractual formalization of social relations and via the capillary network of micro-powers identified by Foucault. But whatever the medium the consequences are the same: people are made to feel less like persons and more like things. When the scope of communicative action is confined in this way, distinctively new conflicts arise around what Habermas calls 'the grammar of forms of life'. These are articulated by new social movements which overlie the supposedly traditional politics of economic, social, and military security and whose concerns revolve around the symbolic reproduction of the life-world itself (Habermas 1987a).

Social struggles, social movements, and the postmodern city

Habermas may well see the life-world as a 'symbolic space', but I have already indicated that, for the most part at any rate, he fails to recognize the materiality of that space: its concretization in place. It is therefore scarcely surprising that he should also fail to grasp that many of the new social movements which he talks about are either territorially based or have important territorial implications. This is the nub of what I take to be Castells's (1983) parallel discussion. It would be absurd to assimilate Castells to critical theory, but his recent writings indicate some congruence between the two. He seeks to explain the emergence of new, essentially urban social movements concerned, among other things, with the creation of landscapes organized around use value rather than exchange value ('collective consumption'), which diffuse power to local administrations rather than condense it around the central state, and which – crucially for Habermas's thesis – safeguard local cultural identities: which defend 'communication between people, autonomously defined social meaning, and face-to-face interaction, against the monopoly of messages by the media, the predominance of one-way information flows and the standardization of culture'.

These are Habermas's main interests too. Like Castells, he thinks the disintegration of the public sphere has been aggravated by 'the centralization of organizations which privilege vertical and one-way flows of second- and third-hand information, privately consumed.' These new media have substituted images for words, Habermas argues, and blurred the boundaries between advertising, entertainment, information, and politics. Visually, the centres of our large cities have

> absorbed elements of surrealism in an ironic way, and promoted the neon-lit enchantment of a de-realized reality. The banal coalesces with the unreal, hellenistically de-differentiated customs [*sic*] blend with high-tech style, and the ruins of popular cultures with the highly personalized, consumeristically polished bizarre. The refuse dumps of civilization are camouflaged with plastic (Habermas, cited in Dews 1986, p. 178).

The rhetoric may be rebarbative, but it is entirely consonant with Habermas's hostility to postmodernism.

Other writers sharpen similar points. Dear (1986) is plainly not so very far from Habermas when he diagnoses the postmodern city in general and Los Angeles in particular as the products of state penetration and corporate commodification of the urban process: 'a mutant money machine' wired to 'a political economy of social dislocation.' In the same vein, though further removed from critical theory, Davis (1985, 1987) dismisses postmodernism – 'at least in its architectural incarnations and sensibilities' (and, as I will show shortly, the qualification is significant) – as 'a decadent trope of a massified modernism'. In raising the 'sumptuary towers of the speculators' and 'large vivariums for the upper middle classes' over and against the burgeoning enclosures of the poor, postmodernism polarizes the city into radically antagonistic spaces. And Castells (1983) himself speaks of 'dualized supercities' whose different worlds display 'the contradictory co-existence of different social, cultural and economic logics within the same spatial structure' (see Cooke 1988).

But postmodernism is Janus-faced: there is also a postmodernism of resistance. In architecture, in aesthetics, in literary theory, in the human sciences, and in philosophy postmodernism can also be read as a critique of those same deformations. Its rejection of totalization and its sensitivity to difference can be made to open a critical distance between its motifs and those of modernism and, perhaps most important, in seeking to recover an attachment to place, postmodernism offers the prospect of subverting the silent spaces of modernity. Habermas (1987a) has a more one-dimensional view of all this, of course, but it is nevertheless inside the pulverized matrix of the postmodern city (and beyond) that he too detects the formation of 'locally fragmented, sub-cultural resistances' which 'have become the core of autonomous counter-public spheres' (Ley 1986).

If these counter-public spheres subsume Castells's urban social movements, however, Castells offers a different account of their constitution and one which is peculiarly attentive to the spatiality of social life. He is even more sceptical than Habermas about the ability of an unreconstructed Marxism to make much sense of them (though for much the same reason), and he makes three conceptual moves which distance him still further from its formulations.

First, Castells draws a distinction between modes of production and 'modes of development', and cross-cuts them to produce the grid shown in Figure 14.6. The theoretical status of this scheme is far from clear, but Castells is certainly not alone in seeking to distinguish capitalism from industrialism nor, for that matter, in accentuating the contemporary significance of the accumulation of information: in particular, the lower left-hand cell of the grid.

Second, more novel is the suggestion that all four cells are territorially differentiated and integrated in an asymmetrical way across different levels. Castells argues that modern social life takes place in a three-dimensional space which cannot be conceptualized as a conventional totality because its dynamics are disjointed. The first two dimensions are **production**, which is now organized at the level of the world economy (the new international division of labour), and **power**, which is focused at the level of the nation–state. The differential mobilization of these allocative and authoritative resources across time and space now depends, so Castells argues, upon 'the twin revolution in communication systems and micro-electronics' – upon the informational mode of development – and this has effectively transformed 'places into flows and channels'. In consequence, the third dimension, **experience**, is dislocated:

> [W]hat tends to disappear is the meaning of places for people. Each place, each city, will receive its social meaning from its location in the hierarchy of a network whose control and rhythm will escape from each place and, even more, from the people in each place . . . . The new space of a world capitalist system, combining the informational and industrial modes of development,

		MODE OF PRODUCTION	
		CAPITALISM	STATE SOCIALISM
MODE OF DEVELOPMENT	*INDUSTRIAL*	Profit maximization: increasing the appropriation of surplus value *Economic growth:* *increasing output*	Power maximization: increasing the military capacity of the political apparatus *Economic growth:* *increasing output*
	INFORMATIONAL	Profit maximization: increasing the appropriation of surplus value *Technological development:* *accumulation of knowledge*	Power maximization: increasing the military capacity of the political apparatus *Technological development:* *accumulation of knowledge*

Figure 14.6 Modes of production and modes of development (after Castells).

> is a space of variable geometry, formed by locations hierarchically ordered in a continuously changing network of flows . . . . The new source of power relies on the control of the entire network of information. Space is dissolved into flows . . . . It is the space of collective alienation and individual violence . . . . Life is transformed into abstraction, cities into shadows (Castells 1983, p. 314).

In effect, Castells suggests that any analysis of the deformation of the life-world – of the disassociation between the 'space of experience' and the 'space of organizations' – *depends upon a conjoint analysis of the spatiality of system integration*. Since the theoretical scaffolding which supports these claims is by no means secure, it might help to translate them into other vocabularies. In Hägerstrand's terms, for example, they refer to the ways in which late 20th-century changes in capability constraints have radically transformed coupling and steering constraints; in Giddens's terms, they confirm the close connections between time–space routinization and time–space distanciations; and in Habermas's terms, they prefigure a reduction of the 'public sphere' (and public space) and a delocalization of the 'private sphere'.

Third, Castells (1983) insists that these transformations are not predetermined: they are the product of social struggles. New social movements arise, therefore, which – for all their differences – seek to subvert and, ultimately, to reverse the hierarchy:

> to establish the command of experience over production and power, instead of adapting experience in the best possible conditions to the structural framework created by production and enforced by power.

And Castells claims that this is not a utopian project, because the technological basis of the informational mode of development allows for the automation of production and depends on communication and co-operation: in principle, and at the limit, it contains the seeds of its own dialectical transcendence.

The emphasis on information is a recurrent theme in much post-Marxist writing, of course, and Lyotard (1984) identifies the comodification of knowledge as a strategic moment in the production of a postmodern network of social relations that is supposedly more complex, more heterogeneous and more mobile than ever before. Although Habermas also accentuates the importance of a more general paradigm of communication – he and Lyotard have at least this much in common – he insists, against Lyotard, that this does not entail the abandonment of historical materialism. The pathologies of modernity are still to be explained 'in terms of a capitalist accumulation process which is largely disconnected from orientations towards use value' (Habermas, cited in Dews 1986, p. 177).[23]

While I think it wrong to reduce the circulation of information to the circulation of capital, the intersection of these two circuits is a prominent feature of the late 20th-century world. Davis (1985, 1987) suggests that the 'great construction bubble' of the postmodern city has been inflated by the rise of new international circuits of financial capital whose hypertrophic expansion is symptomatic not of some new phase of capital accumulation, still less of the grand finale of modernity, but of chronic global crisis. The volatility of these circuits of capital transfer depends, in part, on the speed of information transmission, and they stretch far beyond the orbit of civilian production. In the United States in particular, still the epicentre of the world economy, the diffusion of high technology – the propulsive dynamic of the space of flows – is closely connected to the geography of military expenditure: to the rise of the 'warfare state' and, by extension, the collapse of the welfare state. Not surprisingly, the shock-waves of the new spatial division of labour are registered with a special force in the inner cities. Davis shows with exceptional clarity how the so-called urban renaissance is also a direct consequence of the abandonment of urban reform as part of the new class polarization taking place in advanced capitalist societies.[24]

Habermas passes over these matters partly because he develops the theory of communicative action in terms of the constitution of a single society. While he is evidently not unaware of 'the conflicts generated by the international division of labour' (North–South) or 'the significance of the strategic balance between the superpowers' (East–West), he offers no detailed analysis of them. But it is also partly because he is simply more interested in those pathologies which, as he says himself, 'jeopardize social and cultural integration' (Habermas, cited in Dews 1986, pp. 39–43).

Habermas's sensitivity to these pathologies is unimpeachable, but the wider silences inevitably compromise his reconstruction of historical materialism. They mean that social movements become, at best, purely defensive: 'the temptation becomes stronger to withdraw even further in order to regain individuality and subjectivity in tiny corners of social life not yet colonized by capital and the state administration.' Habermas himself agrees that these new social movements 'do not seek to conquer new territory'. On the contrary, they are sub-institutional, para-political, principally concerned with beating back encroachments on the life-world (O'Connor 1987). It is for precisely this reason that Castells describes them, in their present form, as 'symptoms of resistance' rather than agents of structural change. They are aimed at 'transforming the meaning of the city without being able to transform society', 'calling for a new depth of existence without being able to create that new depth' (Castells 1983, p. 329). And it is for the same reason that Harvey (1985, p. 127) regards their interventions as essentially functional: as the 'disciplining arm of uneven accumulation and uneven class struggle in geographic space'. It is not necessary

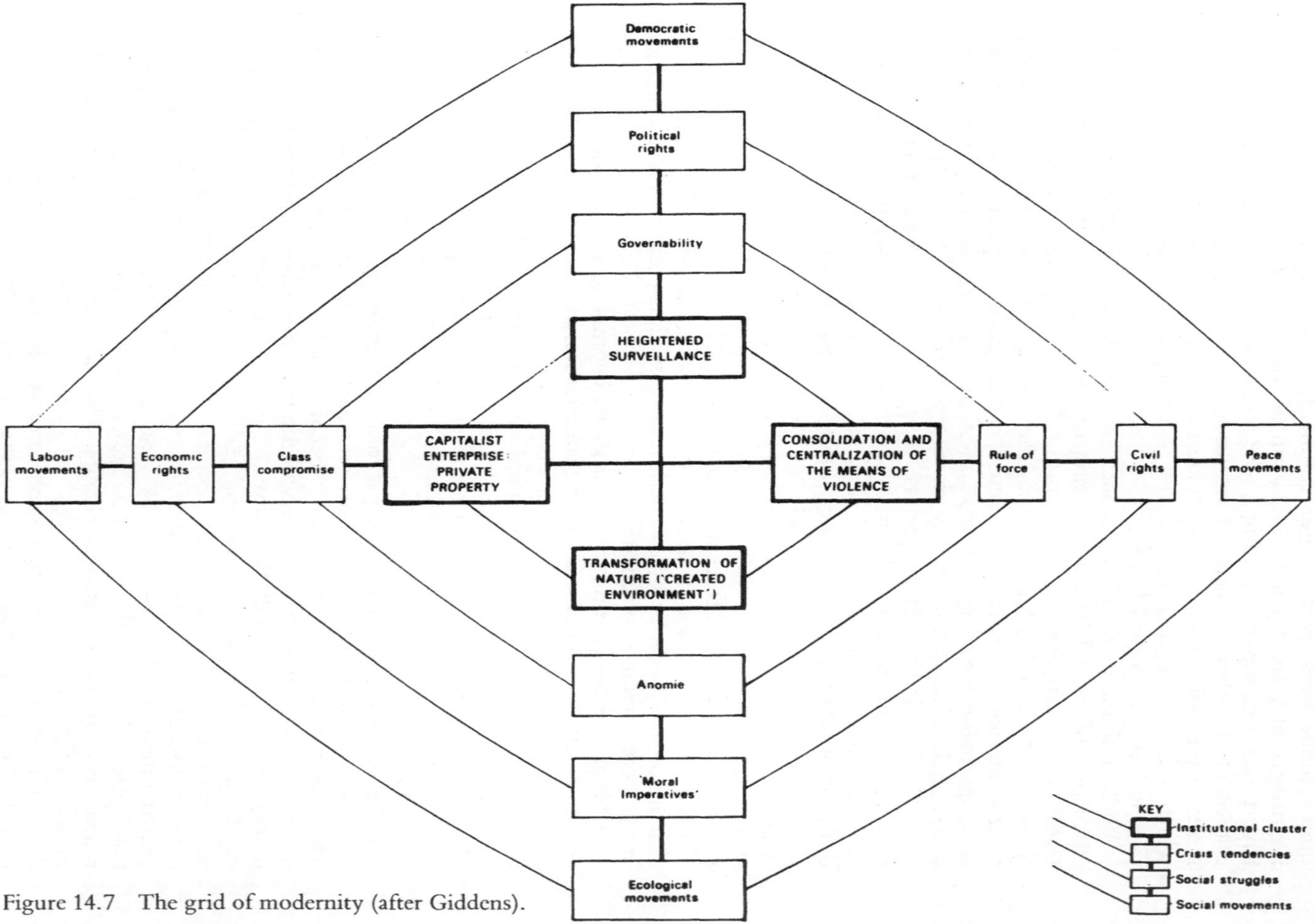

Figure 14.7 The grid of modernity (after Giddens).

to accept the economism of Harvey's argument to see that questions of class struggle must be marginalized in any account which fails to foreground the historical geography of capital accumulation. To be sure, both Habermas and Castells emphasize that these new social movements cannot be *reduced* to the conventional notations of class; that is, after all, what is supposedly new about them. But if they are not located within a wider conceptual grid which illuminates the structural matrix within which they are set, then the possibilities for collective action will be short-circuited; as Castells fears, 'urban darkness [will] . . . overshadow city lights'.[25]

Can structuration theory provide such a grid?

Struturation theory and critical theory

In one sense, clearly it can. Where Habermas focuses on social integration and its cultural pathologies, Giddens views modernity through a wide-angle lens. Structuration theory is developed across more dimensions than the theory of communicative action and, unlike Habermas, Giddens accentuates the time–space distanciation of each of them.

In his recent writings, Giddens claims that the modern world has been shaped by the intersection of capitalism, industrialism, and the system of nation–states. He unfolds their conjunction along four principal axes, none of which is wholly reducible to any of the others, to reveal the conceptual map which I have sketched in Figure 14.7. Working out from the centre of the diagram, four **institutional clusters** are supposed to define the structural grid of modernity and these are connected to characteristic **crisis tendencies**. Giddens argues that those crises which typically confound capitalist societies cluster in the upper-left quadrant, whereas those which confound state socialist societies cluster in the upper right; both types of society are affected by crises which derive more or less directly from industrialism and which arc through the lower-left and lower-right quadrants. These different crises engender distinctive **social struggles** which, in turn, are given shape and substance by particular **social movements**. His gazetteer of them is obviously far more complete, but Giddens (1985) maintains that all modern social movements can, in principle, be located within this co-ordinate system and that their analysis is focal to the development of a properly constituted critical theory.

Although Giddens insists that the spatiality of social life has to have a central place in any such theory, some parts of his map are more thoroughly surveyed than others. He has much of value to say about the geography of administrative power, the territoriality of the nation–state, and the geopolitics of the modern world system, but his discussion of the uneven development of capitalism remains at best perfunctory and he offers no sustained account of the geography of crisis tendencies and, in particular, of the chronic tension between the immobility of spatial structures and their capacity to stretch across even wider spans of time and space.[26]

Still more important for my present purpose, he consistently theorizes the distanciation of this grid in terms of the resources made available by structures of domination. The impact of signification and legitimation, supposedly co-equal axes of structuration, is blunted. This plainly marginalizes the questions which are highest on Habermas's agenda. If culture is not altogether absent from the lexicon of structuration theory, it is certainly not elaborated in any systematic fashion. Even when Giddens (1987c, pp. 28–9) acknowledges, as he now does, that the debates surrounding postmodernism might be 'the first real initiatives in the ambitious task of charting the cultural universe resulting from the ever-more complete disintegration of the traditional world', they remain, for the most part, voyages into the unknown. And, as I must now show, this rebounds directly on his conception of the transformative capacities of human agency.

Time-space distanciation, signification, and subjectivity

I want to substantiate two claims. The first is that conceptions of space are not mirrors but *media* of time–space distanciation: they enter fully into the time–space constitution of societies. Once it is accepted that the stretching of social relations across time and space depends upon the differential construction of these 'geographies of social knowing and unknowing', as Thrift (1985b) calls them more generally, then my second claim follows directly from the first: time–space distanciation is structurally implicated in the time–space constitution of knowledgeable and capable human subjects.

The first of these propositions is unexceptional. The production of printed maps – to take only the most obvious example – was a pivotal moment in the passage of modernity. If by the early 17th century as Eisenstein (1979) suggests, 'navigators and explorers *could be seen* to be moving in a different direction', then the burden of the phrase falls exactly where I have marked it. The geographical imagination was enlarged and enhanced by the codification of these astonishing discoveries in spatial form, and a revitalized history of cartography is now beginning to reveal the ways in which these new world pictures were intrinsically politico-economic visions which helped shape the constitution of the modern world system and the consolidation of nation–states. 'Pictures of Europe's political geography became a familiar sight not just to those who administered governments and dominated economic life,' Mukerji (1983) remarks, 'but also to those subject to them.'

But maps are ever vulnerable to competing interpretations. Saxton's county maps of Elizabethan England were designed to convey the image of a bureaucratic and centralized state, and yet the county gentry had little difficulty in converting them into the mirror image of an insistent regionalism in which the claims of the county community were reflected with a special clarity. Even so, these paper landscapes represented the adminstrative apparatus of the state with an apparently indelible authority, revealing the steady encroachment of surveillance into still more spheres of social life, and in this sense, Harley (1988) reminds us, 'as a regulator of human affairs', 'the map is rather like a clock.' Unlike the clock, however, the map is not a mechanism. It is, in the fullest of senses, a discourse, and one which articulates that abstracted conception of space which Sack (1980) and others (Helgerson 1986) see as characteristic of both class-divided and class societies. Indeed, many of these maps concerned representations of property rights and from the 16th century they became vital instruments in that progressive commodification of space which Giddens portrays as focal to the emergence and expansion of European capitalism. There can be no doubt, therefore, that the graphical representation of space was immensely important in its own right.[27]

If the critical reading of maps poses problems which cannot be directly assimilated to the semiotics of written texts, they cannot be altogether separated from them. Cartographic conventions, for all their seeming 'naturalness', were part of a wider cultural universe whose deconstruction requires an interrogation of multiple, intersecting and overlapping discourses. As Cosgrove (1984, 1985) has shown, there were numerous attempts in 16th- and 17th-century Europe, across the whole field of the visual arts, to clarify 'a new conception of space as a coherent visual structure into which the actions of human life could be inserted in a controlled and orderly fashion.' Cosgrove claims that these experiments were not simply framed by capitalism: they were part of its dynamic. Their conventions, incorporated in landscape painting and in the landscape itself, enunciated and endorsed the alienation and objectification which were hallmarks of the new social order.

All of this may be granted; my comments can be connected to the episodes which Giddens singles out for special attention without in any way compromising his formulation of structuration theory. But the previous paragraphs all relate to the *knowledgeability* of human agents and, in seeking to theorize time–space distanciation without making these connections explicit, Giddens risks obscuring the modalities through which power enters into the constitution of human subjects.

These are, of course, Foucauldian concerns, but Giddens misreads (or at any rate misrepresents) them. While he evidently does not conceive of the human subject as somehow 'pre-formed', he undoubtedly offers an account of the constitution of human subjects which is, at bottom, ahistorical. He treats subjectivity in strictly developmental terms, drawing on the ideas of Erikson, Freud, Lacan and others to establish the transformation of the body into an instrument of acting-in-the-world, and subsequently using Goffman, Hägerstrand, and others to emphasize the significance of time–space routinization for sustaining the stratification of personality. Habermas's account depends on a developmental logic too, of course, but in an altogether different way: his theory of social evolution, whatever its demerits, is at least designed to draw attention to the *different* ways in which human subjects are constituted in *different* types of society. It is this insight which is missing from structuration theory. To be sure, Habermas locates his argument within the sociological rather than the geographical imagination and a more satisfactory account of the time–space constitution of human subjects – of biography formation – would have to incorporate differences in the spatiality of social life. But a project of this kind cannot be confined to the study of time–space routinization (as it usually is). An historical geography of the person is, so I shall claim, strategically incomplete without a conjoint analysis of time–space distanciation.

In fact, the consequences of stretching social relations across time and space for the constitution of human subjects can be discerned in each of the episodes which Giddens identifies. All I can do is list some examples, but I should emphasize that none of them took place on an undifferentiated plane and that a fuller discussion would have to be much more sensitive to the specificities of context and the contours of 'local knowledge' than is possible here.

While it would be absurd to privilege the spoken over the written word as somehow more real and more personal, as what Hirst & Woolley (1982) call 'the true evocation of the speaking subject' (since it is precisely this notion of transhistorical authenticity which is being contested), it seems likely that the intervention of writing marked the inauguration of new concepts of human being, and that the invention of printing provided the basis for the development of still other conceptions of personal identity and, in particular, of the status of 'person-as-proprietor'. Indeed, Ong (1982) claims that typography enforced a doubly spatial coding, fixing words in space and fragmenting the oral world into 'privately claimed freeholdings'. The systematic collation of information about individuals lowered the threshold of description even further. As Foucault remarks,

> For a long time ordinary individuality – the everyday individuality of everybody – remained below the threshold of description. To be looked at, observed, described in detail, followed from day to day by an uninterrupted writing, was a privilege . . . . The disciplinary methods reversed this relation, lowered the threshold of describable individuality and made of this description a means of control and a method of domination (Foucault 1979, p. 191).

Foucault (1979, pp. 191–3) shows in some considerable detail how this depended on a spatial analytics in which 'each individual has [their] own place, and each place its individual'.[28] But the introduction of electronic media, which has dramatically refined these systems of surveillance, has also, more generally and less formally, blurred what Meyrowitz (1985) calls the 'situational geography' of everyday life. Electronic media have overriden the boundaries and definitions of situations supported by physical settings, Meyrowitz claims, and in consequence conceptions of the self and of interpersonal relations are being radically revised: so much so, indeed, that Lash & Urry (1987) connect these developments to a postmodern 'decentring of identity' symptomatic of the disorganization of contemporary capitalism.

Monetization and commodification have had equally powerful effects within the sphere of 'organized' capitalism. On the one hand, as Simmel (1978, p. 477) noted, these processes became embedded in

> the extension of means of transport which . . . progressed from the infrequency of the mailcoach to the almost uninterrupted connections between the most important places and to the telegraph and telephone which [made] communication possible at any time.

This declining distance in external relationships was counterpointed by a 'growing distance in genuine inner relationships'. 'The most remote [came] closer at the price of increasing the distance to what was originally nearer', Simmel (1978) wrote, so that 'an inner barrier develop[ed] between people' and 'penetrate[d] even more deeply into the individual human subject'. On the other hand, therefore, Simmel observed how the enlargement of the money economy 'places us at a distance from the substance of things; they speak to us "as from afar"; reality is touched not with direct confidence but with fingertips that are immediately withdrawn.' Here, surely, are the clearest intimations of Castells's cities 'transformed into shadows' (and, of course, Simmel had much to say about social life in the modern metropolis too).[29]

All of these examples could be qualified in various ways, and I have drawn them from diverse and by no means unassailable theoretical traditions. But even as they stand, they suggest that a purely abstract account of human subjectivity is inadequate for any critical theory with practical intent. For, as Callinicos (1985) argues, such an account must necessarily fail to clarify the different 'modalities of resistance' – the spaces for collective human agency – in *different* types of society. Giddens has so far not shown how the transformative capacities of human beings vary according to the specific circumstances in which they find themselves *and through which they are constituted as knowledgeable and capable human subjects*. His preoccupation with an abstract account of human subjectivity, Callinicos (1987) concludes, 'prevents him from following through the consequences of his own insight into the way in which social structures enable as well as constrain'.

Prospect

Let me now try to summarize the implications of my argument. It is not an exhaustive account of the relations between human geography and critical social theory – far from it – but I want to make three closely connected claims about the prospect of a critical human geography.

First, it is impossible to understand postmodernism and postmodernity without a detailed analysis of the historical geography of modernity. This cannot afford to be an essentialist account, in which everything of any consequence is reduced to the logic of capital accumulation, and it must take seriously the spatiality of both social integration and system integration. Some critics might object that historical materialism needs no lessons in pluralism, since there has always been a plurality of different Marxian traditions. But I think that postmodernism provides a particularly vibrant statement of the polyphony that characterizes (and ought to characterize) contemporary social theory.

Second, the crises of modernity must be seen as moments of opportunity. By this, I mean that any critical theory with practical intent must incorporate both an explanatory–diagnostic perspective *and* an anticipatory–utopian perspective: it must empower people to make their own (and better) human geographies. Yet over the last ten years and across the discipline as a whole, so it seems to me, the normative has been marginalized. There are the clearest of signs that human geography is retreating from some of the most signal advances made during its 'relevance revolution'. A concern with moral and political questions was not confined to the 1970s, I realize, but it was during that decade that some of the most sustained and thought-provoking commentaries were written. They derived their power not just from the substantive areas which they opened up for investigation, important though these were, but from their willingness to reflect upon the process of intellectual enquiry and its intrinsically *practical* implications.

My third point follows directly from this. The debate over modernism and postmodernism is a serious one and deserves to be treated as such. Postmodernism needs to be saved from both its antagonists and its advocates. Its claims need to be approached openly, scrupulously, and vigilantly. For to do otherwise is to abandon any prospect for a critical human geography.

Notes

1 For charges of careerism see Sangren (1988) and the subsequent commentary; for reductions of postmodernism to the logic of capitalism see Jameson (1984), Eagleton (1986), and Harvey (1987). My own reading of postmodernism is indebted to Bauman (1988).
2 For general discussions see Dews (1987) and Vattimo (1988).
3 For superb discussions of the practical imperative, of particular relevance to the present chapter, see Habermas (1974) and Benhabib (1986).
4 In refusing to treat modernism and postmodernism as antinomies, I follow Wellmer (1985) and Hutcheon (1988); see also Gregory (1987).
5 Other maps are possible, of course, and the exclusion of physical geography from this one reflects my limited competence and not any judgement of its importance: quite the reverse. Indeed, I do not pretend that this is anything other than a severe limitation. Only a fool would think that the crisis of modernity could be discussed without the most searching discussion of the dynamics of the natural environment.
6 Bradbury & Macfarlane (1976) offer 'A geography of modernism' and Huyssen (1986) provides one way of 'Mapping the postmodern'. It should be noted that Hutcheon (1988, p. 209) objects to Habermas generalizing about postmodernism on the basis of 'one *particular* and *local* experience of it' (emphasis added).
7 This is not to say that there are no differences between spatial science and its predecessors, of course, though I think some of the most important are to do with textual strategies rather than philosophical bases. Traditional geography often depended upon a conventional narrative – which is why denudation chronology and historical geography were its twin pillars, but the narrative logic resided in the processes being described. Spatial science depended upon a narrative too, but its logic resided not in the processes being analyzed but in the processes *of* analysis: 'literature review'; 'hypothesis construction'; 'data collection'; 'hypothesis testing'; 'interpretation of results'. For a fuller discussion, see Gregory (1989b).
8 Cf. Quaini (1982), who claims that questions of space and spatial structure are 'indelibly present' in Marx's writings (p. 26).
9 Giddens's early writings said virtually nothing about time-space relations, but cf. Giddens (1981) and (1984).
10 Darby (1962) provided the classic statement of the problem, but in doing so drew upon Lessing's *Laocoon* which was the Enlightenment's most comprehensive statement on the semiotics of painting and poetry. According to Lessing, painting expressed subjects which exist side by side in space (bodies) and poetry expressed subjects which succeed one another in time (actions). This mimics Kant's distinction between geography as 'a report of phenomena that follow one another in time' and which was the basis for Hartshorne's (1939) account of the nature of geography (see also Harvey (1969), pp. 215–16). I have provided a fuller (and sharply critical) discussion in Gregory (1989b).
11 This is not to exempt Geertz from criticism, of course: see Shankman (1984) and the subsequent commentary.
12 See Tyler (1987). I have provided a summary account in Gregory (1988).
13 This is also a central focus of Olsson's astonishing linguistic experiments.
14 The suspicion of metanarratives is developed in detail in Rorty (1979). But cf. Norris (1985, pp. 139–66), who claims that there is 'a sharp distinction between Rorty's avowed "post-modernist" attitude and the kind of story he tells by way of backing it up . . . . It is a tale which disclaims meta-narrative authority in the interest of conveying the pragmatist outcome to which all its episodes lead up. But this means adopting an implied teleology, a grasp of its unfolding narrative logic which does, in effect, lay claim to superior insight.'
15 Mead is usually taken to be the principal architect of symbolic interactionism, but

for a scrupulously careful exegesis of his contribution – on which Habermas also relies – see Joas (1985).

16 The account which follows is derived from Gregory (1989b) and it draws in particular upon Foucault (1977, 1980). See also Dreyfus & Rabinow (1982).

17 For Habermas on Foucault, see Habermas (1987b, pp. 238–92).

18 I discuss Habermas's subsequent formulations below. For fuller discussions of the development of Habermas's project, see Roderick (1986), Ingram (1987), and White (1988).

19 The account which follows is derived from Giddens (1981, 1984). I discuss Giddens's subsequent formulations below.

20 Giddens never refers to this particular account, but he provides a critique of interpretative sociologies more generally in Giddens (1976).

21 The focus on modernity (rather than capitalism alone) is a relatively recent feature of the writings of both Habermas and Giddens.

22 For an example in human geography, see Harvey & Scott (1989). Although they do not refer explicitly to Habermas's work the theory which they describe as being overtaken by events is, to all intents and purposes, Habermas's model of legitimation crisis. They do not, however, register the development of Habermas's views since then. For a dramatically different perspective, see Bauman (1988). He rejects the notion of legitimation crisis because the state, so he claims, now dispenses with legitimation altogether. Instead, it relies on seduction (in the marketplace) and repression (via the police, the judiciary, and the military). It is for precisely this reason that Bauman treats postmodernism as symptomatic of a crisis of intellectual activity: as he reads the present situation, it is the traditional (Enlightenment) conception of the rôle of the intellectual that has been overtaken by events.

23 It should also be noted that Habermas does not consider 'the retention of the labour theory of value to be feasible' (1987a, p. 88). For a comparison between Lyotard and Habermas, see Rorty (1985).

24 See also Castells (1985) and Harvey (1987).

25 Harvey is particularly exercised by Castells's 'apparent defection from the Marxist fold' (1985, p. 125).

26 I have provided a more detailed discussion in Gregory (1989a).

27 Harley treats the map as 'pre-eminently a language of power', but insists that this derives as much from 'its representational force as a symbol as through its overt representations'. In some measure following Foucault, he therefore draws attention to 'the "hidden rules" of cartographic discourse whose contours can be traced in the subliminal geometries, the silences and the representational hierarchies of maps' (Harley 1988).

28 Foucault emphasizes that the human sciences were themselves profoundly implicated in these developments and in the formation of a 'disciplinary society' (cf. Bauman 1988).

29 For a commentary, see Frisby (1985).

References

Anderson, P. 1983. *In the tracks of historical materialism.* London: Verso.

Bauman, Z. 1988. Is there a postmodern sociology? *Theory, Culture and Society* **5**, 217–37.

Benhabib, S. 1986. *Critique, norm and utopia: a study of the foundations of critical theory.* New York: Columbia University Press.

Berdoulay, V. 1978. The Vidal-Durkheim debate. In *Humanistic geography: prospects and problems,* D. Ley & M. Samuels (eds), 77–90. Chicago: Maaroufa Press.

Bernstein, R. 1972. *Praxis and action*. London: Duckworth.
Bernstein, R. 1983. *Beyond objectivism and relativism: science, hermeneutics and praxis*. Oxford: Basil Blackwell.
Bernstein, R. (ed.) 1985. *Habermas and modernity*. Cambridge: Polity Press.
Bhaskar, R. 1979. *The possibility of naturalism: a philosophical critique of the contemporary human sciences*. Brighton: Harvester.
Bradbury, M. & J. McFarlane (eds) 1976. *Modernism 1890–1930*. Harmondsworth: Penguin.
Bunge, W. 1962. *Theoretical geography*. Lund: Gleerup.
Callinicos, A. 1985. Anthony Giddens: a contemporary critique. *Theory and Society* **14**, 133–66.
Callinicos, A. 1987. *Making history: agency, structure and change in social theory*. Cambridge: Polity Press.
Castells, M. 1983. *The city and the grassroots: a cross-cultural theory of urban social movements*. London: Edward Arnold.
Castells, M. 1985. High technology, economic restructuring and the urban-regional process in the United States.' In *High technology, space and society*, M. Castells (ed.), 11–40. Beverly Hills: Sage Publications.
Chorley, R. J. & P. Haggett (eds) 1967. *Models in geography*. London Methuen.
Clifford, J. & G. Marcus, (eds) 1986. *Writing cultures: the poetics and politics of ethnography*. Berkeley: University of California Press.
Cooke, P. 1988. Modernity, postmodernity and the city. *Theory, Culture and Society* **5**, 475–92.
Cosgrove, D. 1984. *Social formation and symbolic landscape*. London: Croom Helm.
Cosgrove, D. 1985. Prospect, perspective and the evolution of the landscape idea. *Transactions of the Institute of British Geographers* **1**, 45–62.
Cosgrove, D. & P. Jackson 1987. New directions in cultural geography. *Area* **19**, 95–101.
Darby, H. C. 1962. The problem of geographical description. *Transactions of the Institute of British Geographers* **30**, 1–14.
Davis, M. 1985. Urban renaissance and the spirit of postmodernism. *New Left Review* **151**, 106–13.
Davis, M. 1987. *Chinatown* part two? The 'internationalization' of downtown Los Angeles. *New Left Review* **164**, 65–86.
Dear, M. 1986, Postmodernism and planning. *Environment and Planning D, Society and Space* **4**, 367–84.
Dear, M. 1988. The postmodern challenge: reconstructing human geography. *Transactions of the Institute of British Geographers* **13**, 262–74.
Dews, P. (ed.) 1986. *Habermas: autonomy and solidarity. Interviews with Jurgen Habermas*. London: Verso.
Dews, P. 1987. *Logics of disintegration: post-structuralist thought and the claims of critical theory*. London: Verso.
Dreyfus, H. L. & P. Rabinow 1982. *Michel Foucault: beyond structuralism and hermeneutics*. Brighton: Harvester.
Eagleton, T. 1986. Capitalism, modernism and postmodernism. In *Against the grain: selected essays*, T. Eagleton, 131–48. London: Verso.
Eisenstein, E. L. 1979. *The printing press as an agent of change: communications and cultural transformations in early modern Europe*. Cambridge: Cambridge University Press.
Entrikin, J. N. 1980. Robert Park's human ecology and human geography. *Annals of the Association of American Geographers*.
Foucault, M. 1979. *Discipline and punish: the birth of the prison*. Harmondsworth: Penguin.
Foucault, M. 1980. *Power/knowledge*. Brighton: Harvester.
Fraser, N. 1985. What's critical about critical theory? The case of Habermas and gender. *New German Critique* **35**, 97–131.

Frisby, D. 1985. *Fragments of modernity: theories of modernity in the work of Simmel, Kracauer and Benjamin*. Cambridge: Polity Press.

Geertz, C. 1973. *The interpretation of cultures: selected essays*. New York: Basic Books.

Geertz, C. 1983. *Local knowledge: further essays in interpretive anthropology*. New York: Basic Books.

Geertz, C. 1984. Anti anti-relativism. *American Anthropologist* **86**, 263–78.

Giddens, A. 1976. *New rules of sociological method: a positive critique of interpretative sociologies*. London: Hutchinson.

Giddens, A. 1981. *A contemporary critique of historical materialism*. Vol. 1: *Power, property and the state*. London: Macmillan.

Giddens, A. 1984. *The constitution of society: outline of the theory of structuration*. Cambridge: Polity Press.

Giddens, A. 1985. *A contemporary critique of historical materialism*. Vol. 2: *The nation–state and violence*. Cambridge: Polity Press.

Giddens, A. 1987a. Erving Goffman as a systematic social theorist. In *Social theory and modern sociology*, A. Giddens, 109–39. Cambridge: Polity Press.

Giddens, A. 1987b. Reason without revolution? Habermas's *Theory of communicative action*. In *Social theory and modern sociology*, A. Giddens, 225–52. Cambridge: Polity Press.

Giddens, A. 1987c. Nine theses on the future of sociology. In *Social theory and modern sociology*, A. Giddens, 22–51. Cambridge: Polity Press.

Goody, J. 1986. *The logic of writing and the organization of society*. Cambridge: Cambridge University Press.

Gregory, D. 1985. Suspended animation: the stasis of diffusion theory. In *Social relations and spatial structures*, D. Gregory & J. Urry (eds), 296–336. London: Macmillan.

Gregory, D. 1987. Postmodernism and the politics of social theory. *Environment and Planning D, Society and Space* **5**, 245–8.

Gregory, D. 1988. Areal differentiation and post-modern human geography. In *Horizons in human geography*, D. Gregory & R. Walford (eds), 67–96. London: Macmillan.

Gregory, D. 1989a. Presences and absences; time-space relations and structuration theory. In *Critical theory of the industrial societies*, D. Held & J. Thompson (eds). Cambridge: Cambridge University Press.

Gregory, D. 1989b. *The geographical imagination: social theory and human geography*. London: Unwin Hyman.

Habermas, J. 1974. *Theory and practice*. London: Heinemann.

Habermas, J. 1975. *Legitimation crisis*. London: Heinemann.

Habermas, J. 1979. *Communication and the evolution of society*. London: Heinemann.

Habermas, J. 1982. A reply to my critics. *Habermas: critical debates*, J. Thompson & D. Held (eds), 219–83. London: Macmillan.

Habermas, J. 1984. *The theory of communicative action*. Vol. 1: *Reason and the rationalization of society*. Cambridge: Polity Press.

Habermas, J. 1985. Questions and counterquestions. In *Habermas and modernity*, R. Bernstein (ed.), 192–216. Cambridge: Polity Press.

Habermas, J. 1987a. *The theory of communicative action*. Vol. 2: *Lifeworld and system. A critique of functionalist reason*. Cambridge: Polity Press.

Habermas, J. 1987b. *The philosophical discourse of modernity*. Cambridge: Polity Press.

Habermas, J. 1987c. Philosophy as stand-in and interpreter. In *After philosophy. End or transformation?*, K. Baynes, J. Bohmann & T. McCarthy (eds), 296–315. Cambridge, Mass.: MIT Press.

Hägerstrand, T. 1982. Diorama, path and project. *Tijdshrift voor Economische en Sociale Geographie* **73**, 323–39.

Harley, J. B. 1988. Maps, knowledge and power. In *The iconography of landscape*, D. Cosgrove & S. Daniels (eds), 277–312. Cambridge: Cambridge University Press.

Hartshorne, R. 1939. *The nature of geography*. Lancaster, Penn.: Association of American Geographers.

Harvey, D. 1969. *Explanation in geography*. London: Edward Arnold.

Harvey, D. 1982. *The limits to capital*. Oxford: Basil Blackwell.

Harvey, D. 1985. *The urbanization of capital*. Oxford: Basil Blackwell.

Harvey, D. 1987. Flexible accumulation through urbanisation: reflections on 'post-modernism' in the American city. *Antipode* **19**, 260–86.

Harvey, D. & A. Scott 1989. The practice of human geography: theory and empirical specificity in the transition from Fordism to flexible accumulation. *Remodelling geography*, W. Macmillan (ed.). Oxford: Basil Blackwell.

Helgerson, R. 1986. The land speaks: cartography, chorography and subversion in Renaissance England. *Representations* **16**, 50–85.

Hirst, P. & J. Woolley 1982. *Social relations and human attributes*. London: Tavistock Publications.

Hutcheon, L. 1988. *A poetics of postmodernism: history, theory, fiction*. New York: Routledge.

Huyssen, A. 1986. *After the great divide: modernism, mass culture and postmodernism*. London: Macmillan.

Ingram, D. 1987. *Habermas and the dialectic of reason*. New Haven, Conn.: Yale University Press.

Jameson, F. 1980. *The political unconscious: narrative as a socially symbolic act*. Ithaca, NJ: Cornell University Press.

Jameson, F. 1984. Postmodernism, or the cultural logic of late capitalism. *New Left Review* **146**, 53–92.

Jay, M. 1984. *Marxism and totality*. Cambridge: Polity Press.

Joas, H. 1985. *G. H. Mead: a contemporary re-examination of his thought*. Cambridge: Polity Press.

Lash, S & J. Urry 1986. The dissolution of the social? In *Sociological theory in transition*, M. Wardell & S. Turner (eds), 95–109. Boston: Allen & Unwin.

Lash, S. & J. Urry 1987. *The end of organized capitalism*. Cambridge: Polity Press.

Lawson, H. 1985. *Reflexivity: the post-modern predicament*. London: Hutchinson.

Ley, D. 1986. Modernism, post-modernism and the struggle for place. Paper presented to the seminar series on The power of place, Syracuse University, NY.

Lyotard, J.-F. 1984. *The postmodern condition: a report on knowledge*. Manchester: Manchester University Press.

Mann, M. 1986. *The sources of social power*. Vol. 1: *A history of power from the beginning to AD 1760*. Cambridge: Cambridge University Press.

Marcus, G. & M. Fischer 1986. *Anthropology as cultural critique: the experimental moment in the human sciences*. Chicago: University of Chicago Press.

Meyrowitz, J. 1985. *No sense of place: the impact of electronic media on social behavior*. New York: Oxford University Press.

Mukerji, C. 1983. *From graven images: patterns of modern materialism*. New York: Columbia University Press.

Norris, C. 1985. *The contest of faculties: philosophy and theory after deconstruction*. London: Methuen.

O'Connor, J. 1987. *The meaning of crisis: a theoretical introduction*. Oxford: Basil Blackwell.

Ong, W. J. 1982. *Orality and literacy: the technologizing of the word*. London: Methuen.

Pickles, J. 1985. *Phenomenology, science and geography: spatiality and the human sciences*. Cambridge: Cambridge University Press.

Pred, A. 1981. Social reproduction and the time-geography of everyday life. *Geografisker Annaler* **63B**, 5–22.

Quaini, M. 1982. *Geography and Marxism*. Oxford: Basil Blackwell.

Relph, E. 1976. *Place and placelessness*. London: Pion.

Relph, E. 1981. *Rational landscapes and humanistic geography*. London: Croom Helm.
Roderick, R. 1986. *Habermas and the foundations of critical theory*. London: Macmillan.
Rorty, R. 1979. *Philosophy and the mirror of nature*. Princeton, NJ: Princeton University Press.
Rorty, R. 1982. *Consequences of pragmatism*. Minneapolis: University of Minnesota Press.
Rorty, R. 1985. Habermas and Lyotard on modernity. In *Habermas and modernity*, R. Bernstein (ed.), 161–75. Cambridge: Polity Press.
Sack, R. D. 1980. *Conceptions of space in social thought*. London: Macmillan.
Sangren, P. 1988. Rhetoric and the authority of ethnography: 'postmodernism' and the social reproduction of texts. *Current Anthropology* **29**, 405–24.
Sauer, C. 1963. Folkways of social science. In *Land and life: a selection from the writings of Carl Ortwin Sauer*, J. Leighly (ed.), 380–8. Berkeley: University of California Press.
Schutz, A. & T. Luckmann, 1974. *The structures of the lifeworld*. London: Heinemann.
Shankman, P. 1984. The thick and the thin: on the interpretive theoretical program of Clifford Geertz. *Current Anthropology* **25**, 261–70.
Simmel, G. 1978. *The philosophy of money*. London: Routledge & Kegan Paul.
Skinner, Q. (ed.) 1985. *The return of grand theory in the human sciences*. Cambridge: Cambridge University Press.
Stoddart, D. R. 1986. *On geography and its history*. Oxford: Basil Blackwell.
Storper, M. 1985. The spatial and temporal constitution of social action: a critical reading of Giddens. *Environment and Planning D, Society and Space* **3**, 407–24.
Thompson, J. B. 1984. *Studies in the theory of ideology*. Cambridge: Polity Press.
Thrift, N. 1985a. Bear and mouse or bear and tree? Anthony Giddens's reconstitution of social theory. *Sociology* **19**, 609–23.
Thrift, N. 1985b. Flies and germs: a geography of knowledge. In *Social relations and spatial structures*, D. Gregory & J. Urry (eds), 366–403. London: Macmillan.
Tyler, S. 1987. *The unspeakable; discourse, dialogue and rhetoric in the postmodern world*. Madison: University of Wisconsin Press.
Vattimo, G. 1988. *The end of modernity: nihilism and hermeneutics in post-modern culture*. Cambridge: Polity Press.
Wellmer, A. 1985. On the dialectics of modernism and postmodernism. *Praxis International* **4**, 337–62.
White, H. 1987. *The content of the form: narrative discourse and historical representation*. Baltimore: Johns Hopkins University Press.
White, S. 1988. *The recent work of Jurgen Habermas: reason, justice and modernity*. Cambridge: Cambridge University Press.
Wolf, E. 1982. *Europe and the people without history*. Berkeley: University of California Press.
Wright, E. O. 1985. *Classes*. London: Verso.

Index